THE
SHAPE
OF
INNER
SPACE

The
SHAPE
of
INNER
SPACE

STRING THEORY AND THE
GEOMETRY OF THE UNIVERSE'S
HIDDEN DIMENSIONS

Shing-Tung Yau
and
Steve Nadis

Illustrations by
Xianfeng (David) Gu and Xiaotian (Tim) Yin

BASIC BOOKS

A MEMBER OF THE PERSEUS BOOKS GROUP
New York

Books published by Basic Books are available at special discounts for bulk
purchases in the United States by corporations, institutions, and other
organizations. For more information, please contact the Special Markets
Department at the Perseus Books Group, 2300 Chestnut Street, Suite 200,
Philadelphia, PA 19103, or call (800) 810-4145, ext. 5000, or e-mail
special.markets@perseusbooks.com.

Designed by Timm Bryson

Library of Congress Cataloging-in-Publication Data
Yau, Shing-Tung, 1949-
  The shape of inner space : string theory and the geometry of the universe's
hidden dimensions / Shing-tung Yau and Steve Nadis ; illustrations by Xianfeng
(David) Gu and Xiaotian (Tim) Yin.
    p. cm.
  Includes bibliographical references and index.
  ISBN 978-0-465-02023-2 (alk. paper)
  1. Hyperspace. 2. String models. 3. Fourth dimension. I. Nadis, Steven J. II.
Title.
  QA691.Y38 2010
  530.1—dc22
                        2010009956

10 9 8 7 6 5 4 3 2 1

CONTENTS

"Space/Time" (Poem) vii
Preface ix
Prelude: The Shapes of Things to Come xvii

1 A UNIVERSE IN THE MARGINS 1

2 GEOMETRY IN THE NATURAL ORDER 17

3 A NEW KIND OF HAMMER 39

4 TOO GOOD TO BE TRUE 77

5 PROVING CALABI 103

6 THE DNA OF STRING THEORY 121

7 THROUGH THE LOOKING GLASS 151

8 KINKS IN SPACETIME 183

9 BACK TO THE REAL WORLD 199

10 BEYOND CALABI-YAU 227

11 THE UNIVERSE UNRAVELS 253

12   THE SEARCH FOR EXTRA DIMENSIONS   269

13   TRUTH, BEAUTY, AND MATHEMATICS   289

14   THE END OF GEOMETRY?   307

EPILOGUE   ANOTHER DAY, ANOTHER DONUT   321

POSTLUDE   ENTERING THE SANCTUM   325

"A Flash in the Middle of a Long Night" (Poem)   329
Notes   331
Glossary   345
Index   359

SPACE/TIME

Time, time
  why does it vanish?
All manner of things
  what infinite variety.
Three thousand rivers
  all from one source.
Time, space
  mind, matter, reciprocal.
Time, time
  it never returns.
Space, space
  how much can it hold?
In constant motion
  always in flux.
Black holes lurking
  mysteries afoot.
Space and time
  one without bounds.
Infinite, infinite
  the secrets of the universe.
Inexhaustible, lovely
  in every detail.
Measure time, measure space
  no one can do it.
Watched through a straw
  what's to be learned has no end.

SHING-TUNG YAU
BEIJING, 2002

PREFACE

Mathematics is often called the language of science, or at least the language of the physical sciences, and that is certainly true: Our physical laws can only be stated precisely in terms of mathematical equations rather than through the written or spoken word. Yet regarding mathematics as merely a language doesn't do justice to the subject at all, as the word leaves the erroneous impression that, save for some minor tweaks here and there, the whole business has been pretty well sorted out.

In fact, nothing could be further from the truth. Although scholars have built a strong foundation over the course of hundreds—and indeed thousands—of years, mathematics is still very much a thriving and dynamic enterprise. Rather than being a static body of knowledge (not to suggest that languages themselves are set in stone), mathematics is actually a dynamic, evolving science, with new insights and discoveries made every day rivaling those made in other branches of science, though mathematical discoveries don't capture the headlines in the same way that the discovery of a new elementary particle, a new planet, or a new cure for cancer does. In fact, save for the proof of a centuries-old problem from time to time, they rarely capture headlines at all.

Yet for those who appreciate the sheer force of mathematics, it can be viewed as not just a language but as the surest path to the truth—the bedrock upon which the whole edifice of physical science rests. The strength of this discipline, again, lies not simply in its ability to explain physical reality or to reveal it, because to a mathematician, mathematics is reality. The geometric figures and spaces, whose existence we prove, are just as real to us as are the elementary particles of physics that make up all matter. But we consider mathematical structures even more fundamental than the particles of nature because mathematical structures can be used not only to understand such particles but also to understand the phenomena of everyday life, such as the contours of a human face or the symmetry of flowers. What excites geometers perhaps most of all is the power and beauty of the abstract principles that underlie the familiar forms and shapes of our contemporary world.

For me, the study of mathematics and my specialty, geometry, has truly been an adventure. I still recall the thrill I felt during my first year of graduate school, when—as a twenty-year-old fresh off the boat, so to speak—I first learned about Einstein's theory of gravity. I was struck by the notion that gravity and curvature could be regarded as one and the same, as I'd already become fascinated with curved surfaces during my undergraduate years in Hong Kong. Something about these shapes appealed to me on a visceral level. I don't know why, but I couldn't stop thinking about them. Hearing that curvature lay at the heart of Einstein's theory of general relativity gave me hope that someday, and in some way, I might be able to contribute to our understanding of the universe.

The Shape of Inner Space describes my explorations in the field of mathematics, focusing on one discovery in particular that has helped some scientists build models of the universe. No one can say for sure whether these models will ultimately prove correct. But the theory underlying these models, nevertheless, possesses a beauty that I find undeniable.

Taking on a book of this nature has been challenging, to say the least, for someone like me who's more comfortable with geometry and nonlinear differential equations than writing in the English language, which is not my native tongue. I find it frustrating because there's a great clarity, as well as a kind of elegance, in mathematical equations that is difficult, if not impossible, to express in words. It's a bit like trying to convey the majesty of Mount Everest or Niagara Falls without any pictures.

Fortunately, I've gotten some well-needed help on this front. Although this narrative is told through my eyes and in my voice, my coauthor has been responsible for translating the abstract and abstruse mathematics into (hopefully) lucid prose.

When I proved the Calabi conjecture—an effort that lies at the heart of this book—I dedicated the paper containing that proof to my late father, Chen Ying

Chiu, an educator and philosopher who instilled in me a respect for the power of abstract thought. I dedicate this book to him and to my late mother, Leung Yeuk Lam, both of whom had a profound influence on my intellectual growth. In addition, I want to pay tribute to my wife, Yu-Yun, who has been so tolerant of my rather excessive (and perhaps obsessive) research and travel schedule, and to my sons, Isaac and Michael, of whom I'm very proud.

I also dedicate this book to Eugenio Calabi, the author of the aforementioned conjecture, whom I've known for nearly forty years. Calabi were an enormously original mathematician with whom I've been linked for more than a quarter century through a class of geometric objects, Calabi-Yau manifolds, which serve as the principal subject of this book. The term *Calabi-Yau* has been so widely used since it was coined in 1984 that I almost feel as if Calabi is my first name. And if it is to be my first name—at least in the public's mind—it's one I'm proud to have.

The work that I do, much of which takes place along the borders between mathematics and theoretical physics, is rarely done in isolation, and I have benefited greatly from interactions with friends and colleagues. I'll mention a few people, among many, who have collaborated with me directly or inspired me in various ways.

First, I'd like to pay tribute to my teachers and mentors, a long line of illustrious people that includes S. S. Chern, Charles Morrey, Blaine Lawson, Isadore Singer, Louis Nirenberg, and the aforementioned Calabi. I'm pleased that Singer invited Robert Geroch to speak at a 1973 Stanford conference, where Geroch inspired my work with Richard Schoen on the positive mass conjecture. My subsequent interest in physics-related mathematics has always been encouraged by Singer.

I'm grateful for the conversations I had on general relativity while visiting Stephen Hawking and Gary Gibbons at Cambridge University. I learned about quantum field theory from one of the masters of the subject, David Gross. I remember in 1981, when I was a professor at the Institute for Advanced Study, the time Freeman Dyson brought a fellow physicist, who had just arrived in Princeton, into my office. The newcomer, Edward Witten, told me about his soon-to-be-published proof of the positive mass conjecture—a conjecture I had previously proved with a colleague using a very different technique. I was struck, for the first of many times to come, by the sheer force of Witten's mathematics.

Over the years, I've enjoyed close collaborations with a number of people, including Schoen (mentioned above), S. Y. Cheng, Richard Hamilton, Peter Li, Bill Meeks, Leon Simon, and Karen Uhlenbeck. Other friends and colleagues who have added to this adventure in many ways include Simon Donaldson, Robert Greene, Robert Osserman, Duong Hong Phong, and Hung-Hsi Wu.

I consider myself especially lucky to have spent the past twenty-plus years at Harvard, which has provided an ideal environment for interactions with both mathematicians and physicists. During my time here, I've gained many insights from talking to Harvard math colleagues—such as Joseph Bernstein, Noam Elkies, Dennis Gaitsgory, Dick Gross, Joe Harris, Heisuke Hironaka, Arthur Jaffe (also a physicist), David Kazdhan, Peter Kronheimer, Barry Mazur, Curtis McMullen, David Mumford, Wilfried Schmid, Yum-Tong Siu, Shlomo Sternberg, John Tate, Cliff Taubes, Richard Taylor, H. T. Yau, and the late Raoul Bott and George Mackey—while having memorable exchanges with MIT math colleagues as well. On the physics side, I've had countless rewarding conversations with Andy Strominger and Cumrun Vafa.

In the past ten years, I was twice an Eilenberg visiting professor at Columbia, where I had many stimulating conversations with faculty members, especially with Dorian Goldfeld, Richard Hamilton, Duong Hong Phong, and S. W. Zhang. I was also a Fairchild visiting professor and Moore visiting professor at Caltech, where I learned a lot from Kip Thorne and John Schwarz.

Over the last twenty-three years, I have been supported by the U.S. government through the National Science Foundation, the Department of Energy, and DARPA in my research related to physics. Most of my postdoctoral fellows received their Ph.D.s in physics, which is somewhat unusual in our discipline of mathematics. But the arrangement has been mutually beneficial, as they have learned some mathematics from me and I have learned some physics from them. I am glad that many of these postdoctoral fellows with physics backgrounds later became prominent professors in mathematics departments at Brandeis, Columbia, Northwestern, Oxford, Tokyo, and other universities. Some of my postdocs have done important work on Calabi-Yau manifolds, and many of them have also helped on this book: Mboyo Esole, Brian Greene, Gary Horowitz, Shinobu Hosono, Tristan Hubsch, Albrecht Klemm, Bong Lian, James Sparks, Li-Sheng Tseng, Satoshi Yamaguchi, and Eric Zaslow. Finally, my former graduate students—including Jun Li, Kefeng Liu, Melissa Liu, Dragon

Wang, and Mu-Tao Wang—have made noteworthy contributions in this area as well, some of which will be described in the pages to come.

—SHING-TUNG YAU, CAMBRIDGE, MASSACHUSETTS, MARCH 2010

Odds are I never would have known about this project were it not for Henry Tye, a Cornell physicist (and a friend of Yau's), who suggested that my coauthor-to-be might steer me to an interesting tale or two. Henry was right about this, as he has been about many other things. I'm grateful to him for helping to launch me on this unexpected journey and for assisting me at many junctures along the way.

As Yau has often said, when you venture down a path in mathematics, you never know where it will end up. The same has been true on the writing end of things. The two of us pretty much agreed during our very first meeting to write a book together, though it took a long while for us to know what the book would be about. In some ways, you might say we didn't really know that until the book was finished.

Now a few words about the product of this collaboration in an attempt to keep any confusion to a minimum. My coauthor is, of course, a mathematician whose work is central to much of the story related here. Sections of the book in which he was an active participant are generally written in the first person, with the "I" in this case referring to him and him alone. However, even though the book has its fair share of personal narrative, this work should probably not be characterized as Yau's autobiography or biography. That's because part of the discussion relates to people Yau doesn't know (or who died long before he was born), and some of the subject matter described—such as experimental physics and cosmology—lies outside his areas of expertise. These sections, which are written in a third-person voice, are largely based on interviews and other research I conducted.

While the book is, admittedly, an unusual blend of our different backgrounds and perspectives, it seemed to be the best way for the two of us to recount a story that we both considered worth telling. The task of actually getting this tale down on paper relied heavily on my coauthor's extraordinary grasp of numbers and hopefully profited as well from his collaborator's facility with words.

One other point on the issue of whether this ought to be regarded as an autobiography: Although the book certainly revolves around Yau's work, I would suggest that the main character is not Yau himself but rather the class of geometric shapes—so-called Calabi-Yau manifolds—that he helped invent.

Broadly speaking, this book is about understanding the universe through geometry. General relativity, a geometry-based description of gravity that has achieved stunning success in the past century, offers one example. String theory represents an ambitious attempt to go even further, and geometry is vital to this quest, with six-dimensional Calabi-Yau shapes assuming a special place in this theory. The book tries to present some of the ideas from geometry and physics needed to understand where Calabi-Yau manifolds came from and why some physicists and mathematicians consider them important. The book focuses on various aspects of these manifolds—their defining features, the mathematics that led to their discovery, the reasons string theorists find them intriguing, and the question of whether these shapes hold the key to our universe (and perhaps to other universes as well).

That, at least, is what *The Shape of Inner Space* is supposed to be about. Whether it lives up to that billing may be open to debate. But there is no doubt in my mind that this book would never have come to fruition without technical, editorial, and emotional support from many people—too many, I'm afraid, to list in full, but I will mention as many as I can.

I received a tremendous amount of help from people already singled out by my coauthor. These include Eugenio Calabi, Simon Donaldson, Brian Greene, Tristan Hubsch, Andrew Strominger, Li-Sheng Tseng, Cumrun Vafa, Edward Witten, and, most of all, Robert Greene, Bong Lian, and Li-Sheng Tseng. The latter three provided me with math and physics tutorials throughout the writing process, exhibiting expository skills and levels of patience that boggle the mind. Robert Greene, in particular, spoke with me a couple of times a week during busy stretches to guide me through thorny bits of differential geometry. Without him, I would have been sunk—many times over. Lian got me started in thinking about geometric analysis, and Tseng helped out immensely with last-minute changes in our ever-evolving manuscript.

The physicists Allan Adams, Chris Beasley, Shamit Kachru, Liam McAllister, and Burt Ovrut fielded questions from me at various times of day and night, carrying me through many a rough patch. Other individuals who were exceedingly generous with their time include Paul Aspinwall, Melanie Becker, Lydia

Bieri, Volker Braun, David Cox, Frederik Denef, Robbert Dijkgraaf, Ron Donagi, Mike Douglas, Steve Giddings, Mark Gross, Arthur Hebecker, Petr Horava, Matt Kleban, Igor Klebanov, Albion Lawrence, Andrei Linde, Juan Maldacena, Dave Morrison, Lubos Motl, Hirosi Ooguri, Tony Pantev, Ronen Plesser, Joe Polchinski, Gary Shui, Aaron Simons, Raman Sundrum, Wati Taylor, Bret Underwood, Deane Yang, and Xi Yin.

That is merely the tip of the iceberg, as I've also received help from Eric Adelberger, Saleem Ali, Bruce Allen, Nima Arkani-Hamed, Michael Atiyah, John Baez, Thomas Banchoff, Katrin Becker, George Bergman, Vincent Bouchard, Philip Candelas, John Coates, Andrea Cross, Lance Dixon, David Durlach, Dirk Ferus, Felix Finster, Dan Freed, Ben Freivogel, Andrew Frey, Andreas Gathmann, Doron Gepner, Robert Geroch, Susan Gilbert, Cameron Gordon, Michael Green, Paul Green, Arthur Greenspoon, Marcus Grisaru, Dick Gross, Monica Guica, Sergei Gukov, Alan Guth, Robert S. Harris, Matt Headrick, Jonathan Heckman, Dan Hooper, Gary Horowitz, Stanislaw Janeczko, Lizhen Ji, Sheldon Katz, Steve Kleiman, Max Kreuzer, Peter Kronheimer, Mary Levin, Avi Loeb, Feng Luo, Erwin Lutwak, Joe Lykken, Barry Mazur, William McCallum, John McGreevy, Stephen Miller, Cliff Moore, Steve Nahn, Gail Oskin, Rahul Pandharipande, Joaquín Pérez, Roger Penrose, Miles Reid, Nicolai Reshetikhin, Kirill Saraikin, Karen Schaffner, Michael Schulz, John Schwarz, Ashoke Sen, Kris Snibbe, Paul Shellard, Eva Silverstein, Joel Smoller, Steve Strogatz, Leonard Susskind, Yan Soibelman, Erik Swanson, Max Tegmark, Ravi Vakil, Fernando Rodriguez Villegas, Dwight Vincent, Dan Waldram, Devin Walker, Brian Wecht, Toby Wiseman, Jeff Wu, Chen Ning Yang, Donald Zeyl, and others.

Many of the concepts in this book are difficult to illustrate, and we were fortunate to be able to draw on the extraordinary graphic talents of Xiaotian (Tim) Yin and Xianfeng (David) Gu of the Stony Brook Computer Science Department, who were assisted in turn by Huayong Li and Wei Zeng. Additional help on the graphics front was provided by Andrew Hanson (the premier renderer of Calabi-Yau manifolds), John Oprea, and Richard Palais, among others.

I thank my many friends and relatives, including Will Blanchard, John De Lancey, Ross Eatman, Evan Hadingham, Harris McCarter, and John Tibbetts, who read drafts of the book proposal and chapters or otherwise offered advice and encouragement along the way. Both my coauthor and I are grateful for the invaluable administrative assistance provided by Maureen Armstrong, Lily Chan, Hao Xu, and Gena Bursan.

Several books proved to be valuable references. Among them are *The Elegant Universe* by Brian Greene, *Euclid's Window* by Leonard Mlodinow, *Poetry of the Universe* by Robert Osserman, and *The Cosmic Landscape* by Leonard Susskind.

The Shape of Inner Space might never have gotten off the ground were it not for the help of John Brockman, Katinka Matson, Michael Healey, Max Brockman, Russell Weinberger, and others at the Brockman, Inc., literary agency. T. J. Kelleher of Basic Books had faith in our manuscript when others did not, and—with the help of his colleague, Whitney Casser—worked hard to get our book into a presentable form. Kay Mariea, the project editor at Basic Books, shepherded our manuscript through its many stages, and Patricia Boyd provided expert copyediting, teaching me that "the same" and "exactly the same" are exactly the same thing.

Finally, I'm especially grateful for the support from my family members— Melissa, Juliet, and Pauline, along with my parents Lorraine and Marty, my brother Fred, and my sister Sue—who acted as if six-dimensional Calabi-Yau manifolds were the most fascinating thing in the world, not realizing that these manifolds are, in fact, out of this world.

—STEVE NADIS, CAMBRIDGE, MASSACHUSETTS, MARCH 2010

PRELUDE
THE SHAPES OF THINGS TO COME

God ever geometrizes.
—PLATO

In the year 360 B.C. or thereabouts, Plato published *Timaeus*—a creation story told in the form of a dialogue between his mentor, Socrates, and three others: Timaeus, Hermocrates, and Critias. Timaeus, likely a fictitious character who is said to have come to Athens from the southern Italian city of Locri, is an "expert in astronomy [who] has made it his main business to know the nature of the universe."[1] Through Timaeus, Plato presents his own theory of everything, with geometry playing a central role in those ideas.

Plato was particularly fascinated with a group of convex shapes, a special class of polyhedra that have since come to be known as the Platonic solids. The faces of each solid consist of identical polygons. The tetrahedron, for example, has four faces, each a triangle. The hexahedron, or cube, is made up of six squares. The octahedron consists of eight triangles, the dodecahedron of twelve pentagons, and the icosahedron of twenty triangles.

Plato did not invent the solids that bear his name, and no one knows who did. It is generally believed, however, that one of his contemporaries, Theaetetus, was the first to prove that five, and only five, such solids—or *convex regular*

0.1—The five *Platonic solids*, named for the Greek philosopher Plato: the tetrahedron, hexahedron (or cube), octahedron, dodecahedron, and icosahedron. The prefixes derive from the number of faces: four, six, eight, twelve, and twenty, respectively. One feature of these solids that no other convex polyhedra satisfy is that all their faces, edges, and angles (between two edges) are congruent.

polyhedra, as they're called—exist. Euclid gave a complete mathematical description of these geometric forms in *The Elements.*

The Platonic solids have several intriguing properties, some of which turn out to be equivalent ways of describing them. For each type of solid, the same number of faces meet at each of the corner points, or vertices. One can draw a sphere around the solid that touches every one of those vertices—something that's not possible for polyhedra in general. Moreover, the angle of each vertex, where two edges meet, is always the same. The number of vertices plus faces equals the number of edges plus two.

Plato attached a metaphysical significance to the solids, which is why his name is forever linked with them. In fact, the convex regular polyhedra, as detailed in *Timaeus,* formed the very essence of his cosmology. In Plato's grand scheme of things, there are four basic elements: earth, air, fire, and water. If we could examine these elements in fine detail, we'd notice that they are composed of minuscule versions of the Platonic solids: Earth would thus consist of tiny cubes, the air of octahedrons, fire of tetrahedrons, and water of icosahedrons. "One other construction, a fifth, still remained," Plato wrote in *Timaeus,* referring to the dodecahedron. "And this one god used for the whole universe, embroidering figures on it."[2]

As seen today, with the benefit of 2,000-plus years of science, Plato's conjecture looks rather dubious. While there is, at present, no ironclad agreement as to the basic building blocks of the universe—be they leptons and quarks, or hypothetical subquarks called preons, or equally hypothetical and even tinier strings—we do know that it's not just earth, air, fire, and water embroidered upon one giant dodecahedron. Nor do we believe that the properties of the elements are governed strictly by the shapes of Platonic solids.

On the other hand, Plato never claimed to have arrived at the definitive theory of nature. He considered *Timaeus* a "likely account," the best he could come up with at the time, while conceding that others who came after him might very well improve on the picture, perhaps in a dramatic way. As Timaeus states midway into his discourse: "If anyone puts this claim to the test and discovers that it isn't so, his be the prize, with our congratulations."[3]

There's no question that Plato got many things wrong, but viewing his thesis in the broadest sense, it's clear that he got some things right as well. The eminent philosopher showed perhaps the greatest wisdom in acknowledging that what he put forth might not be true, but that another theory, perhaps building on some of his ideas, could be true. The solids, for instance, are objects of extraordinary symmetry: The icosahedron and dodecahedron, for instance, can be rotated sixty ways (which, not coincidentally, turns out to be twice the number of edges in each shape) and still look the same. In basing his cosmology on these shapes, Plato correctly surmised that symmetry ought to lie at the heart of any credible description of nature. For if we are ever to produce a real theory of everything—in which all the forces are unified and all the constituents obey a handful (or two) of rules—we'll need to uncover the underlying symmetry, the simplifying principle from which everything else springs.

It hardly bears mentioning that the symmetry of the solids is a direct consequence of their precise shape or geometry. And this is where Plato made his second big contribution: In addition to realizing that mathematics was the key to fathoming our universe, he introduced an approach we now call the geometrization of physics—the same leap that Einstein made. In an act of great prescience, Plato suggested that the elements of nature, their qualities, and the forces that act upon them may all be the result of some hidden geometrical structure that conducts its business behind the scenes. The world we see, in other words, is a mere reflection of the underlying geometry that we might not see. This is a notion dear to my heart, and it relates closely to the mathematical proof for which I am best known—to the extent that I am known at all. Though it may strike some as far-fetched, yet another case of geometric grandstanding, there just might be something to this idea, as we'll see in the pages ahead.

One

A UNIVERSE IN THE MARGINS

The invention of the telescope, and its steady improvement over the years, helped confirm what has become a truism: There's more to the universe than we can see. Indeed, the best available evidence suggests that nearly three-fourths of all the stuff of the cosmos lies in a mysterious, invisible form called dark energy. Most of the rest—excluding only the 4 percent composed of ordinary matter that includes us—is called dark matter. And true to form, it too has proved "dark" in just about every respect: hard to see and equally hard to fathom.

The portion of the cosmos we can see forms a sphere with a radius of about 13.7 billion light-years. This sphere is sometimes referred to as a Hubble volume, but no one believes that's the full extent of the universe. According to the best current data, the universe appears to extend limitlessly, with straight lines literally stretching from here to eternity in every direction we can point.

There's a chance, however, that the universe is ultimately curved and bounded. But even if it is, the allowable curvature is so slight that, according to some analyses, the Hubble volume we see is just one out of at least one thousand such volumes that must exist. And a recently launched space instrument, the Planck telescope, may reveal within a few years that there are at least *one million* Hubble volumes out there in the cosmos, only one of which we'll ever have access to.[1] I'm trusting the astrophysicists on this one, realizing that some may quarrel with the exact numbers cited above. One fact, however, appears to be unassailable: What we see is just the tip of the iceberg.

At the other extreme, microscopes, particle accelerators, and various imaging devices continue to reveal the universe on a miniature scale, illuminating a previously inaccessible world of cells, molecules, atoms, and smaller entities. By now, none of this should be all that surprising. We fully expect our telescopes to probe ever deeper into space, just as our microscopes and other tools bring more of the invisible to light.

But in the last few decades—owing to developments in theoretical physics, plus some advances in geometry that I've been fortunate enough to participate in—there has been another realization that is even more startling: Not only is there more to the universe than we can see, but there may even be more dimensions, and possibly quite a few more than the three spatial dimensions we're intimately acquainted with.

That's a tough proposition to swallow, because if there's one thing we know about our world—if there's one thing our senses have told us from our first conscious moments and first groping explorations—it's the number of dimensions. And that number is three. Not three, give or take a dimension or so, but exactly three. Or so it seemed for the longest time. But maybe, just maybe, there are additional dimensions so small that we haven't noticed them yet. And despite their modest size, they could be crucial in ways we could not have possibly appreciated from our entrenched, three-dimensional perspective.

While this may be hard to accept, we've learned in the past century that whenever we stray far from the realm of everyday experience, our intuition can fail us. If we travel extremely fast, special relativity tells us that time slows down, not something you're likely to intuit from common sense. If we make an object extremely small, according to the dictates of quantum mechanics, we can't say exactly where it is. When we do experiments to determine whether the object has ended up behind Door A or Door B, we find it's neither here nor there, in the sense that it has no absolute position. (And it sometimes may appear to be in both places at once!) Strange things, in other words, can and will happen, and it's possible that tiny, hidden dimensions are one of them.

If this idea is true, then there might be a kind of universe in the margins—a critical chunk of real estate tucked off to the side, just beyond the reach of our senses. This would be revolutionary in two ways. The mere existence of extra dimensions—a staple of science fiction for more than a hundred years—would be startling enough on its own, surely ranking among the greatest findings in the history of physics. But such a discovery would really be a starting point rather than an end unto itself. For just as a general might obtain a clearer perspective on the battlefield by observing the proceedings from a hilltop or tower and thereby gaining the benefit of a vertical dimension, so too may our laws of physics become more apparent, and hence more readily discerned, when viewed from a higher-dimensional vantage point.

We're familiar with travel in three basic directions: north or south, east or west, and up or down. (Or, equivalently, left or right, backward or forward, and up or down.) Wherever we go—whether it's driving to the grocery store or flying to Tahiti—we move in some combination of those three independent directions. So familiar are we with these dimensions that trying to conceive of an additional dimension—and figuring out exactly where it would point—might seem impossible. For a long while, it seemed as if what you see is what you get. In fact, more than two thousand years ago, Aristotle argued as much in his treatise *On the Heavens*: "A magnitude if divisible one way is a line, if two ways a surface, and if three a body. Beyond these there is no other magnitude, because the three dimensions are all that there are."[2] In A.D. 150, the astronomer and mathematician Ptolemy tried to prove that four dimensions are impossible, insisting that you cannot draw four mutually perpendicular lines. A fourth perpendicular, he contended, would be "entirely without measure and without definition."[3] His argument, however, was less a rigorous proof than a reflection of our inability both to visualize and to draw in four dimensions.

To a mathematician, a dimension is a "degree of freedom"—an independent way of moving in space. A fly buzzing around over our heads is free to move in any direction the skies permit. Assuming there are no obstacles, it has three degrees of freedom. Suppose that fly lands on a parking lot and gets stuck in a patch of fresh tar. While it is temporarily immobilized, the fly has zero degrees of freedom and is effectively confined to a single spot—a zero-dimensional world. But this creature is persistent and, after some struggle, wrests itself free from the tar, though injuring its wing in the process. Unable to fly, it has two degrees of freedom and can roam the surface of the parking lot at will. Sensing a predator—a ravenous frog, perhaps—our hero seeks refuge in a rusted tailpipe lying in the lot. The fly thus has one degree of freedom, trapped at least for now in the one-dimensional or linear world of this narrow pipe.

But is that all there is? Does a fly buzzing through the air, stuck in tar, crawling on the asphalt, or making its way through a pipe include all the possibilities imaginable? Aristotle or Ptolemy would have said yes, but while this may be the case for a not terribly enterprising fly, it is not the end of the story for contemporary mathematicians, who typically find no compelling reason to stop at three dimensions. On the contrary, we believe that to truly understand a concept in geometry, such as curvature or distance, we need to understand it in all possible dimensions, from zero to n, where n can be a very big number indeed. Our grasp of that concept will be incomplete if we stop at three dimensions—the point being that if a rule or law of nature works in a space of any dimension, it's more powerful, and seemingly more fundamental, than a statement that only applies in a particular setting.

Even if the problem you're grappling with pertains to just two or three dimensions, you might still secure helpful clues by studying it in a variety of dimensions. Let's return to our example of the fly flitting about in three-dimensional space, which has three directions in which to move, or three degrees of freedom. Yet let's suppose another fly is moving freely in that same space; it too has three degrees of freedom, and the system as a whole has suddenly gone from three to six dimensions—with six independent ways of moving. With more flies zigzagging through the space—all moving on their own without regard to the other—the complexity of the system goes up, as does the dimensionality.

One advantage in looking at higher-dimensional systems is that we can divine patterns that might be impossible to perceive in a simpler setting. In the next chapter, for instance, we'll discuss the fact that on a spherical planet, hypothetically covered by a giant ocean, all the water cannot flow in the same direction—say, from west to east—at every point. There have to be some spots where the water is not moving at all. Although this rule applies to a two-dimensional surface, it can only be derived by looking at a much higher-dimensional system in which all possible configurations—all possible movements of tiny bits of water on the surface—are considered. That's why we continually push to higher dimensions to see what it might lead to and what we might learn.

One thing that higher dimensions lead to is greater complexity. In topology, which classifies objects in terms of shape in the most general sense, there are just two kinds of one-dimensional spaces: a line (a curve with two open ends) and a circle (a closed curve with no ends). There aren't any other possibilities. Of course, the line could be squiggly, or the closed curve oblong, but those are questions of geometry, not topology. The difference between geometry and topology is like the difference between looking at the earth's surface with a magnifying glass and going up in a rocket ship and surveying the planet as a whole. The choice comes down to this: Do you insist on knowing every last detail— every ridge, undulation, and crevice in the surface—or will the big picture ("a giant round ball") suffice? Whereas geometers are often concerned with identifying the exact shape and curvature of some object, topologists only care about the overall shape. In that sense, topology is a holistic discipline, which stands in sharp contrast to other areas of mathematics in which advances are typically made by taking complicated objects and breaking them down into smaller and simpler pieces.

As for how this ties into our discussion of dimensions, there are—as we've said—just two basic one-dimensional shapes in topology: A straight line is identical to a wiggly line, and a circle is identical to any "loop"—oblong, squiggly, or even square—that you can imagine. The number of two-dimensional spaces is similarly restricted to two basic types: either a sphere or a donut. A topologist considers any two-dimensional surface without holes in it to be a sphere, and this includes everyday geometric shapes such as cubes, prisms, pyramids, and even watermelon-like objects called ellipsoids.

The presence of the hole in the donut or the lack of the hole in the sphere makes all the difference in this case: No matter how much you manipulate or deform a sphere—without ripping a hole in it, that is—you'll never wind up with a donut, and vice versa. In other words, you cannot create new holes in an object, or otherwise tear it, without changing its topology. Conversely, topologists regard two shapes as functionally equivalent if—supposing they are made out of malleable clay or Play-Doh—one shape can be molded into the other by squeezing and stretching but not ripping.

A donut with one hole is technically called a *torus*, but a donut-like surface could have any number of holes. Two-dimensional surfaces that are both compact (closed up and finite in extent) and orientable (double-sided) can be classified

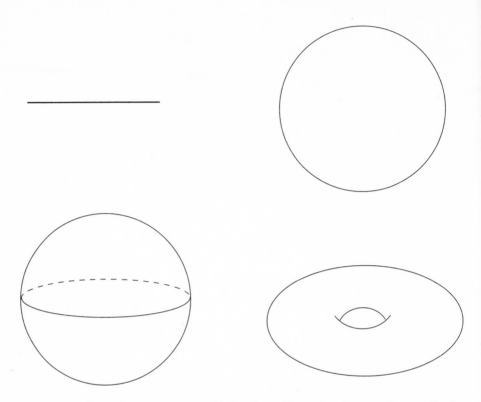

1.1—In topology, there are just two kinds of one-dimensional spaces that are fundamentally distinct from each other: a line and a circle. You can make a circle into all kinds of loops, but you can't turn it into a line without cutting it.

Two-dimensional surfaces, which are *orientable*—meaning they have two sides like a beach ball, rather than just one side like a Möbius strip—can be classified by their *genus*, which can be thought of, in simple terms, as the number of holes. A sphere of genus 0, which has no holes, is therefore fundamentally distinct from a donut of genus 1, which has one hole. As with the circle and line, you can't transform a sphere into a donut without cutting a hole through the middle of it.

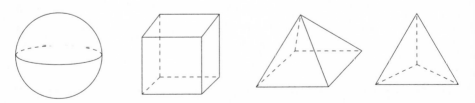

1.2—In topology, a sphere, cube, and tetrahedron—among other shapes—are all considered equivalent because each can be fashioned from the other by bending, stretching, or pushing, without their having to be torn or cut.

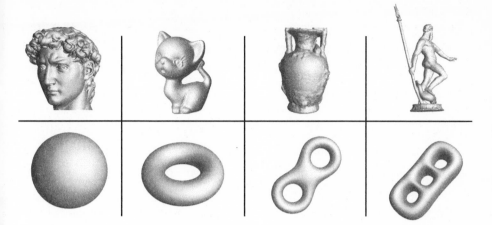

1.3—Surfaces of genus 0, 1, 2, and 3; the term *genus* refers to the number of holes.

by the number of holes they have, which is also known as their *genus*. Objects that look quite different in two dimensions are considered topologically identical if they have the same genus.

The point made earlier about there being just two possible two-dimensional shapes—a donut or a sphere—is only true if we restrict ourselves to orientable surfaces, and those are the surfaces we'll generally be referring to in this book. A beach ball, for example, has two sides, an inside and an outside, and the same goes for a tire's inner tube. There are, however, more complicated cases—single-sided, "nonorientable" surfaces such as the Klein bottle and Möbius strip—where the foregoing is not true.

In dimensions three and beyond, the number of possible shapes widens dramatically. In contemplating higher-dimensional spaces, we must allow for movements in directions we can't readily imagine. We're not talking about heading somewhere in between north and west like northwest or even "North by Northwest." We're talking about heading off the grid altogether, following arrows in a coordinate system that has yet to be drawn.

One of the first big breakthroughs in charting higher-dimensional space came courtesy of René Descartes, the seventeenth-century French mathematician, philosopher, scientist, and writer, though his work in geometry stands foremost for me. Among other contributions, Descartes taught us that thinking in terms of coordinates rather than pictures can be extremely productive. The labeling system he invented, which is now called the Cartesian coordinate system, united algebra and geometry. In a narrow sense, Descartes showed that by drawing x, y, and z axes that intersect in a point and are all perpendicular to each other, one can pin down any spot in three-dimensional space with just three numbers—the x, y, and z coordinates. But his contribution was much broader than that, as he vastly enlarged the scope of geometry and did so in one brilliant stroke. For with his coordinate system in hand, it became possible to use algebraic equations to describe complex, higher-dimensional geometric figures that are not readily visualized.

Using this approach, you can think about any dimension you want—not just (x, y, z) but (a, b, c, d, e, f) or $(j, k, l, m, n, o, p, q, r, s)$—the dimension of a given space being the number of coordinates you need to determine the location of a given point. Armed with this system, one could contemplate higher-dimensional spaces of any order—and do various calculations concerning them—without having to worry about trying to draw them.

The great German mathematician Georg Friedrich Bernhard Riemann took off with this idea two centuries later and carried it far. In the 1850s, while working on the geometry of curved (non-Euclidean) spaces—a subject that will be taken up in the next chapter—Riemann realized that these spaces were not restricted in terms of the number of dimensions. He showed how distance, curvature, and other properties in such spaces could be precisely computed. And in an 1854 inaugural lecture in which he presented principles that have since come to be known as Riemannian geometry, he speculated on the dimensionality and geometry of the universe itself. While still in his twenties, Riemann also began work on a mathematical theory that attempted to tie together electricity, magnetism, light, and gravity—thereby anticipating a task that continues to occupy scientists to this day.

Although Riemann helped free up space from the limitations of Euclidean flatness and three dimensions, physicists did not do much with that idea for decades. Their lack of interest may have stemmed from the absence of experimental evidence to suggest that space was curved or that any dimensions beyond three existed. What it came down to was that Riemann's advanced mathematics had simply outpaced the physics of his era, and it would take time—another fifty years or so—for the physicists, or at least one physicist in particular, to catch up. The one who did was Albert Einstein.

In developing his special theory of relativity—which was first presented in 1905 and further advanced in the years after, culminating in the general theory of relativity—Einstein drew on an idea that was also being explored by the German mathematician Hermann Minkowski, namely, that time is inextricably intertwined with the three dimensions of space, forming a new geometrical construct known as *spacetime.* In an unexpected turn, time itself came to be seen as the fourth dimension that Riemann had incorporated decades before in his elegant equations.

Curiously, the British writer H. G. Wells had anticipated this same outcome ten years earlier in his novel *The Time Machine.* As explained by the Time Traveller, the main character of that book, "there are really four dimensions, three which we call the three planes of Space, and a fourth, Time. There is, however, a tendency to draw an unreal distinction between the former three dimensions and the latter."[4]

Minkowski said pretty much the same thing in a 1908 speech—except that in this case, he had the mathematics to back up such an outrageous claim: "Henceforth space by itself, and time by itself, are doomed to fade away into mere shadows, and only a kind of union of the two will preserve an independent reality."[5] The rationale behind the marriage of these two concepts—if, indeed, marriages ever have a rationale—is that an object moves not only through space but through time as well. It thus takes four coordinates, three of space and one of time, to describe an event in four-dimensional spacetime (x, y, z, t).

Although the idea may seem slightly intimidating, it can be expressed in extremely mundane terms. Suppose you make plans to meet somebody at a shopping mall. You note the location of the building—say it's at the corner of First Street and Second Avenue—and decide to meet on the third floor. That takes care of your x, y, and z coordinates. Now all that remains is to fix the fourth coordinate and settle on the time. With those four pieces of information specified, your assignation is all set, barring any unforeseen circumstances that might intervene. But if you want to put it in Einstein's terms, you shouldn't look at it as setting the exact place for this little get-together, while separately agreeing on the time. What you're really spelling out is the location of this event in spacetime itself.

So in a single bound, early in the twentieth century, our conception of space grew from the cozy three-dimensional nook that had nurtured humankind since

1.4—As we don't know how to draw a picture in four dimensions, this is a rather crude, conceptual rendering of four-dimensional *spacetime*. The basic idea of spacetime is that the three spatial dimensions of our world (represented here by the *x-y-z* coordinate axis) have essentially the same status as a fourth dimension—that being time. We think of time as a *continuous variable* that's always changing, and the figure shows snapshots of the coordinate axis at various moments frozen in time: t_1, t_2, t_3, and so forth. In this way, we're trying to show that there are four dimensions overall: three of space plus the additional one labeled by time.

antiquity to the more esoteric realm of four-dimensional spacetime. This conception of spacetime formed the bedrock on which Einstein's theory of gravity, the general theory of relativity, was soon built. But is that the end of the line, as we asked once before? Does the buck stop there, at four dimensions, or can our notion of spacetime grow further still? In 1919, a possible answer to that question arrived unexpectedly in the form of a manuscript sent to Einstein for review by a then-unknown German mathematician, Theodor Kaluza.

In Einstein's theory, it takes ten numbers—or ten *fields*—to precisely describe the workings of gravity in four dimensions. The force can be represented most succinctly by taking those ten numbers and arranging them in a four-by-four matrix more formally known as a *metric tensor*—a square table of numbers that serves as a higher-dimensional analogue of a ruler. In this case, the metric has sixteen entries in all, only ten of which are independent. Six of the numbers repeat because gravity, along with the other fundamental forces, is inherently symmetrical.

In his paper, Kaluza had basically taken Einstein's general theory of relativity and added an extra dimension to it by expanding the four-by-four matrix to a five-by-five one. By expanding spacetime to the fifth dimension, Kaluza was able to take the two forces known at the time, gravity and electromagnetism, and combine them into a single, unified force. To an observer in the five-dimensional world that Kaluza envisioned, those forces would be one and the same, which is what we mean by unification. But in a four-dimensional world, the two can't go together; they would appear to be wholly autonomous. You could say that's the case simply because both forces do not fit into the same four-by-four matrix. The additional dimension, however, provides enough extra elbow room for both of them to occupy the same matrix and hence be part of the same, more all-encompassing force.

I may get in trouble for saying this, but I believe that only a mathematician would have been bold enough to think that higher-dimensional space would afford us special insight into phenomena that we've so far only managed to observe in a lower-dimensional setting. I say that because mathematicians deal with extra dimensions all the time. We're so comfortable with that notion, we don't give it a moment's thought. We could probably manipulate extra dimensions in our sleep without interfering with the REM phase.

Even if I think that only a mathematician would have made such a leap, in this case, remarkably, it was a mathematician building on the work of a physicist, Einstein. (And another physicist, Oskar Klein, whom we'll be discussing momentarily, soon built on that mathematician's work.) That's why I like to position myself at the interface between these two fields, math and physics, where a lot of interesting cross-pollination occurs. I've hovered around that fertile zone since the 1970s and have managed to get wind of many intriguing developments as a result.

But returning to Kaluza's provocative idea, people at the time were puzzled by a question that is equally valid today. And it's one that Kaluza undoubtedly grappled with as well: If there really is a fifth dimension—an entirely new direction to move at every point in our familiar four-dimensional world—how come nobody has seen it?

The obvious explanation is that this dimension is awfully small. But where would it be? One way to get a sense of that is to imagine our four-dimensional universe as a single line that extends endlessly in both directions. The idea here is that the three spatial dimensions are either extremely big or infinitely large. We'll also assume that time, too, can be mapped onto an infinite line— an assumption that may be questionable. At any rate, each point w on what we've thought of as a line actually represents a particular point (x, y, z, t) in four-dimensional spacetime.

In geometry, lines are normally just length, having no breadth whatsoever. But we're going to allow for the possibility that this line, when looked at with an exceedingly powerful magnifying glass, actually has some thickness. When

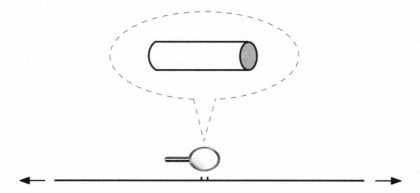

1.5—Let's picture our infinite, four-dimensional spacetime as a line that extends endlessly in both directions. A line, by definition, has no thickness. But if we were to look at that line with a magnifying glass, as suggested in the Kaluza-Klein approach, we might discover that the line has some thickness after all—that it is, in fact, harboring an extra dimension whose size is set by the diameter of the circle hidden within.

seen in this light, our line is not really a line at all but rather an extremely slender cylinder or "garden hose," to choose the standard metaphor. Now, if we slice our hose at each point w, the cross-section of that cut will be a tiny circle, which, as we've said, is a one-dimensional curve. The circle thus represents the extra, fifth dimension that is "attached," in a sense, to every single point in four-dimensional spacetime.

A dimension with that characteristic—being curled up in a tiny circle—is technically referred to as being compact. The word *compact* has a fairly intuitive meaning: Physicists sometimes say that a compact object or space is something you could fit into the trunk of your car. But there's a more precise meaning as well: If you travel in one direction long enough, it is possible to return to the same spot. Kaluza's five-dimensional spacetime includes both extended (infinite) and compact (finite) dimensions.

But if that picture were correct, why wouldn't we notice ourselves moving around in circles in this fifth dimension? The answer to that question came in 1926 from Oskar Klein, the Swedish physicist who carried Kaluza's idea a step further. Drawing on quantum theory, Klein actually calculated the size of the compact dimension, arriving at a number that was tiny indeed—close to the so-called Planck length, which is about as small as you can get—around 10^{-30} cm in circumference.[6] And that is how a fifth dimension could exist, yet remain forever unobservable. There is no foreseeable means by which we could see this minuscule dimension; nor could we detect movements within it.

Kaluza-Klein theory, as the work is now known, was truly remarkable, showing the potential of extra dimensions to demystify the secrets of nature. After sitting on Kaluza's original paper for more than two years, Einstein wrote back saying he liked the idea "enormously."[7] In fact, he liked the idea enough to pursue Kaluza-Klein-inspired approaches (sometimes in collaboration with the physicist Peter Bergmann) off and on over the next twenty years.

But ultimately, Kaluza-Klein theory was discarded. In part this was because it predicted a particle that has never been shown to exist, and in part because attempts to use the theory to compute the ratio of an electron's mass to its charge went badly awry. Furthermore, Kaluza and Klein—as well as Einstein after them—were trying to unify only electromagnetism and gravity, as they didn't know about the weak and strong forces, which were not well understood until the latter half of the twentieth century. So their efforts to unify all the forces were doomed to failure because the deck they were playing with was still missing a couple of important cards. But perhaps the biggest reason that Kaluza-Klein theory was cast aside had to do with timing: It was introduced just as the quantum revolution was beginning to take hold.

Whereas Kaluza and Klein put geometry at the center of their physical model, quantum theory is not only an ungeometric approach, but also one that directly conflicts with conventional geometry (which is the subject of Chapter 14). In the wake of the upheaval that ensued as quantum theory swept over physics in the twentieth century, and the amazingly productive period that followed, it took almost fifty years for the idea of new dimensions to be taken seriously again.

General relativity, the geometry-based theory that encapsulates our current understanding of gravity, has also held up extraordinarily well since Einstein introduced it in 1915, passing every experimental test it has faced. And quantum theory beautifully describes three of the known forces: the electromagnetic, weak, and strong. Indeed, it is the most precise theory we have, and "probably the most accurately tested theory in the history of human thought," as Harvard physicist Andrew Strominger has claimed.[8] Predictions of the behavior of an electron in the presence of an electric field, for example, agree with measurements to ten decimal points.

Unfortunately, these two very robust theories are totally incompatible. If you try to mix general relativity with quantum mechanics, the combination can create a horrific mess. The trouble arises from the quantum world, where things are always moving or fluctuating: The smaller the scale, the bigger those fluctuations get. The result is that on the tiniest scales, the turbulent, ever-changing picture afforded by quantum mechanics is totally at odds with the smooth geometric picture of spacetime upon which the general theory of relativity rests.

Everything in quantum mechanics is based on probabilities, and when general relativity is thrown into the quantum model, calculations often lead to infinite probabilities. When infinities pop up as a matter of course, that's a tipoff that something is amiss in your calculations. It's hardly an ideal state of affairs when your two most successful theories—one describing large objects such as planets and galaxies, and the other describing tiny objects such as electrons and quarks—combine to give you gibberish. Keeping them separate is not a satisfactory solution, either, because there are places, such as black holes, where the very large and very small converge, and neither theory on its own can make sense of them. "There shouldn't be laws of physics," Strominger maintains. "There should be just one law and it ought to be the nicest law around."[9]

Such a sentiment—that the universe can and should be describable by a "unified field theory" that weaves all the forces of nature into a seamless whole—is both aesthetically appealing and tied to the notion that our universe started with an intensely hot Big Bang. At that time, all the forces would have been at the same unimaginably high energy level and would therefore act as if they were a single force. Kaluza and Klein, as well as Einstein, failed to build a theory that could capture everything we knew about physics. But now that we have more pieces of the puzzle in hand, and hopefully all the big pieces, the question remains: Might we try again and this time succeed where the great Einstein failed?

That is the promise of string theory, an intriguing tough unproven approach to unification that replaces the pointlike objects of particle physics with extended (though still quite tiny) objects called *strings*. Like the Kaluza-Klein approaches that preceded it, string theory assumes that extra dimensions beyond our everyday three (or four) are required to combine the forces of nature. Most

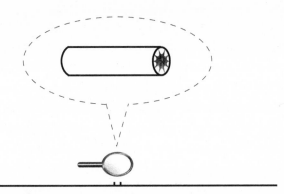

1.6—String theory takes the old Kaluza-Klein idea of one hidden "extra" dimension and expands it considerably. If we were to take a detailed look at our four-dimensional spacetime, as depicted by the line in this figure, we'd see it's actually harboring six extra dimensions, curled up in an intricate though minuscule geometric space known as a Calabi-Yau manifold. (More will be said about these spaces later, as they are the principal subject of this book.) No matter where you slice this line, you will find a hidden Calabi-Yau, and all the Calabi-Yau manifolds exposed in this fashion would be identical.

versions of the theory hold that, altogether, ten or eleven dimensions (including time) are needed to achieve this grand synthesis.

But it isn't just a matter of throwing in some extra dimensions and hoping for the best. These dimensions must conform to a particular size and shape—the right one being an as-of-yet unsettled question—for the theory to have a chance of working. Geometry, in other words, plays a special role in string theory, and many adherents would argue that the geometry of the theory's extra dimensions may largely determine the kind of universe we live in, dictating the properties of all the physical forces and particles we see in nature, and even those we don't see. (Because of our focus on so-called Calabi-Yau manifolds and their potential role in providing the geometry for the universe's hidden dimensions—assuming such dimension exist—this book will not explore loop quantum gravity, an alternative to string theory that does not involve extra dimensions and therefore does not rely on a compact, "internal" geometry such as Calabi-Yau.)

We will explore string theory in depth, starting in Chapter 6. But before plunging into the complex mathematics that underlies that theory, it might be useful to establish a firmer grounding in geometry.

This subject will be explored in depth, starting in Chapter 6. But before plunging into the complex mathematics that underlies that theory, it might be useful to establish a firmer grounding in geometry. (In my admittedly biased experience, that is always a useful tactic.) So we're going to back up a few steps from the twentieth and twenty-first centuries to review the history of this venerable field and thereby gain a sense of its place in the order of things.

And as for that place, geometry has always struck me as a kind of express lane to the truth—the most direct route, you might say, of getting from where we are to where we want to be. That's not surprising, given that a fair chunk of geometry is devoted to the latter problem—finding the distance between two points. Bear with me if the path from the mathematics of ancient Greece to the intricacies of string theory seems a bit convoluted, or tangled, at times. Sometimes, the shortest path is not a straight line, as we shall see.

Two

GEOMETRY IN THE
NATURAL ORDER

Over most of the last two and a half thousand years in the European or Western tradition, geometry has been studied because it has been held to be the most exquisite, perfect, paradigmatic truth available to us outside divine revelation. Studying geometry reveals, in some way, the deepest true essence of the physical world.
—PIERS BURSILL-HALL,
"WHY DO WE STUDY GEOMETRY?"

What is geometry? Many think of it as simply a course they took in high school—a collection of techniques for measuring the angles between lines, calculating the area of triangles, circles, and rectangles, and perhaps establishing some measure of equivalence between disparate objects. Even with such a limited definition, there's no doubt that geometry is a useful tool—one that architects, for instance, make use of every day. Geometry is these things, of course, and much, much more, for it actually concerns architecture in the broadest sense of the term, from the very smallest scales to the largest. And for someone like me, obsessed with understanding the size, shape, curvature, and structure of space, it is the essential tool.

The word *geometry*, which comes from *geo* ("earth") and *metry* ("measure"), originally meant "measuring the earth." But we now put it in more general terms to mean measuring space, where space itself is not a particularly well-defined concept. As Georg Friedrich Bernhard Riemann once said, "geometry presupposes the concept of space, as well as assuming the basic principles for constructions in space," while giving "only nominal definitions of these things."[1]

Odd as it may sound, we find it useful to keep the concept of space rather fuzzy because it can imply many things for which we have no other terms. So there's some convenience to that vagueness. For example, when we contemplate how many dimensions there are in space or ponder the shape of space as a whole, we might just as well be referring to the entire universe. A space could also be more narrowly defined to mean a simple geometric construct such as a point, line, plane, sphere, or donut—the sorts of figures a grade school student might draw—or it could be more abstract, more complex, and immensely more difficult to picture.

Suppose, for instance, you have a bunch of points spread out in some complicated, haphazard arrangement with absolutely no way of determining the distance between them. As far as mathematicians are concerned, that space has no geometry; it's just a random assortment of points. But once you put in some kind of measurement function, technically called a metric, which tells you how to compute the distance between any two points, then your space has suddenly become navigable. It has a well-defined geometry. The metric for a space, in other words, gives you all the information you need to divine its shape. Armed with that measurement capability, you can now determine its flatness to great precision, as well as its deviation from flatness, or curvature, which is the thing I find most interesting of all.

Lest one conclude that geometry is little more than a well-calibrated ruler—and this is no knock against the ruler, which happens to be a technology I admire—geometry is one of the main avenues available to us for probing the universe. Physics and cosmology have been, almost by definition, absolutely crucial for making sense of the universe. Geometry's role in all this may be less obvious, but it is equally vital. I would go so far as to say that geometry not only deserves a place at the table alongside physics and cosmology, but in many ways it *is* the table.

For you see, this entire cosmic drama—a complex dance of particles, atoms, stars, and other entities, constantly shifting, moving, interacting—is played out on a stage, inside a "space," if you will, and it can never be truly understood without grasping the detailed features of that space. More than just a passive backdrop, space actually imbues its constituents with intrinsically vital properties. In fact, as we currently view things, matter or particles sitting (or moving) in a space are actually part of that space or, more precisely, spacetime. Geometry can impose constraints on spacetime and on physical systems in general—constraints that we can deduce purely from the principles of mathematics and logic.

Consider the climate of the earth. Though it may not be obvious, the climate can be profoundly influenced by geometry—in this case by the essentially spherical shape of our host planet. If we resided on a two-dimensional torus, or donut, instead, life—as well as our climate—would be substantially different. On a sphere, winds can't blow in the same direction (say, east), nor can the ocean's waters all flow in the same direction (as mentioned in the final chapter). There will inevitably be places—such as at the north and south poles—where wind or current direction no longer points east, where the whole notion of "east" disappears, and all movement grinds to a halt. This is not the case on the surface of a single-holed donut, where there are no such impasses and everything can flow in the same direction without ever hitting a snag. (That difference would surely affect global circulation patterns, but if you want to know the exact climatological implications—and get a seasonal comparison between spherical and toroidal living—you'd better ask a meteorologist.)

The scope of geometry is even broader still. In concert with Einstein's theory of general relativity, for example, geometry has shown that the mass and energy of the universe are positive and hence that spacetime, the four-dimensional realm we inhabit, is stable. The principles of geometry also tell us that somewhere in the universe, there must be strange places known as *singularities*—thought to lie, for instance, in the center of black holes—where densities approach infinity and known physics breaks down. In string theory, to take another example, the geometry of weird six-dimensional spaces called Calabi-Yau manifolds—where much important physics supposedly takes place—may explain why we have the assortment of elementary particles we do, dictating not only their masses but the forces between them. The study of these higher-dimensional spaces, moreover, has offered possible insights into why gravity appears to be so much weaker than the other forces of nature, while also providing clues about the mechanisms behind the inflationary expansion of the early universe and the dark energy that's now driving the cosmos apart.

So it's not just idle boasting when I say that geometry has been an invaluable tool for unlocking the universe's secrets, right up there with physics and cosmology. Moreover, with the advances in mathematics that we'll be describing here, along with progress in observational cosmology and the advent of string theory, which is attempting a grand synthesis that has never before been realized, all three of these disciplines seem to be converging at the same time. As a result, human knowledge now stands poised and raring to go, on the threshold of remarkable insights, with geometry, in many ways, leading the charge.

It's important to bear in mind that whatever we do in geometry, and wherever we go, we never start from scratch. We're always drawing on what came before—be it conjectures (which are unproven hypotheses), proofs, theorems, or axioms—building from a foundation that, in many cases, was laid down thousands of years before. In that sense, geometry, along with other sciences, is like an elaborate construction project. The foundation is laid down first, and if it's built correctly—placed on firm ground, so to speak—it will last, as will the structures built on top of it, provided they too are engineered according to sound principles.

That, in essence, is the beauty and strength of my elected calling. When it comes to mathematics, we always expect a completely true statement. A mathematical theorem is an exact statement that will remain an eternal truth and is independent of space, time, people's opinions, and authority. This quality sets it apart from empirical science, where you do experiments and, if a result looks good, you accept it after a satisfactory trial period. But the results are always subject to change; you can never expect a finding to be 100 percent, unalterably true.

Of course, we often come across broader and better versions of a mathematical theorem that don't invalidate the original. The foundation of the building is still sound, to continue our construction analogy; we've kept it intact while doing some expansion and remodeling. Sometimes we have to go farther than just remodeling, perhaps even "gutting" the interior and starting afresh. Even though the old theorems are still true, we may need entirely new developments, and a fresh batch of materials, to create the overall picture we seek to achieve.

The most important theorems are usually checked and rechecked many times and in many ways, leaving essentially no chance for error. There may be problems, however, in obscure theorems that have not received such close scrutiny. When a mistake is uncovered, a room of the building—or perhaps a whole wing—might have to be torn down and reassembled. Meanwhile, the rest of the structure—a sturdy edifice that has stood the test of time—remains unaffected.

One of the great architects of geometry is Pythagoras, with the well-known formula attributed to him being one of the sturdiest edifices ever erected in mathematics. The Pythagorean theorem, as it's called, states that for a right triangle (a triangle, that is, with one 90-degree angle), the length of the longest side (or hypotenuse) squared equals the sum of the squares of the two shorter sides. Or as schoolchildren, former and present, may recall: $a^2 + b^2 = c^2$. It's a simple, yet very powerful statement that amazingly is as relevant now as it was when formulated some 2,500 years ago. The theorem is not just restricted to elementary school mathematics. Indeed, I use the theorem just about every day, almost without thinking about it, because it has become so central and so ingrained.

To my mind, the Pythagorean theorem is the most important statement in geometry, as crucial for advanced, higher-dimensional math—such as for working out distances in Calabi-Yau spaces and solving Einstein's equations of motion—as it is for calculations on a two-dimensional plane (such as the sheet of a homework assignment) or in a three-dimensional grade school classroom. The theorem's importance stems from the fact that we can use it to figure out distances between two points in spaces of *any* dimension. And, as I said at the outset of this chapter, geometry has a lot to do with distance, which is why this formula is central to practically everything we do.

I find the theorem, moreover, to be extremely beautiful, although beauty, admittedly, is in the eye of the beholder. We tend to like things that we know—things that have become so familiar, so natural, that we take them for granted, just like the rising and setting of the sun. Then there's the great economy of it all, just three simple letters raised to the second power, $a^2 + b^2 = c^2$, almost as

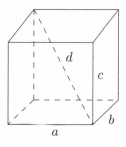

2.1—The Pythagorean theorem is often pictured in two dimensions in terms of a right triangle with the sum of the lengths of the sides squared equaling the length of the hypotenuse squared: $a^2 + b^2 = c^2$. But, as shown here, the theorem also works in three dimensions ($a^2 + b^2 + c^2 = d^2$) and higher.

$$a^2 + b^2 + c^2 = d^2$$

terse as other famous laws like $F = ma$ or $E = mc^2$. For me, the beauty stems from the elegance of a simple statement that sits so comfortably within nature.

In addition to the theorem itself, which is without a doubt a cornerstone of geometry, equally important is the fact that it was *proved* to be true and appears to be the first documented proof in all of mathematics. Egyptian and Babylonian mathematicians had used the relation between the sides of a right triangle and its hypotenuse long before Pythagoras was even born. But neither the Egyptians nor the Babylonians ever *proved* the idea, nor do they seem to have considered the abstract notion of a proof. This, according to the mathematician E. T. Bell, was where Pythagoras made his greatest contribution:

> Before him, geometry had been largely a collection of rules of thumb empirically arrived at without any clear indication of the mutual connections of the rules. Proof is now so commonly taken for granted as the very spirit of mathematics that we find it difficult to imagine the primitive thing which must have preceded mathematical reasoning.[2]

Well, maybe Pythagoras is responsible for the proof, though you might have noticed I said the theorem was "attributed" to him, as if there were some doubt as to the authorship. There is. Pythagoras was a cultlike figure, and many of the contributions of his math-crazed disciples, the so-called Pythagoreans, were attributed to him retroactively. So it's possible that the proof of the Pythagorean theorem originated with one of his followers a generation or two later. Odds are we'll never know: Pythagoras lived primarily in the sixth century B.C. and left behind little, if anything, in the way of written records.

Fortunately, that's not the case with Euclid, one of the most famous geometers of all time and the man most responsible for turning geometry into a precise, rigorous discipline. In stark contrast to Pythagoras, Euclid left behind reams of documents, the most illustrious of them being *The Elements* (published around 300 B.C.)—a thirteen-volume treatise, of which eight volumes are devoted to the geometry of two and three dimensions. *The Elements* has been called one of the most influential textbooks ever penned, "a work of beauty whose impact rivaled that of the Bible."[3]

In his celebrated tome, Euclid laid the groundwork not just for geometry but also for all of mathematics, which depends inextricably on a manner of reasoning we now call Euclidean: Starting with clearly defined terms and a set of explicitly stated axioms, or postulates (the two words being synonymous), one can then employ cool logic to prove theorems that, in turn, can be used to prove other assertions. Euclid did just that, proving more than four hundred theorems in all, thereby encapsulating virtually all of the geometric knowledge of his era.

Stanford mathematician Robert Osserman explained the enduring appeal of Euclid's manifesto this way: "First there is the sense of certainty—that in a world full of irrational beliefs and shaky speculations, the statements found in *The Elements* were proven true beyond a shadow of a doubt." Edna St. Vincent Millay expressed similar appreciation in her poem "Euclid Alone Has Looked on Beauty Bare."[4]

The next crucial contribution for the purposes of our narrative—with no slight intended to the many worthy mathematicians whose contributions are being overlooked—comes from René Descartes. As discussed in the previous chapter, Descartes greatly enlarged the scope of geometry by introducing a coordinate system that enabled mathematicians to think about spaces of any dimension and to bring algebra to bear on geometric problems. Before he rewrote the field, geometry was pretty much limited to straight lines, circles, and *conic sections*—the shapes and curves, such as parabolas and hyperbolas, that you get by slicing an infinitely long cone at different angles. With a coordinate system in place, we could suddenly describe very complicated figures, which we otherwise would not know how to draw, by means of equations. Take the equation $x^n + y^n = 1$, for example. Using Cartesian coordinates, one can solve the equation and trace out a curve. Before we had a coordinate system, we didn't know how to draw such a figure. Where we had been stuck before, Descartes offered us a way to proceed.

And that way became even stronger when, about fifty years after Descartes shared his ideas on analytic geometry, Isaac Newton and Gottfried Leibniz invented calculus. Over the coming decades and centuries, the tools of calculus were eventually incorporated in geometry by mathematicians like Leonhard Euler, Joseph Lagrange, Gaspard Monge, and perhaps most notably Carl Friedrich Gauss, under whose guidance the field of *differential geometry* finally came of age in the 1820s. The approach used Descartes' system of coordinates to describe surfaces that could then be analyzed in detail by applying the techniques of differential calculus—differentiation being a technique for finding the slope of any smooth curve.

The development of differential geometry, which has continued to evolve since Gauss's era, was a major achievement. With the tools of calculus in their grasp, geometers could characterize the properties of curves and surfaces with far greater clarity than had been possible before. Geometers obtain such information through differentiation or, equivalently, by taking *derivatives*, which measure how functions change in response to changing inputs. One can think of a function as an algorithm or formula that takes a number as an input and produces a number as an output: $y = x^2$ is an example, where values for x go in and values for y come out. A function is consistent: If you feed it the same input, it will always produce the same output; if you put 2 in our example, you will always get 4. A derivative is how we describe the changes in output given incremental changes in input; the value of the derivative reflects the sensitivity of the output to slight changes in the input.

The derivative is not just some abstract notion; it's an actual number that can be computed and tells us the slope of a curve, or of a surface, at a given point. In the above example, for instance, we can determine the derivative at a point $(x = 2)$ on our function, which in this case happens to be a parabola. If we move a little bit away from that point to, say, $x = 2.001$, what happens to the output, y? Here, y (if computed to three decimal points) turns out to be 4.004. The derivative here is the ratio of the change in output (0.004) to the change in input (0.001), which is just 4. And that is, in fact, the exact derivative of this function at $x = 2$, which is another way of saying it's the slope of the curve (a parabola) there, too.

The calculations, of course, can get much more involved than the foregoing when we pick more complicated functions and move into higher dimensions. But returning, for a moment, to the same example, we obtained the derivative of $y = x^2$ from the *ratio* of the change in y to the change in x because the derivative of this function tells us its *slope,* or steepness, at a given point—with the slope being a direct measure of how y changes with respect to x.

To picture this another way, let's consider a ball on a surface. If we nudge it to the side a tiny bit, how will that affect its height? If the surface is more or less flat, there will be little variation in height. But if the ball is on the edge of a steep grade, the change in height is more substantial. Derivatives can thus reveal the slope of the surface in the immediate vicinity of the ball.

Of course, there's no reason to limit ourselves to just a single spot on the surface. By taking derivatives that reveal variations in the geometry (or shape) at different points on the surface, we can calculate the precise curvature of the object as a whole. Although the slope at any given point provides local information regarding only the "neighborhood" around that point, we can pool the information gathered at different points to deduce a general function that describes the slope of the object at any point. Then, by means of *integration*, which is a way of adding and averaging in calculus, we can deduce the function that describes the object as a whole. In so doing, one can learn about the structure of the entire object. This is, in fact, the central idea of differential geometry— namely, that you can obtain a global picture of an entire surface, or manifold, strictly from local information, drawn from derivatives, that reveals the geometry (or metric) at each point.

Gauss made many other noteworthy contributions in math and physics in addition to his work on differential geometry. Perhaps the most significant contribution for our purposes relates to his startling proposition that objects within a space aren't the only things that can be curved; space itself can be curved. Gauss's view directly challenged the Euclidean concept of flat space—a notion that applied not only to the intuitively flat two-dimensional plane but also to three-dimensional space, where flatness means (among other things) that on

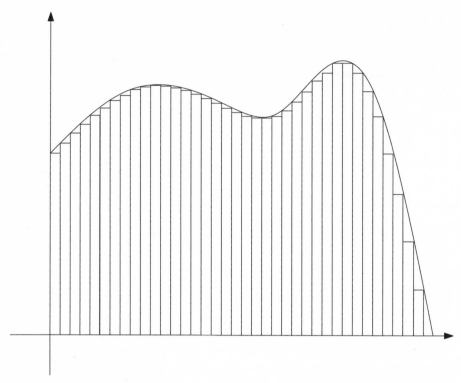

2.2—One can compute the area bounded by a curve by means of a calculus technique, *integration*, which divides the bounded regions into infinitesimally thin rectangles and adds up the area of all the rectangles. As the rectangles become narrower and narrower, the approximation gets better and better. Taken to the limit of the infinitesimally small, the approximation becomes as good as you can get.

very large scales, parallel lines never cross and the sum of the angles of a triangle always add up to 180 degrees.

These principles, which are essential features of Euclidean geometry, do not hold in curved spaces. Take a spherical space like the surface of a globe. When viewed from the equator, the longitudinal lines appear to be parallel because they are both perpendicular to the equator. But if you follow them in either direction, they eventually converge at the north and south poles. That doesn't happen in (flat) Euclidean space—such as on a Mercator projection map—where two lines that are perpendicular to the same line are truly parallel and never intersect.

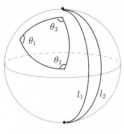 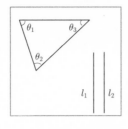 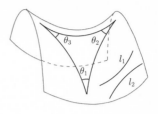

$\theta_1 + \theta_2 + \theta_3 > 180°$ $\theta_1 + \theta_2 + \theta_3 = 180°$ $\theta_1 + \theta_2 + \theta_3 < 180°$

Spherical (positive curvature)   Euclidean (zero curvature)   Hyperbolic (negative curvature)

2.3—On a surface with positive curvature such as a sphere, the sum of the angles of a triangle is greater than 180 degrees, and lines that appear to be parallel (such as longitudinal lines) can intersect (at the north and south poles, for instance). On a flat planar surface (of zero curvature), which is the principal setting of Euclidean geometry, the sum of the angles of a triangle equals 180 degrees, and parallel lines never intersect. On a surface with negative curvature such as a saddle, the sum of the angles of a triangle is less than 180 degrees, and seemingly parallel lines diverge.

In non-Euclidean space, the angles of a triangle can either add up to more than 180 degrees or to less than 180 degrees depending on how space is curved. If it is *positively* curved like a sphere, the angles of a triangle always add up to more than 180 degrees. Conversely, if the space has *negative* curvature, like the middle part of a horse's saddle, the angles of a triangle always add up to less than 180 degrees. One can obtain a measure of a space's curvature by determining the extent to which the angles of a triangle add up to more than, less than, or equal to 180 degrees.

Gauss also advanced the concept of *intrinsic geometry*—the idea that an object or surface has its own curvature (the so-called Gauss curvature) that is independent of how it may be sitting in space. Let's start, for example, with a piece of paper. You'd expect its overall curvature to be zero, and it is. But now let's roll that sheet up into a cylinder. A two-dimensional surface like this, according to Gauss, has two principal curvatures running in directions that are orthogonal to each other: One curvature relates to the circle and has the value of $1/r$, where r is the radius. If r is 1, then this curvature is 1. The other curvature runs along the length of the cylinder, which happens to be a straight line. The curvature of a straight line is obviously zero, since it doesn't curve at all. The Gauss curvature of this object—or any two-dimensional object—equals the product of those

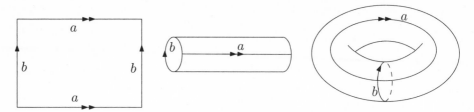

2.4—A torus, or donut-shaped, surface can be entirely "flat" (zero Gauss curvature), because it can be made, in principle, by rolling up a piece of paper into a tube or cylinder and then attaching the ends of the tube to each other.

two curvatures, which in this case is $1 \times 0 = 0$. So in terms of its intrinsic curvature, the cylinder is the same as the sheet of paper it can be constructed from: perfectly flat. The zero intrinsic curvature of the cylinder is a result of the fact that one can form it from a sheet of paper without any stretching or distortion. To put it another way, the distance measurements between any two points on the surface of a sheet—whether the sheet is flat on a table or rolled into a tube— remain unchanged. That means that the geometry, and hence the intrinsic curvature, of the sheet stays intact regardless of whether it's flat or curled up.

Similarly, if we could create a donut or torus by attaching the circular ends of a cylinder together—again doing so without any stretching or distortion— the torus would have the same intrinsic curvature as the cylinder, namely, zero. In practice, however, we cannot actually construct this so-called flat torus—at least not in two dimensions where folds or wrinkles will inevitably be introduced at the seams. But we can construct such an object (known as an abstract surface) in theory, and it holds just as much importance to mathematics as the objects we call real.

A sphere, on the other hand, is quite different from a cylinder or flat torus. Consider, for example, the curvature of a sphere of radius r. It is defined by the equation $1/r^2$ and is the same everywhere on the surface of the sphere. As a result, every direction looks the same on the surface of a sphere, whereas this is obviously not the case on a cylinder or donut. And that doesn't change, no matter how the sphere is oriented in three-dimensional space, just as a small bug living on that surface is presumably oblivious to how the surface is aligned in three-dimensional space; all it likely cares about, and experiences, is the geometry of its local, two-dimensional abode.

Gauss—in concert with Nikolai Lobachevsky and János Bolyai—made great contributions to our understanding of abstract space, particularly the two-dimensional case, though he personally admitted to some confusion in this area. And ultimately, neither Gauss nor his peers were able to liberate our conception of space entirely from the Euclidean framework. He expressed his puzzlement in an 1817 letter to the astronomer Heinrich Wilhelm Matthäus Olbers: "I am becoming more and more convinced that the necessity of our geometry cannot be proved, at least by human reason and for human reason. It may be that in the next life we shall arrive at views on the nature of space that are now inaccessible to us."[5]

Some answers came not in the "next life," as Gauss had written, but in the next generation through the efforts, and sheer brilliance, of his student Georg Friedrich Bernhard Riemann. Riemann suffered from poor health and died young, but in his forty years on this planet, he helped overturn conventional notions of geometry and, in the process, overturned our picture of the universe as well. Riemann introduced a special kind of field, a set of numbers assigned to each point in space that could reveal the distance along any path connecting two points—information that could be used, in turn, to determine the extent to which that space was curved.

Measuring space is simplest in one dimension. To measure a one-dimensional space, such as a straight line, all we need is a ruler. In a two-dimensional space, such as the floor of a grand ballroom, we'd normally take two perpendicular rulers—one called the x-axis and the other the y-axis—and work out distances between two points by creating right triangles and then using the Pythagorean theorem. Likewise, in three dimensions, we'd need three perpendicular rulers, x, y, and z.

Things get more complicated and interesting, however, in curved, non-Euclidean space, where properly labeled, perpendicular rulers are no longer available. We can rely on Riemannian geometry, instead, to calculate distances in spaces like these. The approach we'll take in computing the length of a curve, which itself is sitting on a curved manifold, will seem familiar: We break the curve down into tangent vectors of infinitesimal size and integrate over the entire curve to get the total length.

The tricky part stems from the fact that in curved space, the measurement of the individual tangent vectors can change as we move from point to point on the manifold. To handle this variability, Riemann introduced a device, known as a *metric tensor*, that provides an algorithm for computing the length of a tangent vector at each point. In two dimensions, the metric tensor is a two-by-two matrix; in *n* dimensions, the metric tensor is an *n*-by-*n* matrix. (It's worth noting that this new measurement approach, despite Riemann's great innovation, still relies heavily on the Pythagorean theorem, adapted to a non-Euclidean setting.)

A space endowed with a Riemannian metric is called a *Riemannian manifold*. Equipped with the metric, we can measure the length of any curve in a manifold of arbitrary dimension. But we're not limited to measuring the length of curves; we can also measure the area of a surface in that space, and a "surface" in this case is not limited to the usual two dimensions.

With the invention of the metric, Riemann showed how a space that was only vaguely defined could instead be granted a well-described geometry, and how curvature, rather than being an imprecise concept, could be encapsulated in a precise number associated with each point in space. And this approach, he showed, could apply to spaces of all dimensions.

Prior to Riemann, a curved object could only be studied from the "outside," like surveying a mountain range from afar or gazing at the surface of Earth from a rocket ship. Up close, everything would seem flat. Riemann showed how we could still detect the fact that we were living in a curved space, even with nothing to compare that space with.[6] This poses a huge question for physicists and astronomers: If Riemann was right, and that one space was all there is, without a bigger structure to fall back, it meant we had to readjust our picture of almost everything. It meant that on the largest scales, the universe need not be confined by the strictures of Euclid. Space was free to roam, free to curve, free to do whatever. It is for this very reason that astronomers and cosmologists are now making meticulous measurements in the hopes of finding out whether our universe is curved or not. Thanks to Riemann, we now know that we don't have to go outside our universe to make these measurements, which would be a difficult feat to pull off. Instead, we should be able to figure this out from right where we're sitting—a fact that could offer comfort to both cosmologists and couch potatoes.

These, in any event, were some of the new geometric ideas circulating when Einstein began drawing together his thoughts on gravity. Early in the twentieth century, Einstein had been struggling for the better part of a decade to combine his special theory of relativity with the principles of Newtonian gravity. He suspected that the answer may lie somewhere in geometry and turned to a friend, the geometer Marcel Grossman, for assistance. Grossman, who had previously helped Einstein get through graduate coursework that he'd found uninspiring, introduced his friend to Riemann's geometry, which was unknown to physics at the time—although the geometer did so with a warning, calling it "a terrible mess which physicists should not be involved with."[7]

Riemann's geometry was the key to solving the puzzle Einstein had been wrestling with all those years. As we saw in the previous chapter, Einstein was grappling with the idea of a curved, four-dimensional spacetime (otherwise known as our universe) that was not part of a bigger space. Fortunately for him, Riemann had already provided such a framework by defining space in exactly that way. "Einstein's genius lay in recognizing that this body of mathematics was tailor-made for implementing his new view of the gravitational force," Brian Greene contends. "He boldly declared that the mathematics of Riemann's geometry aligns perfectly with the physics of gravity."[8]

Einstein recognized not only that spacetime could be described by Riemann's geometry, but also that the geometry of spacetime would influence its physics. Whereas special relativity had already unified space and time through the notion of spacetime, Einstein's subsequent theory of general relativity unified space and time with matter and gravity. This was a conceptual breakthrough. Newtonian physics had treated space as a passive background, not an active participant in the proceedings. The breakthrough was all the more spectacular considering that there was no experimental motivation for this theory at all. The idea literally sprang from one person's head (which is not to say, of course, that it could have sprung from anyone's head).

The physicist C. N. Yang called Einstein's formulation of general relativity an act of "pure creation" that was "unique in human history . . . Einstein was not trying to seize an opportunity that had presented itself. He created the opportunity himself. And then fulfilled it on his own, through deep insight and grand design."[9]

It was a remarkable achievement that might even have surprised Einstein, who hadn't always recognized that basic physics and mathematics could be so intricately intertwined. He would conclude years later, however, that "the creative principle resides in mathematics. In a certain sense, therefore, I hold it true that pure thought can grasp reality, as the ancients dreamed."[10] Einstein's theory of gravitation was arrived at by such a process of pure thought—realized through mathematics without any prompting from the outside world.

Equipped with Riemann's metric tensor, Einstein worked out the shape and other properties—the geometry, in other words—of his newly conceived spacetime. And the resulting synthesis of geometry and physics, culminating in the famous Einstein field equation, illustrates that gravity—the force that shapes the cosmos on the largest scales—can be regarded as a kind of illusion caused by the curvature of space and time. The metric tensor of Riemannian geometry not only described the curvature of spacetime, but also described the gravitational field in Einstein's new theory. Thus, a massive body like the sun warps the fabric of spacetime in the same way that a large man deforms a trampoline. And, just as a small marble thrown onto the trampoline will spiral around the heavier man, ultimately falling into the dip he creates, the geometry of warped spacetime causes Earth to orbit the sun. Gravity, in other words, is geometry. The physicist John Wheeler once explained Einstein's picture of gravity this way: "Mass grips space by telling it how to curve; space grips mass by telling it how to move."[11]

Another example might help drive this point home: Suppose that two people start at different spots on the equator and set out at the same speed toward the north pole, moving along longitudinal lines. As time goes on, they get closer and closer to each other. They may think they are affected by some invisible force that's drawing them together. But another way to think of it is that the assumed force is really a consequence of the geometry of the earth and that there's actually no force at all. And that, in short, gives you some idea as to the force of geometry itself.

The power of that example hit me with full impact when I was a first-semester graduate student learning about general relativity for the first time. It was no secret, of course, that gravity shapes our cosmos and that gravity was, indeed, its principal architect in terms of the big picture. On smaller scales, in the confined venue of most physics apparatus, gravity is extremely feeble compared with the other forces: electromagnetic, strong, and weak. But in the grand scheme of things, gravity is pretty much all there is: It is responsible for the creation of structure in the universe, from individual stars and galaxies up to giant superclusters stretching a billion light-years across. If Einstein was right, and it all came down to geometry, then geometry, too, was a force to be reckoned with.

I was sitting in a lecture class, pondering the implications, when a series of thoughts occurred to me. I had been interested in curvature since college and sensed that in light of Einstein's insights, it may be a key to understanding the universe, as well as an avenue through which I might make my own mark someday. Differential geometry had provided tools for describing how mass moves in a curved spacetime without explaining why spacetime is curved in the first place. Einstein had taken those same tools to explain where that curvature comes from. What had been seen as two separate questions—the shape of a space under the influence of gravity and its shape under the influence of curvature—turned out to be the same problem.

Taking it a step further, the question I pondered was this: If gravity comes from mass telling space how to curve, what happens in a space that has no mass whatsoever—a space we call a vacuum? Who does the talking then? Put in other terms, does the so-called Einstein field equation for the vacuum case have a solution other than the most uninteresting one—that is, a "trivial" spacetime with no matter, no gravity, and no interaction and where absolutely nothing happens? Might there be, I mused, a "nontrivial" space that has no matter, yet whose curvature and gravity are nonzero?

I wasn't yet in any position to answer these questions. Nor did I realize that a fellow named Eugenio Calabi had posed a special case of that very question more than fifteen years before, though he had approached it from a purely mathematical standpoint and wasn't thinking about gravity or Einstein at all. The best I could do then was to marvel, open-mouthed, and wonder: "What if?"

It was a surprising question for me to ask, in many ways, especially given where I'd come from—starting on a trajectory that was as likely to have taken me to the poultry trade as it was to have led me to geometry, general relativity, and string theory.

I was born in mainland China in 1949, but my family moved within a year to Hong Kong. My father was a university professor with a modest salary and a wife and eight children to feed. Despite his taking three teaching jobs at three universities, his total earnings were meager, affording neither enough money nor food to go around. We grew up poor, without electricity or running water, taking our baths in a river nearby. Enrichment, however, came in other forms. Being a philosopher, my father inspired me to try to perceive the world through a more abstract lens. I remember as a young child overhearing the conversations he had with students and peers; I could feel the excitement of their words even though I couldn't grasp their meaning.

My father always encouraged me in mathematics, despite my not getting off to the most promising start. When I was five, I took an entrance exam for a top-notch public school but failed the mathematics part because I wrote 57 instead of 75 and 69 instead of 96—a mistake, I now tell myself, that's easier to make in Chinese than in English. As a result, I was forced to go to an inferior rural school populated by a lot of rough kids who had little patience for formal education. I had to be rough to survive, so rough that I dropped out of school for a time in my preteen years and headed a gang of youths who, like me, wandered the streets looking for trouble and, more often than not, found it.

Personal tragedy turned that around. My father died unexpectedly when I was fourteen, leaving our family not only grief-stricken but destitute, with a slew of debts to pay off and virtually no income. As I needed to earn some money to support the family, an uncle advised me to leave school and raise ducks instead. But I had a different idea: teaching mathematics to other students. Given our financial circumstances, I knew there was just one chance for me to succeed and I placed my bets on math, double or nothing. If I didn't do well, my whole future was done, leaving nothing to fall back on (other than fowl husbandry, perhaps) and no second chances. In situations like that, I've found, people tend to work harder. And though I may have my shortcomings, no one has ever accused me of being lazy.

I wasn't the best student in high school but tried to make up for that in college. While I was a reasonably good student in my first year, though by no means exceptional, things really picked up for me in the second year when Stephen Salaff, a young geometer from Berkeley, came to teach at our school, the Chinese University of Hong Kong. Through Salaff I got my first taste of what real mathematics was all about. We taught a course together on ordinary differential equations and later wrote a textbook together on that same subject. Salaff

2.5—The geometer S. S. Chern (Photo by George M. Bergman)

introduced me to Donald Sarason, a distinguished Berkeley mathematician who paved the way for me to come to the university as a graduate student after I had completed just three years of undergraduate work. Nothing I'd encountered in mathematics up to that point rivaled the bureaucratic challenges we overcame— with the help of S. S. Chern, the great Chinese geometer, also based at Berkeley— in order to secure my early admittance.

Arriving in California at the age of twenty, with the full range of mathematics lying before me, I had no idea of what direction to pursue. I was initially inclined toward operator algebra, one of the more abstract areas of algebra, owing to my vague sense that the more abstract a theory was, the better.

Although Berkeley was strong in many branches of math, it happened to be a world center—if not *the* world center—for geometry at the time, and the presence of many impressive scholars like Chern began to exert an inexorable tug on me. That, coupled with a growing recognition that geometry constituted a large, rich subject ripe with possibilities, slowly lured me into the fold.

Nevertheless, I continued to expose myself to as many subjects as possible, enrolling in six graduate courses and auditing many others on subjects including geometry, topology, differential equations, Lie groups, combinatorials, number theory, and probability theory. That kept me in the classroom from 8 A.M. to 5 P.M. every day, barely leaving time for lunch. When I wasn't in the classroom, I was in the math library, my second home, reading as many books as I could lay my hands on. As I couldn't afford to buy books when I was younger, I was now like the proverbial kid in the candy shop, literally working my way through the stacks from one end to the other. Having nothing better to do, I often stayed until closing time, regularly qualifying as the last man sitting. Confucius once said: "I have spent a whole day without eating and a whole night without sleeping in order to think, but I got nothing out of it. Thinking cannot compare with studying." While I may not have been aware of that quotation at the time, I embraced the philosophy nevertheless.

So why did geometry, of all the areas of mathematics, come to occupy center stage for me, both in my waking thoughts and in my dreams? Primarily because it struck me as the field closest to nature and therefore closest to answering the kinds of questions I cared about most. Besides, I find it helpful to look at pictures when grappling with difficult concepts, and pictures are few and far between in the more abstruse realms of algebra and number theory. Then there was this fantastic group of people doing geometry at Berkeley (including Professors Chern and Charles Morrey, as well as some young faculty members like Blaine Lawson and fellow graduate students like future Fields Medal winner William Thurston) that made me want to be part of that excitement and, hopefully, add to it.

On top of that, there was a much larger community of people, not only at other campuses, but throughout the world, and—as we've seen in this chapter, throughout history as well—who had paved the way for the fertile period I was now fortunate enough to step into. It's kind of like what Isaac Newton said about "standing on the shoulders of giants," though Newton himself is one of the foremost giants upon whose shoulders we now stand.

Around the time that I first began thinking about Einstein's general theory of relativity and the curvature of space in a vacuum, my adviser, Chern, returned from a trip to the East Coast, excited because he'd just heard from the renowned Princeton mathematician André Weil that the "Riemann hypothesis," a problem posed more than a century before, might soon be solved. The hypothesis relates to the distribution of prime numbers, which don't appear to follow any pattern. Yet Riemann proposed that the frequency of these numbers was, in fact, related to a complex function since named the Riemann zeta function. More specifically, he suggested that the frequency of prime numbers corresponded to the location of the zeros of his zeta function. Riemann's assertion has been confirmed for the location of more than a billion zeros, but it still has not been proved as a general matter.

Although this was one of the most prominent problems in all of mathematics—and if I were lucky enough to solve it, it would bring in countless job offers and seal my fame for life—I couldn't muster much enthusiasm for Chern's proposition. The Riemann hypothesis just didn't excite me, and you have to be excited if you're going to set off on an ambitious project, which had thwarted so many talented people before and would, at a minimum, take years to complete. The lack of passion for this problem would unquestionably have hurt my chances of solving it, which meant there was a real possibility that I could work on the Riemann hypothesis for years and have nothing to show for it. What's more, I like pictures too much. I like mathematical structures you can look at in some fashion, which is why I like geometry. Plus, I already knew of some areas in geometry where I could achieve some results—albeit not nearly so spectacular.

It's like going out fishing. If you're content to bring back small fish, you're likely to catch something. But if you only care about bringing back the biggest fish ever caught—a mythical creature that is the stuff of legends—you're likely to come home empty-handed. Thirty-five years later, the Riemann hypothesis remains an open problem. As we say in mathematics, there's no such thing as 90 percent proved.

So that was part of my thinking when I turned down Chern's request. But there was more to it than that. At the time, as I've said, I was already getting intrigued by general relativity, trying to sort out how many of the features of our universe emerge from the interplay of gravity, curvature, and geometry. I didn't know where this line of inquiry might take me, yet I had an inkling, nevertheless, that I was embarking on a great adventure, harnessing the powers of geometry to go after the truth.

As a child born of modest circumstances, I never had the opportunity to see much of the world. Yet my passion for geometry was piqued at an early age and grew out of desire to map a land as great as China and travel a sea without knowing the end. I've journeyed farther since then, yet geometry still serves that same purpose for me. Only now, the land has become the whole earth, and the sea, the universe. And the little straw bag I used to carry around has been replaced by a small briefcase containing a ruler, compass, and protractor.

Three

A NEW KIND OF HAMMER

Despite the rich history of geometry and the spectacular accomplishments to date, we should keep in mind that this is an evolving story rather than a static thing, with the subject constantly reinventing itself. One of the more recent transformations, which has made some contributions to string theory, is *geometric analysis*, an approach that has swept over the field only in the last few decades. The goal of this approach, broadly stated, is to exploit the powerful methods of analysis, an advanced form of differential calculus, to understand geometric phenomena and, conversely, to use geometric intuition to understand analysis. While this won't be the last transformation in geometry—with other revolutions being plotted as we speak—geometric analysis has already racked up a number of impressive successes.

My personal involvement in this area began in 1969, during my first semester of graduate studies at Berkeley. I needed a book to read during Christmas break. Rather than selecting *Portnoy's Complaint, The Godfather, The Love Machine,* or *The Andromeda Strain*—four top-selling books of that year—I opted for a less popular title, *Morse Theory*, by the American mathematician John Milnor. I was especially intrigued by Milnor's section on topology and curvature, which explored the notion that local curvature has a great influence on geometry and topology. This is a theme I've pursued ever since, because the local curvature of a surface is determined by taking the derivatives of that surface, which is another way of saying it is based on analysis. Studying how that curvature influences geometry, therefore, goes to the heart of geometric analysis.

Having no office, I practically lived in Berkeley's math library in those days. Rumor has it that the first thing I did upon arriving in the United States was visit that library, rather than, say, explore San Francisco as others might have done. While I can't remember exactly what I did, forty years hence, I have no reason to doubt the veracity of that rumor. I wandered around the library, as was my habit, reading every journal I could get my hands on. In the course of rummaging through the reference section during winter break, I came across a 1968 article by Milnor, whose book I was still reading. That article, in turn, mentioned a theorem by Alexandre Preissman that caught my interest. As I had little else to do at the time (with most people away for the holidays), I tried to see if I could prove something related to Preissman's theorem.

Preissman looked at two nontrivial loops, A and B, on a given surface. A loop is simply a curve that starts at a particular point on a surface and winds around that surface in some fashion until coming back to the same starting point. *Nontrivial* means the loops cannot be shrunk down to a point while resting on that surface. Some obstruction is preventing that, just as a loop through a donut hole cannot be shrunk indefinitely without slicing clear through the donut (in which case the loop would no longer be *on* the surface and the donut, topologically speaking, would no longer be a donut). If one were to go around loop A and, from there, immediately go around loop B, the combined path would trace out a new loop called $B \times A$. Conversely, one could just as well go around loop B first and then loop A, thus tracing out a loop called $A \times B$. Preissman proved that in a space whose curvature is everywhere negative—one that slopes inward like the inside of a saddle—the resulting loops $B \times A$ and $A \times B$ can never be smoothly deformed to the other simply by bending, stretching, or shrinking except in one special case: If a multiple of A (a loop made by going around A once or an integer number of times) can be smoothly deformed to a multiple of B, then the combined loop of $B \times A$ can be smoothly deformed to $A \times B$ and vice versa. In this single, exceptional case, loops A and B are said to be *commuting,* just as the operations of addition and multiplication are commutative ($2 + 3 = 3 + 2$, and $2 \times 3 = 3 \times 2$), whereas subtraction and division are noncommutative ($2-3 \neq 3-2$, and $2/3 \neq 3/2$).

My theorem was somewhat more general than Preissman's. It applied to a space whose curvature is not positive (that is, it can either be negative or, in some places, zero). To prove the more general case, I had to make use of some mathematics that had not previously been linked to topology or differential geometry: group theory. A group is a set of elements to which specific rules apply: There is an identity element (e.g., the number 1) and an inverse element (e.g., $1/x$ for every x). A group is *closed*, meaning that when two elements are combined through a specific operation (such as addition or multiplication), the result will also be a member of the group. Furthermore, the operations must obey the associative law—that is, $a \times (b \times c) = (a \times b) \times c$.

The elements of the group I considered (which is known as the fundamental group) consisted of loops one could draw on the surface, such as the aforementioned A and B. A space with nontrivial loops has a nontrivial fundamental group. (Conversely, if every loop can be shrunk to a point, we say the space has a trivial fundamental group.) I proved that if those two elements are commuting, if $A \times B = B \times A$, then there must be a "subsurface" of lower dimension—specifically a torus—sitting somewhere inside the surface.

In two dimensions, we can think of a torus as the "product" of two circles. We start with one circle—the one that goes around the donut hole—and imagine that each point on this circle is the center of an identical circle. When you put all those identical circles together, you get a torus. (One could similarly make a donut shape by stringing Cheerios onto a cord and tying the ends together into a tight circle.) And that's what we mean by saying the product of these two circles is a torus. In the case of my theorem, which built upon Preissman's work, those two circles are represented by the loops A and B.

Both Preissman's efforts and mine were rather technical and may seem obscure. But the important point is that both arguments show how the global topology of a surface can affect its overall geometry, not just the local geometry. That's true because the loops in this example define a fundamental group, which is a global rather than local feature of a space. To demonstrate that one loop can be continuously deformed to another, you may have to move over the entire surface, which makes it a global property of that space. This is, in fact, one of the major themes in geometry today—to see what kind of global geometric structures a given topology can support. We know, for instance, that the average

3.1—The geometer Charles Morrey (Photo by George M. Bergman)

curvature of a surface topologically equivalent to a sphere cannot be negative. Mathematicians have complied a long list of statements like that.

As far as I could tell, my proof looked OK, and when the vacation was over, I showed it to one of my instructors, Blaine Lawson, who was then a young lecturer at the university. Lawson thought it looked OK, too, and together we used some ideas from that paper to prove a different theorem on a similar subject that related curvature to topology. I was pleased to have finally contributed something to the great body of mathematics, but I didn't feel as if what I'd done was especially noteworthy. I was still searching for a way to truly make my mark.

It dawned on me that the answer may lie in a class I was taking on nonlinear partial differential equations. The professor, Charles Morrey, impressed me greatly. His course, on a subject that was anything but fashionable, was very demanding, drawing on a textbook that Morrey himself had written, which was extremely difficult to read. Before long, everyone dropped out of the course but me, with many students off protesting the bombing of Cambodia. Yet Morrey still continued his lectures, apparently putting a great deal of effort into their preparation, even with just one student in the class.

Morrey was a master of partial differential equations, and the techniques he'd developed were very deep. It's fair to say that Morrey's course laid the foundation for the rest of my mathematics career.

Differential equations pertain to just about anything that takes place or changes on infinitesimal scales, including the laws of physics. Some of the most useful and difficult of these equations, called partial differential equations, describe how something changes with respect to multiple variables. With partial differential equations, we can ask not only how something changes with respect to, say, time but also how it changes with respect to other variables, such as space (as in moving along the x-axis, y-axis, and z-axis). These equations provide a means of peering into the future and seeing how a system might evolve; without them, physics would have no predictive power.

Geometry needs differential equations, too. We use such equations to measure the curvature of objects and how it changes. And this makes geometry essential to physics. To give a simple example, the question of whether a rolling ball will accelerate—whether its velocity will change over time—is strictly governed by the curvature of the ball's trajectory. That's one reason curvature is so closely linked to physics. That's also why geometry—the "science of space" that is all about curvature—is instrumental to so many areas of physics.

The fundamental laws of physics are *local*, meaning that they can describe behavior in particular (or localized) regions but not everywhere at once. This is even true in general relativity, which attempts to describe the curvature of all spacetime. The differential equations that describe that curvature, after all, are derivatives taken at single points. And that poses a problem for physicists. "So you'd like to go from local information like curvature to figuring out the structure of the whole thing," says UCLA mathematician Robert Greene. "The question is how."[1]

Let's start by thinking about the curvature of the earth. As it's hard to measure the entire globe at once, Greene suggests the following picture instead: Imagine a dog attached by leash to a stake in the front yard. If the dog can move a little bit, it can learn something about the curvature of the tiny patch to which it's confined. In this case, we'll assume that patch has positive curvature. Imagine that every yard all over the world has a dog attached to a post, and every single

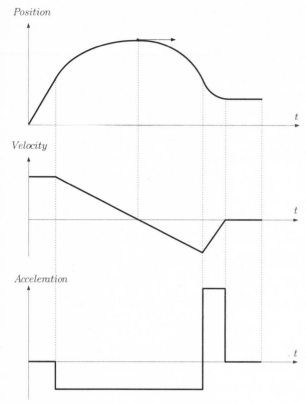

Position

Velocity

Acceleration

3.2—Think of an object moving along a particular path. The velocity, which reflects how the object's position changes with time, can be obtained by taking the derivative of the position curve. The derivative yields the slope of that curve at a given point in time, which also represents the velocity. The acceleration, which reflects how the object's velocity changes with time, can be obtained by taking the derivative of the velocity curve. The value of the acceleration at a given point in time is given by the slope of the velocity curve.

patch around those posts has positive curvature. From all these local curvature measurements, one can infer that topologically, the world must be a sphere.

There are, of course, more rigorous ways of determining the curvature of a patch than basing it on how the surface feels to a dog. For example, if a leash has length r, as a dog walks around the post with the leash fully extended, the animal will trace out a circle whose circumference is exactly $2\pi r$, assuming the space (or ground) is perfectly flat. The circumference will be somewhat smaller than $2\pi r$ on the surface of a sphere (with positive curvature) that slopes "downward" in all directions, and it will be larger than $2\pi r$ if the post is sitting in a dip or saddle point (with negative curvature) that slopes downward in some directions and upward in others. So we can determine the curvature of each patch by measuring the longest round-trip journey each dog can make and then combining the results from the various patches.

That's pretty much what differential geometers do. We measure the curvature locally, at a particular point, but attempt to use that knowledge to understand the entire space. "Curvature governs topology" is the basic slogan that we geometers embrace. And the tools we use to effect that aim are differential equations.

Geometric analysis—a relatively recent development that we'll be taking up in a moment—carries this idea further, but the general approach of including differential equations in geometry has been going on for centuries, dating back nearly to the invention of calculus itself. Leonhard Euler, the great eighteenth-century Swiss mathematician, was one of the earliest practitioners in this area. Among his many accomplishments was the use of partial differential equations to systematically study surfaces in three-dimensional space. More than two hundred years later, we are, in many ways, still following in Euler's footsteps. In fact, Euler was one of the first to look at nonlinear equations, and those equations lie at the heart of geometric analysis today.

Nonlinear equations are notoriously difficult to solve, partly because the situations they describe are more complicated. For one thing, nonlinear systems are inherently less predictable than linear systems—the weather being a familiar example—because small changes in the initial conditions can lead to wildly different results. Perhaps the best-known statement of this is the so-called butterfly effect of chaos theory, which fancifully refers to the possibility that air currents generated by the flapping of a butterfly's wings in one part of the world might conceivably cause a tornado to sprout up elsewhere.

Linear systems, by contrast, hold far fewer surprises and are consequently much easier to comprehend. An algebraic equation like $y = 2x$ is called linear because when you graph it, you literally get a straight line. Any value of x you pick automatically yields a single value for y. Doubling the value of x doubles the value of y, and vice versa. Change, when it comes, is always proportional; a small change in one parameter will never lead to a fantastically large change in another. Our world would be much easier to understand—though far less interesting—if nature worked that way. But it does not, which is why we have to talk about nonlinear equations.

We do have some methods, however, to make things a bit more manageable. For one thing, we draw on linear theory as much as we can when confronting a nonlinear problem. To analyze a wiggly (nonlinear) curve, for example, we can take the derivative of that curve (or the function defining it) to get the tangents—which are, essentially, *linear elements* or straight lines—at any point on the curve we want.

Approximating the nonlinear world with linear mathematics is a common practice, but, of course, it does nothing to change the fact that the universe is, at its heart, nonlinear. To truly make sense of it, we need techniques that merge geometry with nonlinear differential equations. That's what we mean by geometric analysis, an approach that has been useful in string theory and in recent mathematics as well.

I don't want to give the impression that geometric analysis started in the early 1970s, when I cast my lot with this approach. In mathematics, no one can claim to have started anything from scratch. The idea of geometric analysis, in a sense, dates back to the nineteenth century—to the work of the French mathematician Henri Poincaré, who had in turn built upon the efforts of Riemann and those before him.

Many of my immediate predecessors in mathematics made further critical contributions, so that by the time I came on the scene, the field of nonlinear analysis was already reaching a maturity of sorts. The theory of two-dimensional, nonlinear partial differential equations (of the sort we call elliptic and which will be discussed in Chapter 5) were worked out previously by Morrey, Aleksei Pogorelov, and others. In the 1950s, Ennio De Giorgi and John Nash paved the way for dealing with such equations of higher dimensions and indeed of *any* dimension. Additional progress on the higher-dimensional theories was subsequently made by people like Morrey and Louis Nirenberg, which meant that I had entered the field at almost the perfect time for applying these techniques to geometric problems.

Nevertheless, while the approach that my colleagues and I were taking in the 1970s was not brand new, our emphasis was rather different. To someone of Morrey's bent, partial differential equations were fundamental in their own right—a thing of beauty to be studied, rather than a means to an end. And while he was interested in geometry, he saw it primarily as a source of interesting differential equations, which is also how he viewed many areas of physics. Although both of us shared an awe for the power of these equations, our objectives were almost opposite: Instead of trying to extract nonlinear equations from geometric examples, I wanted to use those equations to solve problems in geometry that had previously been intractable.

Up to the 1970s, most geometers had shied away from nonlinear equations, but my contemporaries and I tried not to be intimidated. We vowed to learn what we could to manage these equations and then exploit them in a systematic way. At the risk of sounding immodest, I can say the strategy has paid off, going far beyond what I'd initially imagined. For over the years, we've managed to solve through geometric analysis many outstanding problems that have yet to be solved by any other means. "The blending of geometry with [partial differential equation] theory," notes Imperial College mathematician Simon Donaldson, "has set the tone for vast parts of the subject over the past quarter century."[2]

So what do we do in geometric analysis? We'll start first with the simplest example I can think of. Suppose you draw a circle and compare it with an arbitrary loop or closed curve of somewhat smaller circumference—this could be just a rubber band you've carelessly tossed on your desk. The two curves look different and clearly have different shapes. Yet you can also imagine that the rubber band can easily be deformed (or stretched) to make a circle—an identical circle, in fact.

There are many ways of doing so. The question is, what's the best way? Is there a way to do it that will always work so that, in the process, the curve doesn't develop a knot or a kink? Can you find a systematic way of deforming that irregular curve into a circle without resorting to trial and error? Geometric analysis can use the geometry of the arbitrary curve (i.e., the rubber band in our example) to prescribe a way of driving that curve to a circle. The process should not be arbitrary. The geometry of the circle ought to determine a precise way, and preferably a canonical way, of getting to a circle. (For mathematicians, *canonical* is a watered-down way of saying "unique," which is sometimes too strong. Suppose you want to travel from the north pole to the south pole. There are many great circles connecting these points. Each of these paths offers the shortest route but none of them is unique; we call them canonical instead.)

You can ask the same questions in higher dimensions, too. Instead of a circle and a rubber band, let's compare a sphere or fully inflated basketball with a deflated basketball with all kinds of dents and dimples. The trick, again, is to turn

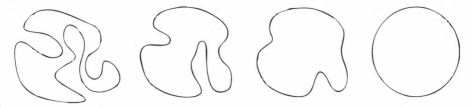

3.3—A technique in geometric analysis, called *curve shortening flow,* can provide a mathematical prescription for turning any non-self-intersecting closed curve into a circle, without running into any complications—such as snags, tangles, or knots—along the way.

that deflated basketball into a perfect sphere. Of course, we can do it with a pump, but how can we do it through math? The mathematical equivalent of a pump in geometric analysis is a differential equation, which is the driving mechanism for the evolution of shape by means of tiny, continuous changes. Once you've determined the starting point—the geometry of the deflated basketball—and identified the proper differential equation, you've solved the problem.

The hard part, of course, is finding the right differential equation for the job and even determining whether there is, in fact, an equation that's up to the task. (Fortunately, Morrey and others have developed tools for analyzing these equations—tools that can tell us whether the problem we're trying to solve has a solution at all and, if so, whether that solution is unique.)

The class of problems I've just described come under the heading of *geometric flow.* Such problems have lately garnered a good deal of attention, as they were used in solving the hundred-year-old Poincaré conjecture, which we'll get to later in this chapter. But I should emphasize that problems of this sort constitute just a fraction of the field we now call geometric analysis, which encompasses a far broader range of applications.

When you're holding a hammer, as the saying goes, every problem looks like a nail. The trick is figuring out which problems are best suited to a particular line of attack. An important class of questions that geometric analysis lets us solve is those involving minimal surfaces. These are nails for which geometric analysis may sometimes be almost the perfect hammer.

Odds are, we've all seen minimal surfaces at one time or another. When we dip the plastic ring from a soap bubble kit into the jar of soapy water, surface tension will force the soap film that forms to be perfectly flat, thereby covering the smallest possible area. A minimal surface, to be more mathematical about it, is the smallest possible surface that can span a given closed-loop boundary.

Minimization has been a foundational concept in both geometry and physics for hundreds of years. In the seventeenth century, for example, the French mathematician Pierre de Fermat showed that light traveling through different media always follows the path that takes the least energy (or "least action"), which was one of the first great physics principles expressed in terms of minimization.

"You often see this phenomenon in nature," explains Stanford mathematician Leon Simon, "because of all the possible configurations you can have, the ones that actually occur have the least energy."[3] The least-area shape corresponds to the lowest energy state, which, other things being equal, tends to be the preferred state. The least-area surface has a surface tension of zero, which is another way of saying its mean curvature is zero. That's why the surface of a liquid tends to be flat (with zero curvature) and why soap films tend to be flat as well.

A confusing aspect of minimal surfaces stems from the fact that the terminology has not changed over the centuries while the mathematics has become increasingly sophisticated. It turns out there is a large class of related surfaces that are sometimes called minimal surfaces even though they are not necessarily area-minimizing. This class includes the surface whose area is smallest compared with all other surfaces bounded by the same border—which might be called a true minimal surface or "ground state"—but it also includes an even greater number of so-called stationary surfaces that minimize the area in small patches (locally) but not necessarily everywhere (globally). Surfaces in this category, which have zero surface tension and zero mean curvature, are of great interest to mathematicians and engineers. We tend to think of minimal surfaces in terms of a family, all of whose members are similar. And while every minimal surface is intriguing, one stands out as truly exceptional.

Finding the shortest path, or the geodesic, is the one-dimensional version of the generally more complex problem of finding minimal surface areas in higher dimensions. The shortest path between any two points—such as a straight line on a plane, or the segment of a great circle connecting two points on the

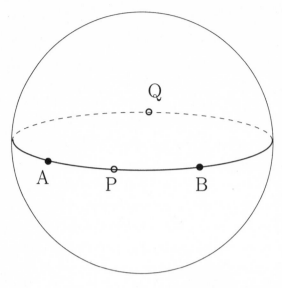

3.4—The shortest distance between *A* and *B* lies along a "great circle" (which in this case happens to be the equator) by way of point *P*. This path is also called a *geodesic*. The path from *A* to *B* via point *Q* is a geodesic too, even though this route obviously does not represent the shortest distance between the two points. (But it is the shortest path compared with all other routes one might take in the vicinity of that arc.)

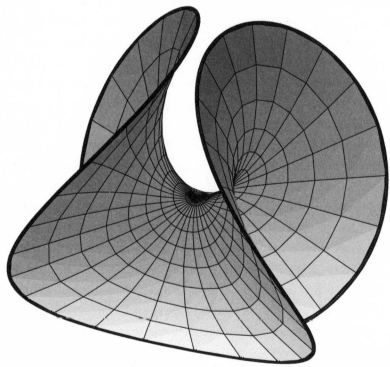

3.5—Joseph Plateau postulated that for any simple closed curve, one could find a *minimal surface*—a surface, in other words, of minimal area—bounded by that curve. The minimal surface spanning the curve (in bold) in this example, called an Enneper surface, is named after the German mathematician Alfred Enneper. (Image courtesy of John F. Oprea)

globe—is sometimes called a *geodesic*, although that term (to confuse matters further) also includes paths that are not necessarily the shortest but are still of considerable importance to geometers and physicists. If you take two points on a great circle that are not on opposite ends of the "globe," there will be two ways of going from one to the other—the short way around and the long way around. Both paths, or arcs, are geodesics, but only one represents the shortest distance between those points. The long way around also minimizes length, but it only does so locally, meaning that among all possible paths one might draw that are close to that geodesic, it alone offers the shortest path. But it is not the shortest path among all possibilities, since one could take the short way around instead. (Things get even more complicated on an ellipsoid—a flattened sphere made by rotating an ellipse around one of its axes—on which many geodesics do not minimize length among all possible paths.)

To find those minimal distances, we need to use differential equations. To find the minimum values, you look for places where the derivative is zero. A surface of minimum area satisfies a particular differential equation—an equation, namely, that expresses the fact that the mean curvature is zero everywhere. Once you've found this specific partial differential equation, you have lots of information to bring to bear on the problem because over the years we've learned a lot about these equations.

"But it's not as if we've plundered a well-developed field and just taken things straight off the shelf. It's been a two-way street, because a lot of information about the behavior of partial differential equations has been developed through geometry," says Robert Greene.[4] To see what we've learned through this marriage of geometric analysis and minimal surfaces, let's resume our discussion of soap films.

In the 1700s, the Belgian physicist Joseph Plateau conducted classic experiments in this area, dipping wires bent into assorted shapes in tubs of soapy water. Plateau concluded that the soap films that formed were always minimal surfaces. He hypothesized, moreover, that for any given closed curve, you can always produce a minimal surface with that same boundary. Sometimes there's just one surface, and we know it's unique. Other times there is more than one surface that minimizes the area, and we don't know how many there are in total.

Plateau's conjecture was not proved until 1930, when Jesse Douglas and Tibor Rado independently arrived at solutions to the so-called Plateau problem.

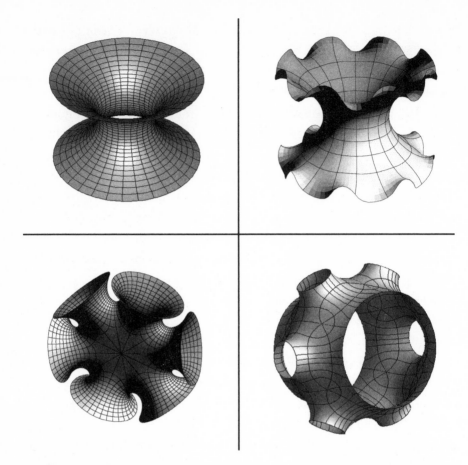

3.6—Although the original version of the Plateau problem related to surfaces spanning simple closed curves, you can also ask—and sometimes answer—more complicated versions of that same question such as this: If your boundary consists of not just a single closed curve but rather several closed curves, such as circles, can you find a minimal surface that connects them all? Here are some examples of minimal surfaces that offer solutions to that more complicated framing of the Plateau problem. (Image courtesy of the 3D-XplorMath Consortium)

Douglas was awarded the Fields Medal for his work in 1936, the first year the award was given.

Not every minimal surface is as simple as a soap film. Some minimal surfaces that mathematicians think about are much more complex—riddled with intricate twists and folds called *singularities*—yet many of these can be found in nature as well. Following up on the Douglas-Rado work a couple of decades later, Robert Osserman of Stanford (author of a masterful book on geometry called

Poetry of the Universe) showed that the minimal surfaces encountered in Plateau-style
experiments only exhibit one kind of singularity of a particularly simple sort, which
looks like disks or planes crossing each other
along a straight line. Then, in the 1970s, a
colleague—William Meeks, a professor at
the University of Massachusetts with whom
I studied at Berkeley—and I carried this a
step further.

We looked at situations in which the minimal surfaces are so-called embedded disks,
meaning that the surface does not fold back
anywhere along its vast extent to cross itself.
(Locally, such a crossing would look like the
intersection of two or more planes.) In particular, we were interested in *convex* bodies,

3.7—The mathematician William
Meeks (Photo courtesy of Joaquín
Pérez)

where a line segment or geodesic connecting any two points in the object always
stays on or in the object. A sphere and a cube are thus convex, while a saddle is
not. Nor is any hollow, dented, or crescent-shaped object convex, because lines
connecting some points will necessarily stray from the object. We proved that
for any closed curve you can draw on the boundary of a convex body, the minimal surface spanning that curve is always embedded: It won't have any of the
folds or crossings that Osserman talked about. In a convex space, we showed,
everything goes down nice and smooth.

We had thus settled a major question in geometry that had been debated for
decades. But that was not the end of the story. To prove that version of the
Plateau problem, Meeks and I drew on something called Dehn's lemma. (A
lemma is a statement proven in the hopes of proving another, more general statement.) This problem was thought to have been proved in 1910 by the German
mathematician Max Dehn, but an error was uncovered more than a decade later.
Dehn stated that if a disk in a three-dimensional space has a singularity, meaning
that it intersects itself in a fold or crisscross, then it can be replaced by a disk
with no singularity but with the same boundary circle. The statement would
be quite useful, if true, because it means that geometers and topologists could

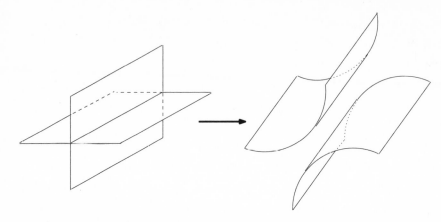

3.8—Dehn's lemma, a geometric version of which was proven by William Meeks and the author (Yau), provides a mathematical technique for simplifying a surface that crosses, or intersects, itself into a surface with no crossings, folds, or other singularities. The lemma is typically framed in terms of topology, but the geometric approach taken by Meeks and Yau offers a more precise solution.

simplify their jobs immensely by replacing a surface that crosses itself with one that has no crossings at all.

The lemma was finally proved in 1956 by the Greek mathematician Christos Papakyriakopoulos—an effort lionized in a limerick penned by John Milnor:

> *The perfidious lemma of Dehn*
> *drove many a good man insane*
> *but Christos Papa-*
> *akyriakop-*
> *oulos proved it without any pain.*

Meeks and I applied Papakyriakopoulos's topology-based approach to the geometry problem inspired by Plateau. We then flipped that around, using geometry to prove stronger versions of both Dehn's lemma and the related *loop theorem* than topologists had been able to achieve. First, we showed that you could find a least-area disk in such a space that was embedded and hence not self-crossing. But in this particular setting (called equivariant), there's not just one disk to consider but all its symmetry pairs—a situation that was like looking in a multiply bent funhouse mirror that has not just one mirror image but many. (The case we considered involved a finite, though arbitrarily large, number of mirror images or symmetry pairs.) We proved that the minimal surface disk would neither intersect itself nor intersect any of the other disks in its symmetry group. You might say that the disks in that group are all "parallel" to each other, with one exception: In cases where the disks do intersect, they must overlap completely.

While that was considered an important problem on its own, it turned out to be even more important than we thought, as it tied into a famous problem in topology known as the Smith conjecture, which dates back to the 1930s. The American topologist Paul Smith was then thinking about rotating an ordinary, three-dimensional space around an infinitely long vertical axis. Smith knew that, if the axis were a perfectly straight line, the rotation could be easily done. But such a rotation would be impossible, he conjectured, if the axis were knotted.

You might wonder why someone would consider such a strange notion, but this is exactly the sort of thing topologists and geometers worry about. "All your intuition tells you the conjecture is obviously true," notes Cameron Gordon of the University of Texas, "for how can you possibly rotate space around a knotted line?" The work Meeks and I had done on Dehn's lemma and the loop theorem contained the last two pieces needed to solve the Smith conjecture. The conjecture was proved by combining our results with those of William Thurston and Hyman Bass. Gordon took on the job of assembling those disparate pieces into a seamless proof that upheld Smith's original assertion that you cannot rotate a three-dimensional space around a knotted line. It turns out, however, despite how ridiculous it might seem, that the statement is false in higher dimensions, where such rotations around knotted lines are indeed possible.[5]

This proof was a nice example of geometers and topologists working together to solve a problem that would surely have taken longer had they been pursuing it entirely on their own. It was also the first time I'm aware of that minimal-surface arguments had been applied to a question in topology. Moreover, it provided some validation of the idea of using geometry to solve problems in topology and physics. Although we've talked about topology, we haven't really said much about physics yet, which leaves the question of whether geometric analysis has had anything to contribute there.

At an international geometry conference held at Stanford in 1973, a problem from general relativity came to my attention that would show how powerful geometric analysis could be for physics, although quite a few years passed before I tried to do anything about it. At that conference, the University of Chicago physicist Robert Geroch spoke of a long-standing riddle called the positive mass conjecture, or positive energy conjecture. It states that in any isolated physical system, the total mass or energy must be positive. (In this case, it's OK to speak of either mass or energy interchangeably, as Einstein showed, most plainly through his famous equation, $E = mc^2$, that the two concepts are equivalent.) Because the universe can be thought of as an isolated system, the conjecture also applies to the universe as a whole. The question was important enough to have warranted its own special session at major general-relativity meetings for years because it related to the stability of spacetime and the consistency of Einstein's theory itself. Simply put, spacetime cannot be stable unless its overall mass is positive.

At the Stanford conference, Geroch laid down the gauntlet, challenging geometers to solve a problem that physicists, up to that point, had been unable to settle on their own. He felt that geometers might help not only because of the fundamental connection between geometry and gravity but also because the statement that the matter density must be positive is equivalent to saying the average curvature of space at each point must be positive.

Geroch was anxious for some resolution of this issue. "It was hard to believe the conjecture was wrong, but it was equally hard to prove it was right," he recently said. Yet one cannot rely on intuition when it comes to matters like this, he added, "because it doesn't always lead us correctly."[6]

His challenge stuck in my mind, and several years later, while I was working on a different question with my former graduate student Richard Schoen (now a Stanford professor), it occurred to us that some of the geometric analysis techniques we'd recently developed might be applied to the positive mass conjecture. The first thing we did, employing a strategy commonly used on big problems, was to break the problem up into smaller pieces, which could then be taken on one at a time. We proved a couple of special cases first, before tackling the full conjecture, which is difficult for a geometer even to comprehend, let alone attempt to prove. Moreover, we didn't believe it was true from a pure geometry standpoint, because it seemed to be too strong a statement.

We weren't alone. Misha Gromov, a famous geometer now at New York University and the Institut des Hautes Études Scientifiques in France, told us that based on his geometric intuition, the general case was clearly wrong, and many geometers agreed. On the other hand, most physicists thought it was true (as they kept bringing it up at their conferences, year after year). That was enough to inspire us to take a closer look at the idea and see if it made any sense.

The approach we took involved minimal surfaces. This was the first time anyone had applied that strategy to the positive mass conjecture, probably because minimal surfaces had no obvious connection to the problem. Nevertheless, Schoen and I sensed that this avenue might pay off. Just as in engineering, you need the right tools to solve a problem (although, after a proof is com-

3.9—Stanford mathematician Richard Schoen

plete, we often find there's more than one way of arriving at the solution). If the local matter density was in fact positive, as postulated in general relativity, then the geometry had to behave in a manner congruent with that fact. Schoen and I decided that minimal surfaces might offer the best way of determining how the local matter density affects the global geometry and curvature.

The argument is difficult to explain mainly because the Einstein field equation, which relates the physics in this situation to geometry, is a complicated, nonlinear formulation that is not intuitive. Basically, we started off by assuming that the mass of a particular space was *not* positive. Next we showed that you could construct an area-minimizing surface in such a space whose average curvature was non-negative. The surface, in other words, could have *zero* average curvature. That would be impossible, however, if the space in which the surface sat was our universe, where the observed matter density is positive. And assuming that general relativity is correct, a positive matter density implies positive curvature.

While this argument might seem circular, it actually is not. The matter density can be positive in a particular space, such as our universe, even though the total mass is not positive. That's because there are two contributions to total mass—one coming from matter and the other coming from gravity. Even though the matter contribution may be positive, as we assumed in our argument, the gravity contribution could be negative, which means the total mass could be negative, too.

Put in other terms, starting from the premise that the total mass was not positive, we proved that an area-minimizing "soap film" could be found, while at the same time we showed that in a universe like ours, such a film could not exist, because its curvature would be all wrong. The supposition of nonpositive mass had thus led to a major contradiction, pointing to the conclusion that the mass and energy must be positive. We proved this in 1979, thereby resolving the issue as the physicist Geroch had hoped someone might.

That discovery was just the first stage of our work, which Schoen and I broke down into two parts, because the problem Geroch had proposed was really a special case, what physicists call the *time-symmetric case*. Schoen and I had taken on the special case first, and the argument that brought us to a contradiction was based on that same assumption. To prove the more general case, we needed to solve an equation proposed by P. S. Jang, who had been Geroch's student. Jang did not try to solve the equation himself, because he believed it had no global solution. Strictly speaking, that was true, but Schoen and I felt we could solve the equation if we made one assumption, which allowed the solution to blow up to infinity at the boundary of a black hole. With that simplifying assumption, we were able to reduce the general case to the special case that we'd already proved.

Our work on this problem received important guidance, as well as motivation, from the physics community. Although our proof rested on pure mathematics— built upon nonlinear arguments few physicists are comfortable with—the intuition of physicists, nevertheless, gave us hope that the conjecture might be true, or was at least worth expending the time and energy to find out. Relying on our geometric intuition in turn, Schoen and I then managed to succeed where physicists had previously failed.

The dominion of geometers in this area did not last long, however. Two years later, the physicist Edward Witten of the Institute for Advanced Study in Princeton proved the positive mass conjecture in an entirely different way—a way that depends on linear (as opposed to nonlinear) equations, which certainly made the argument more accessible to physicists.

Yet both proofs affirmed the stability of spacetime, which was comforting to say the least. "Had the positive mass theorem been untrue, this would have had drastic implications for theoretical physics, since it would mean that conventional spacetime is unstable in general relativity," Witten explains.[7]

Although the average citizen has not lost sleep over this issue, the implications concern more than theoretical physicists only, as the concepts extend to the universe as a whole. I say this because the energy of any system tends to drop to the lowest energy level allowable. If the energy is positive, then there is a floor, set at zero, that it must stay above. But if the overall energy can be negative, there is no bottom. The ground state of general relativity theory—the vacuum—would keep dropping to lower and lower energy levels. Spacetime, itself, would keep degenerating and deteriorating until the universe as a whole disappeared. Fortunately, that is not the case. Our universe is still here, and it appears that spacetime has been saved—at least for now. (More on its possible demise later.)

Despite those rather sweeping implications, one might think that the two proofs of the positive mass conjecture were somewhat beside the point. After all, many physicists had been simply operating under the assumption that the positive mass conjecture was true. Did the proofs really change anything? Well, to my mind, there is an important difference between knowing that something is true and assuming it is true. To some extent, that is the difference between science and belief. In this case, we did not know the conjecture was true until it was a proven fact. As Witten stated in his 1981 paper that presented the proof, "it is far from obvious that the total energy is always positive."[8]

Beyond those more philosophical issues surrounding the proofs of the positive mass conjecture, the theorem also offers some clues for thinking about mass, which turns out to be a subtle and surprisingly elusive concept in general relativity. The complications stem in part from the intrinsic nonlinearity of the theory itself. That nonlinearity means that gravity, too, is nonlinear. And being nonlinear, gravity can interact with itself and, in the process, create mass—the kind of mass that is especially confusing to deal with.

In general relativity, mass can only be defined globally. In other words, we think in terms of the mass of an entire system, enclosed in a figurative box, as measured from far, far away (from infinity, actually). In the case of "local" mass—the mass of a given body, for instance—there is no clear definition yet, even though this may seem like a simpler issue to the layperson. (Mass density is a similarly ill-defined concept in general relativity.) The question of where mass comes from and how you define it has fascinated me for decades, and when time permits, I continue to work on it with math colleagues like Melissa Liu and Mu-Tao Wang from Columbia. I now feel that we're finally narrowing in on a definition of local mass, incorporating ideas from various physicists and geometers, and we may even have the problem in hand. But we couldn't have begun to even think about this issue without first having established as a baseline that the total mass is positive.

In addition, the positive mass theorem led Schoen and me to another general relativity–related proof of some note, this time concerning black holes. When most people think of exotic astrophysical entities like black holes, geometry is the farthest thing from their mind. Yet geometry has a lot to say about black holes, and through geometry, people were able to say something about the existence of these objects before there was strong astronomical evidence for them. This was a major triumph of the geometry of general relativity.

In the 1960s, Stephen Hawking and Roger Penrose proved, through geometry (though a different kind of geometry than we've been discussing here) and the laws of general relativity, that if a *trapped surface* (i.e., an extremely curved surface from which light cannot escape) exists, then the surface would eventually evolve, or devolve, into the kind of singularity thought to lie in the center of a black hole—a place at which we believe the curvature of spacetime approaches infinity. If you find yourself in such a place, spacetime curvature will continue to increase as you move toward the center. And if there's no cap to the curvature—no upper limit—then the curvature will keep getting bigger until you reach the center, where its value will be infinite.

That's the funny thing about curvature. When we walk on the surface of Earth, which has a huge radius (about 4,000 miles) compared with us (normally not much more than 6 feet tall), we can't detect its curvature at all. But if we were to walk on a planet with just a 10- or 20-foot radius (like that inhabited by Saint-Exupéry's Little Prince), we could not ignore its curvature. Because the curvature of a sphere is inversely proportional to the radius squared, as the radius goes to infinity, the curvature goes to zero. Conversely, as the radius goes to zero, the curvature blows up, so to speak, and goes to infinity.

3.10a—Cambridge University physicist Stephen Hawking (Photo by Philip Waterson, LBIPP, LRPS)

3.10b—Oxford University mathematician Roger Penrose (© Robert S. Harris [London])

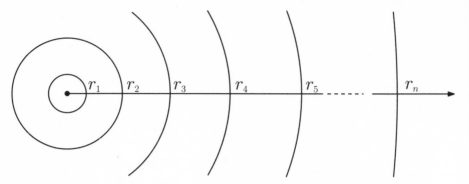

3.11—The smaller the sphere, the more sharply it's curved. Conversely, as the radius of a sphere increases to infinity, its curvature decreases to zero.

Imagine, then, a flash of light emitted simultaneously over the surface of an ordinary two-dimensional sphere. Light will move from this surface in two directions, both inwardly and outwardly. The inward-moving flash will form a surface of rapidly decreasing area that converges toward the center, whereas the surface area of the outgoing flash steadily increases. A trapped surface is different from a typical sphere in that the surface area decreases regardless of whether you move inward or outward.[9] You're trapped no matter which direction you head. In other words, there's no way out.

How can this be possible? Well, it's possible, in part, because that's the definition of a trapped surface. But the explanation also stems from the fact that trapped surfaces have what's called *positive mean curvature* taken to the extreme. Even outward-going rays of light get wrapped around by this intense curvature—as if the roof and walls were closing in on them—and, as a result, they end up converging toward the center. "If the surface area is initially decreasing, it will continue to decrease because there's a focusing effect," my colleague Schoen explains. "You can also think of great circles on the globe that start at the north pole and separate but because the curvature is positive on a sphere, the lines start to converge and eventually come together at the south pole. Positive curvature gives you this focusing effect."[10]

Penrose and Hawking had proved that once formed, trapped surfaces would degenerate into objects from which light cannot escape—objects that we call black holes. But what does it take, exactly, to make a trapped surface? Before

Schoen and I began our work, people had generally asserted that if the matter density in a given region were high enough, a black hole would inevitably form, but these arguments were rather vague and involved a lot of hand-waving. No one had ever formulated the statement in a clear, rigorous manner. This is the problem that Schoen and I attacked, again using minimal-surface approaches that came directly out of our work on the positive mass theorem.

We wanted to know the precise conditions under which you'd produce a trapped surface, and in 1979 we proved that when the density of a region reaches twice that of a neutron star (an environment already 100 trillion times denser than water), the curvature will be high enough that a trapped surface will invariably form. Our argument, coupled with that of Hawking and Penrose, spells out the circumstances under which black holes must exist. More specifically, we showed that when a celestial object has a matter density greater than that of a neutron star, it will collapse directly to a black hole and not to another state. This was a purely mathematical discovery about objects whose existence would soon be confirmed by observation. (A few years ago, Demetrios Christodolou of ETH Zurich developed another mechanism for the formation of trapped surfaces through gravitational collapse.)[11]

More recently, Felix Finster, Niky Kamran, Joel Smoller, and I studied the question of whether spinning black holes are stable in the face of certain perturbations. That is, you could "kick" these objects in various ways, so to speak, but they will not split into two or spin out of control or otherwise fall apart. Although this work looks robust, it is not yet complete and we cannot rule out the possibility of other, more general kinds of kicks that might be destabilizing.

Two years later, Finster, Kamran, Smoller, and I offered what we believe to be the first rigorous mathematical proof of a long-standing black hole problem posed by Roger Penrose. In 1969, Penrose suggested a mechanism for extracting energy from a rotating black hole by drawing down its angular momentum. In this scenario, a piece of matter spiraling toward a black hole can split into two fragments—one that crosses the event horizon and plunges into the hole, and another that is catapulted out with even more energy than the original lump of in-falling matter. Rather than looking at a particle, my colleagues and I considered its analogue—a wave traveling toward the black hole—proving that the mathematics of the Penrose process, as it is known, is entirely sound. While discussing our proof at a 2008 Harvard conference on geometric analysis, Smoller joked that someday, we might use this mechanism to solve our energy crisis.

Although geometers have helped to penetrate some of the enigmas of black holes, the study of these objects now lies primarily in the hands of astrophysicists, who are presently making observations almost to the edge of the event horizon—the point beyond which no observations are possible because nothing (including light) can make it back from the "other side." Nevertheless, had it not been for the work of theorists like Hawking, Penrose, John Wheeler, Kip Thorne, and others, it's doubtful that astronomers would have had the confidence to look for such things in the first place.

Despite these great successes, I don't want to give the impression that this is all there is to geometric analysis. I've focused on the developments I know best, namely, those I was directly or indirectly involved in. But the field is much bigger than that, having involved the efforts of more than one hundred top scholars from all over the world, and I've given just a small taste of that overall effort. We've also managed to get through the bulk of our chapter on geometric analysis without mentioning some of the discipline's biggest achievements. I cannot describe them all; a mere outline of these topics that I wrote in 2006 filled seventy-five single-spaced pages, but we will discuss three that I consider to be among the most important.

The first of these milestones lies in the realm of four-dimensional topology. The principal goal of a topologist is not unlike that of a taxonomist: to classify the kind of spaces or manifolds that are possible in a given dimension. (A *manifold* is a space or surface of any dimension, and we will use these words interchangeably. In the next chapter, however, we'll describe manifolds in greater detail.) Topologists try to lump together objects that have the same basic structure even though there may be wild differences in their outward appearance and detailed structure. Two-dimensional surfaces—with the insistence that they be compact (i.e., bounded and noninfinite) and orientable (having both an inside and an outside)—can be classified by the number of holes they have: Tori, or donut-like surfaces, have at least one hole, whereas the surfaces of topological spheres have none. If two such surfaces have the same number of holes, they are equivalent to a topologist, regardless of how different they may appear.

(Thus, both a coffee mug and the donut being dipped in it are tori of genus one. If you prefer milk with your donut, the glass you're drinking out of will be the topological equivalent of a sphere—made, for instance, by pushing the north pole toward the south pole and then modifying the shape a bit.)

While the two-dimensional situation has been understood for more than a century, higher dimensions have proved more challenging. "Remarkably, the classification is easier in five dimensions and higher," notes University of Warwick mathematician John D. S. Jones. "Three and four dimensions are by far the most difficult."[12] Coincidentally, these happen to be the dimensions deemed most important in physics. William Thurston worked out a classification scheme in 1982 that carved up three-dimensional space into eight basic types of geometries. This hypothesis, known as Thurston's geometrization conjecture, was proved about two decades later (as will be discussed shortly).

The assault on the fourth dimension began at about the same time that Thurston advanced his bold proposition. Four-dimensional spaces are not only harder to visualize but also harder to describe mathematically. One example of a four-dimensional object is a three-dimensional object, like a bouncing basketball, whose shape changes over time as it smashes against the ground, recoils, and then expands. The detailed geometry of such shapes is confusing, to say the least, yet essential to understand if we are ever to truly make sense of the four-dimensional spacetime we supposedly inhabit.

Some clues came in 1982, when Simon Donaldson, then a second-year graduate student at Oxford, published the first of several papers on the structure of four-dimensional space. To gain a window into the fourth dimension, Donaldson drew on nonlinear partial differential equations developed in the 1950s by the physicists Chen Ning Yang and Robert Mills. The Yang-Mills equations—which describe the strong forces that bind quarks and gluons inside the atomic nucleus, the weak forces associated with radioactive decay, and the electromagnetic forces that act on charged particles—operate within the context of four-dimensional space. Rather than just trying to solve the equations, which one would normally attempt by drawing on the geometric and topological features of the underlying space, Donaldson turned the problem on its head: The solution to those equations, he reasoned, should yield information about the four-dimensional space in which they operate. More specifically, the solutions should point to key identifying features—what mathematicians call invariants—that

3.12 —The geometer Simon Donaldson

can be used to determine whether four-dimensional shapes are different or the same.

Donaldson's work shed light on the invariants he was hoping to find, but also turned up unexpected and mysterious phenomena—a new class of "exotic" spaces—that only appear in four dimensions. To explain what *exotic* means, we first have to explain what it means to call two surfaces or manifolds the same. Mathematicians have different ways of comparing manifolds. One is the notion of topological equivalence. Here we can borrow the example from earlier in this chapter of two basketballs, one fully inflated and the other deflated. We say the two objects are effectively the same (or homeomorphic) if we can go from one to the other by folding, bending, squishing, or stretching but not cutting. Going from one manifold to another in this fashion is called *continuous mapping*. It's a one-to-one mapping, meaning that a single point on one surface corresponds to a single point on the other. What's more, any two points that are near each other on one surface will end up being near each other on the other surface as well.

But another way of comparing manifolds is a bit more subtle and more stringent. In this case, the question is whether you can go from one manifold to another *smoothly*, without introducing what mathematicians call singularities, such as sharp corners or spikes on the surface. Manifolds that are equivalent in this sense are called *diffeomorphic*. To qualify, a function that takes you from one manifold to another—transferring one set of coordinates in one space to a different set of coordinates in the other space—has to be a smooth function that is differentiable, which of course means you can take the derivative of this function at all times and at all places. A graph of the function would not appear jagged in any sense: There are no hard edges or steep vertical jumps or rises that would render the whole notion of a derivative meaningless.

As an example, let's place a sphere inside a large ellipsoid, or watermelon-shaped surface, so that the centers of those objects coincide. Drawing radial lines extending outward in all directions from the center will literally match a point on the sphere with a point on the watermelon. You can do this, moreover, for every single point on both the sphere and watermelon. The mapping in this place is not only continuous and one-to-one, but also smooth. There's nothing especially tricky about the function linking these two objects, as it's literally a straight line with no zigzags, sharp turns, or anything out of the ordinary. The two objects in this case—a sphere and an ellipsoid—are both homeomorphic and diffeomorphic.

The so-called exotic sphere offers a contrary example. An exotic sphere is a seven-dimensional manifold that is everywhere smooth, yet it cannot be smoothly deformed into a regular, round (seven-dimensional) sphere, even though it can be continuously deformed into one. We say these two surfaces are thus homeomorphic but not diffeomorphic. John Milnor, who was discussed earlier in this chapter, won a Fields Medal, largely based on his work establishing the fact that exotic spaces exist. People didn't believe such spaces were possible before, which is why the spaces were called exotic.

In two dimensions, flat Euclidean space is about the simplest space you can imagine—just a smooth plane, like a tabletop, that stretches endlessly in all directions. Is a flat two-dimensional disk, which is a subset of that plane, both homeomorphic and diffeomorphic to that plane? Yes it is. You can imagine a bunch of people standing on the plane and grabbing an end of the disk and then walking in an outward direction without stopping. As they march out toward infinity, they'll cover the plane in a nice, continuous, one-to-one fashion. They're topologically identical. It's also pretty easy to imagine that this stretching process, which involves moving a point radially outward, can be done smoothly.

This same basic result holds for three dimensions and every other dimension you might pick except for four dimensions, where you can have manifolds that are homeomorphic to a plane (or flat Euclidean space) without being diffeomorphic. In fact, there are infinitely many four-dimensional manifolds that are homeomorphic but not diffeomorphic to four-dimensional Euclidean space— what we call R^4 (to indicate a real, as opposed to complex, coordinate space in four dimensions).

This is a peculiar and puzzling fact about four dimensions. In a spacetime of 3 + 1 dimensions (three spatial dimensions and one of time), for instance, "electric fields and magnetic fields look similar," Donaldson says. "But in other dimensions, they are geometrically distinct objects. One is a tensor [which is a kind of matrix] and the other is a vector, and you can't really compare them. Four dimensions is a special case in which both are vectors. Symmetries appear there that you don't see in other dimensions."[13]

No one yet knows, from a fundamental standpoint, exactly what makes four dimensions so special, Donaldson admits. Prior to his work, we knew virtually nothing about "smooth equivalence" (diffeomorphism) in four dimensions, although the mathematician Michael Freeman (formerly at the University of California, San Diego) had provided insights on topological equivalence (homeomorphism). In fact, Freeman topologically classified all four-dimensional manifolds, building on the prior work of Andrew Casson (now at Yale).

Donaldson provided fresh insights that could be applied to the very difficult problem of classifying smooth (diffeomorphic) four-dimensional manifolds, thereby opening a door that had previously been closed. Before his efforts, these manifolds were almost totally impenetrable. And though the mysteries largely remain, at least we now know where to start. On the other hand, Donaldson's approach was exceedingly difficult to implement in practice. "We worked like dogs trying to extract information from it," explained Harvard geometer Clifford Taubes.[14]

In 1994, Edward Witten and his physics colleague Nathan Seiberg came up with a much simpler method for studying four-dimensional geometry, despite the fact that their solution sprang from a theory in particle physics called *supersymmetry*, whereas Donaldson's technique sprang from geometry itself. "This new equation had all of the information of the old one," says Taubes, "but it's probably 1000 times easier to get all the information out."[15] Taubes has used the Seiberg-Witten approach, as have many others, to further our understanding of geometric structures in four dimensions—a grasp that is still rather tentative but is nevertheless indispensable for pondering questions about spacetime in general relativity.

For most four-dimensional manifolds, Witten showed that the number of solutions to the Seiberg-Witten equation depends solely on the topology of the manifold in question. Taubes then proved that the number of solutions to those equations, which is dictated by topology, is the same as the number of subspaces or curves of a certain type (or *family*) that can fit within the manifold. Knowing how many curves of this sort can fit in the manifold enables you to deduce the geometry of the manifold, while providing other information as well. So it's fair to say that Taubes's theorem has greatly advanced the study of such manifolds.

This whole excursion into the four-dimensional realm, going back to the work of the physicists Yang and Mills in the 1950s, represents a strange episode that has yet to run its course in which physics has influenced math, which has influenced physics. Though its origins were in physics, Yang-Mills theory was aided by geometry, which helped us better understand the forces that bind elementary particles together. That process was turned around by the geometer Donaldson, who exploited Yang-Mills theory to gain insights into the topology and geometry of four-dimensional space. The same pattern, this give-and-take between physics and math, has continued with the work of the physicists Seiberg and Witten and beyond. Taubes summed up the dynamic history this way: "Once upon a time a Martian arrived, gave us the Yang-Mills equations and left. We studied them and out came Donaldson theory. Years later the Martian has returned and given us the Seiberg-Witten equations."[16] While I can't guarantee that Taubes is right, that's about as plausible an explanation as I've heard.

The second major accomplishment of geometric analysis—and many would place this at the very top—relates to the proof of the famous conjecture formulated in 1904 by Henri Poincaré, which for more than a century stood as the central problem of three-dimensional topology. One reason I consider it so beautiful is that the conjecture can be summed up in a single sentence that nevertheless kept people busy for one hundred years. Stated in simple terms, the conjecture says that a compact three-dimensional space is topologically equivalent to a sphere if every possible loop you can draw in that space can be shrunk to a point without tearing either that loop or the space in the process. We say that a space satisfying this requirement, as discussed earlier in this chapter, has a trivial fundamental group.

The Poincaré conjecture sounds simple enough, but it's not entirely obvious. Let's take a two-dimensional analogue, even though the actual problem—and the hardest one to solve—is in three dimensions. Start with a sphere, say a globe, and place a rubber band on the equator. Now let's gently nudge that rubber band up toward the north pole, keeping it on the surface at all times. If the rubber band is taut enough, when it reaches the north pole, it will shrink down virtually to a point. That's not the case with a torus. Suppose there's a rubber band that runs through the hole and around the other side. There's no way to shrink that rubber band to a point without cutting right through. A rubber band running along the outside of the donut can be nudged to the top of the donut and from there moved down to the donut's inner ring. But it cannot shrink to a point while maintaining contact with the donut. To a topologist, therefore, a sphere is fundamentally different from a donut or any other manifold with a hole (or multiple holes, for that matter). The Poincaré conjecture is essentially a question about what a sphere really means in topology.

Before getting to the proof itself, I'm going to back up a few decades to the year 1979, when I was still at the Institute for Advanced Study. I had invited more than a dozen researchers from all over the world working in geometric analysis to come to Princeton and try to lay out a foundation for our field. I identified 120 outstanding questions in geometry, about half of which have since been completely solved. The Poincaré conjecture was not on that list. In part this was because there was no need to draw attention to the problem, given that it was arguably the most renowned in all of mathematics. It also didn't make the list because I was looking for more narrowly defined problems that I felt could be answered in the end—and hopefully within a reasonable time frame. Although we usually learn something through our struggles, we make the most progress by solving problems; that's what guides mathematicians more than anything else. At the time, however, no one knew exactly how to proceed with Poincaré.

One person who did not participate in our discussions was the mathematician Richard Hamilton, who was then at Cornell and has since settled down in the Columbia math department. Hamilton had just embarked on an ambitious project to find a good dynamical way of changing a complicated, unsmooth metric to a much smoother metric. This effort showed no sign of a short-term payoff, which was apparently how he liked it. He was interested in an extremely difficult set of equations related to the Ricci flow—an example of a geometric flow problem that we touched on earlier in this chapter. Essentially it's a technique for smoothing out bumps and other irregularities to give convoluted spaces more uniform curvature and geometry so that their essential, underlying shapes might be more readily discerned. Hamilton's project did not make my list of geometry's 120 top problems, either, because he hadn't published anything on it yet. He was still toying around with the idea without trying to make a big splash.

I found out what he was up to in 1979, when I gave a talk at Cornell. Hamilton didn't think his equations could be used to solve the Poincaré conjecture; he just thought it was an interesting thing to explore. And I must admit that when I first saw the equations, I was skeptical about their utility, too. They looked too difficult to work with. But work with them he did, and in 1983, he published a paper revealing solutions to what are now called the Hamilton equations. In that paper, Hamilton had solved a special case of the Poincaré conjecture— namely, the case in which the Ricci curvature is positive. (We'll say more about Ricci curvature, which has close ties to physics, in the next chapter.)

My initial skepticism prompted me to go through Hamilton's paper, line by line, before I could believe it. But his argument quickly won me over—so much so, in fact, that the next thing I did was to get three of my graduate students from Princeton to work on the Hamilton equations right away. I immediately suggested to him that his approach could be used to solve Thurston's geometrization conjecture about classifying three-dimensional space into specific geometries, which by extension would imply a general proof of Poincaré itself. At the time, I wasn't aware of any other tools that were up to the job. To my surprise, Hamilton took up the problem with great vigor, pushing ahead with his investigation of Ricci flow over the next twenty years, mostly on his own but with some interactions with me and my students. (Those interactions picked up considerably in 1984, when both Hamilton and I moved to the University of California, San Diego, where we occupied adjacent offices. His seminars on Ricci flow were attended by all my students. We learned a lot from him, though I hope he might have picked up a useful tip or two from me as well. One of the things I missed most upon relocating to Harvard in 1987 was working in such close proximity to Hamilton.)

Regardless of who was around him, Hamilton stuck to his program with steadfast determination. All told, he published a half dozen or so long, important papers—about ninety pages each—and in the end, none of his arguments were wasted. All were ultimately used in the coming ascent of Mount Poincaré.

He showed, for instance, how roundish geometric objects would invariably evolve to spheres—in accordance with Poincaré—as space deformed under the influence of Ricci flow. But more complicated objects, he realized, would inevitably run into snags, producing folds and other singularities. There was no way around it, so he needed to know exactly what kind of singularities could crop up. To catalog the full range of possibilities that might occur, he drew on the work I'd done with Peter Li, which I'd brought to Hamilton's attention some years before, though he generalized our results in impressive ways.

My contribution to this effort dates back to 1973, when I began using a new technique I had developed for harmonic analysis—a centuries-old area of mathematics that is used to describe equilibrium situations. My method was based on an approach called the *maximum principle,* which basically involves looking at worst-case scenarios. Suppose, for instance, you want to prove the inequality $A < 0$. You then ask: What's the biggest value A can possibly assume? If you can show that even in the worst case—at its largest conceivable value—A is still less than zero, then you've finished the job and have my permission to take the rest of the day off. I applied this maximum principle to a variety of nonlinear problems, sometimes in collaboration with my former Hong Kong classmate S. Y. Cheng. Our work concerned questions arising in geometry and physics that are mathematically classified as *elliptic.* Although problems of this sort can be incredibly difficult, they are simplified by the fact that they do not involve any variation in time and can therefore be considered static and unchanging.

In 1978, Peter Li and I took on the more complicated, time-dependent or *dynamic* situation. In particular, we studied equations that describe how heat propagates in a body or manifold. We considered situations where a particular variable like entropy—which measures the randomness of a system—changes in time. Our best-known contribution in this area, the Li-Yau inequality, provides a mathematical description of how a variable like heat may change over time. Hamilton looked at the change in a different variable, entropy, which measures the randomness of a system. The Li-Yau relation is called an inequality because something— in this case, the heat or entropy at one point in time—is bigger or smaller than something else, the heat or entropy at another time.

Our approach provided a quantitative way of seeing how a singularity may develop in a nonlinear system, which was done by charting the distance between two points over time. If the two points collided, with the distance between them vanishing to zero, you got a singularity, and understanding those singularities was the key to understanding almost everything about how heat moves. In particular, our technique offered a way of getting as close to the singularity as possible, showing what happened just before the collision occurred—such as how fast the points were moving—which is kind of like trying to reconstruct what happened before a car crash.

To obtain a close-up view of the singularity—or resolve it, as we say in mathematics—we developed a special kind of magnifying glass. In essence, we zoomed in on the region at which space pinches down to a single point. Then we enlarged that region, smoothing out the creases or pinch points in the process. We did this not once or twice but an infinite number of times. Not only did we enlarge the space, so that we could see the whole picture, but we also enlarged time, in a sense, which effectively meant slowing it down. The next step was to compare that description of the point of singularity—or, equivalently, at the *limit* after an infinite number of blow-ups—with descriptions of the system before the two points collide. The Li-Yau inequality provides an actual measure of the changes in the "before" and "after" shots.

Hamilton took advantage of our approach to get a more detailed look at the Ricci flow, probing the structure of the singularities that might form therein. Incorporating our inequality into his Ricci flow model was a difficult task, which took him nearly five years, because the setting in which his equations resided was far more nonlinear—and hence more complex—than ours.

One of Hamilton's approaches was to focus on a special class of solutions that appear stationary in a particular frame of reference. In the same way, you can find a rotating reference frame in general relativity where the people and objects on a spinning carousel will not be moving, which makes the situation much simpler to analyze. By picking stationary solutions that were easier to understand, Hamilton figured out the best way of incorporating the Li-Yau estimation methods into his equations. This, in turn, afforded him a clearer picture of Ricci flow dynamics—that is, of how things move and evolve. In particular, he was interested in how singularities arise through these complex motions in

spacetime. Ultimately, he was able to describe the structure of all possible singularities that might occur (although he was unable to show that all of these
singularities would occur). Of the singularities that Hamilton identified, all but
one were manageable—they could be removed by topological "surgery," a notion that he introduced and studied extensively in four dimensions. The surgical
procedure is quite complicated, but if it can be performed successfully, one can
show that the space under study is indeed equivalent to a sphere, just as Poincaré had posited.

But there was one kind of singularity—a cigar-shaped protuberance—that
Hamilton could not dispose of in this fashion. If he could show that the "cigar"
does not appear, he would have understood the singularity problem much better, thereby being a big step closer to solving both the Poincaré and the
Thurston conjecture. The key to doing so, Hamilton concluded, was in adapting
the Li-Yau estimate to the more general case where curvature does not have to
be positive. He immediately enlisted me to work with him on this problem,
which turned out to be surprisingly obstinate. Yet we made considerable
progress and felt it was only a matter of time before we'd see the project through.

We were surprised when, in November 2002, the first of three papers on geometric applications of Ricci flow techniques posted on the Internet by Grisha
Perelman, a geometer based in St. Petersburg, Russia. The second and third papers appeared online less than a year later. In these papers, Perelman aimed to
"carry out some details of the Hamilton program" and "give a brief sketch of
the proof of the geometrization conjecture."[17] He too had used the Li-Yau inequality to control the behavior of singularities, though he incorporated these
equations in a different way than Hamilton had, while introducing many inno-
vations of his own.

In a sense, Perelman's papers came out of the blue. No one knew that he'd
even been working on Ricci flow–related problems, as he'd made a name for
himself in an entirely different branch of mathematics called metric geometry,
for solving a famous conjecture put forth by the geometers Jeff Cheeger and
Detlef Gromoll. But in the years prior to his 2002 online publication, Perelman
had largely dropped out of circulation. Occasionally, mathematicians would re-
ceive e-mails in which he inquired about the literature on Ricci flow. But nobody guessed that Perelman was seriously working on the Ricci flow as a way

to solve the Poincaré conjecture, as he hadn't told many people (or perhaps *any* people) exactly what he was up to. In fact, he'd been keeping such a low profile that many of his peers weren't sure he was still doing mathematics at all.

Equally surprising were the papers themselves—a scant sixty-eight pages in total—which meant that it took people a long time to digest their contents and flesh out the key arguments outlined in his approach. Among other advances, Perelman showed how to get past the cigar singularity problem that Hamilton had not yet resolved. Indeed, it is now widely acknowledged that the program pioneered by Hamilton and carried through by Perelman has solved the long-standing Poincaré problem and the more recent geometrization conjecture.

If that consensus is correct, the collective efforts of Hamilton and Perelman represent a great triumph for mathematics and perhaps the crowning achievement of geometric analysis. These contributions far exceed the established standards for a Fields Medal, which Perelman was duly awarded and which Hamilton deserved as well were he not ineligible due to the prize's age restriction. (Winners must be no older than 40.) So far as geometric analysis is concerned, I estimate that roughly half of the theorems, lemmas, and other tools developed in this field over the previous three decades were incorporated in the work of Hamilton and Perelman that culminated in proofs of the Poincaré and Thurston geometrization conjectures.

These are some of the nails that the hammer of geometric analysis has helped to drive home. But you may recall that I promised to describe the *three* biggest successes of geometric analysis. Advances in four-dimensional topology and the Poincaré conjecture, along with the Ricci flow methods that led to its proof, constitute the first two. That leaves number three—a matter I've given considerable thought to and which will be taken up next.

Four

TOO GOOD TO BE TRUE

The third major success of our new "hammer"—geometric analysis—relates to a conjecture raised in 1953 by Eugenio Calabi, a mathematician who has been at the University of Pennsylvania since 1964. The conjecture has turned out to be a seminal piece of work for the field, as we shall see, and important for my own career as well. I consider myself fortunate to have stumbled upon Calabi's ideas or perhaps collided with them head-on. (And in those days, no one wore helmets.) Any mathematician with sufficient talent and training is likely to make some contribution to the field, but it can take luck to find the problem especially suited to your talents and style of thought. Although I've gotten lucky a few times in math, coming across Calabi's conjecture was certainly a high point for me in that regard.

The problem takes the form of a proof that links the topology of *complex space* (which we'll talk about shortly) to geometry or curvature. The basic idea is that we start with some raw topological space, which is like a bare patch of land that's been razed for construction. On top of that, we'd like to build some kind of geometric structure that can later be decorated in various ways. The question Calabi asked, though original in its details, is of the form we often ask in geometry—namely, given a general topology, or rough shape, what sort of precise geometric structures are allowed?

4.1—The geometer Eugenio Calabi (Photo by Dirk Ferus)

This may not seem like a statement replete with implications for physics. But let's frame it in a different way. Calabi's conjecture is concerned with spaces that have a specific type of curvature known as Ricci curvature (to be expanded upon shortly). It turns out that the Ricci curvature of a space relates to the distribution of matter within that space. A space that we call Ricci flat, meaning it has zero Ricci curvature, corresponds to a space with no matter. Viewed in those terms, the question Calabi asked was intimately tied to Einstein's theory of general relativity: Could there be gravity in our universe even if space is a vacuum totally devoid of matter? Curvature, if Calabi were right, makes gravity without matter possible. He put it more generally as his question pertained to spaces of any even dimension and was not limited to the four-dimensional spacetime posited in general relativity. Yet for me, that was the most exciting way of framing the conjecture. It resonated strongly with my conviction that the deepest ideas of math, if shown to be true, would almost invariably have consequences for physics and manifest themselves in nature in general.

Calabi maintains that when he first hit upon the idea, "it had nothing to do with physics. It was strictly geometry."[1] And I have no doubt that is true. The conjecture could have been posed in exactly the same way even if Einstein had never come up with the idea of general relativity. And it could have been proved in the way that it was, even if Einstein's theory did not exist. Yet I still believe that by the time Calabi formulated this problem—almost forty years after Einstein published his revolutionary paper—Einstein's theory was already "in the air." No one in mathematics could help thinking about Einsteinian physics, even if it wasn't deliberate. By that time, Einstein's equations, forever tying curvature to gravity, had become firmly embedded in mathematics. General relativity, you might say, had become part of the general consciousness. (Or perhaps the "collective unconscious," as Jung put it.)

Regardless of whether Calabi consciously (or unconsciously) considered physics, the link between his geometric conjecture and gravity was a great motivating factor for me to take up this work. Proving the Calabi conjecture, I sensed, could be an important step toward uncovering some deep secrets.

Questions like Calabi's are frequently framed in terms of the metric, the geometry of a space—the set of functions that enables us to determine the length of every path—which we first encountered in Chapter 1. A given topological space can have many possible shapes and therefore many possible metrics. Thus the same topological space can accommodate a cube, sphere, pyramid, or tetrahedron, all of which are topologically equivalent. So the question the conjecture raises, regarding what kind of metric a space can "support," can be rephrased in an equivalent way: For a given topology, what kind of geometry is possible?

Of course, Calabi didn't put it in precisely those terms. He wanted to know, among other things, whether a certain kind of complex manifold—a space that was compact (or finite in extent) and "Kähler"—that satisfied specific topological conditions (concerning an intrinsic property known as a "vanishing first Chern class") could also satisfy the geometrical condition of having a Ricci-flat metric. All the terms of this conjecture are, admittedly, rather difficult to grasp, and defining the concepts needed to understand Calabi's statement—complex manifolds, Kähler geometry (and metric), first Chern class, and Ricci curvature—will take some doing. We'll work up to an explanation gradually. But the main thrust of the conjecture is that spaces meeting that complicated set of demands are indeed mathematically and geometrically possible.

To me, such spaces are rare like diamonds, and Calabi's conjecture provided a road map for finding them. If you know how to solve the equation for one manifold and can understand the general structure of that equation, you can use the same idea to solve the equation for *all* Kähler manifolds meeting the same requirements. The Calabi conjecture offers a general rule for telling us that the "diamonds" are there—for telling us that the special metric we seek does, in fact, exist. Even if we cannot see it in its full glory, we can be confident nevertheless that it's genuine. Among mathematical theories, therefore, this question stood out as a kind of jewel—or diamond in the rough, you might say.

From this sprang the work I've become most famous for. One might say it was my calling. No matter what our station, we'd all like to find our true calling in life—that special thing we were put on this earth to do. For an actor, it might be playing Stanley Kowalski in *A Streetcar Named Desire*. Or the lead role in *Hamlet*. For a firefighter, it could mean putting out a ten-alarm blaze. For a crime-fighter, it could mean capturing Public Enemy Number One. And in mathematics, it might come down to finding that one problem you're destined to work on. Or maybe destiny has nothing to do with it. Maybe it's just a question of finding a problem you can get lucky with.

To be perfectly honest, I never think about "destiny" when choosing a problem to work on, as I tend to be a bit more pragmatic. I try to seek out a new direction that could bring to light new mathematical problems, some of which might prove interesting in themselves. Or I might pick an existing problem that offers the hope that in the course of trying to understand it better, we will be led to a new horizon.

The Calabi conjecture, having been around a couple of decades, fell into the latter category. I latched on to this problem during my first year of graduate school, though sometimes it seemed as if the problem latched on to me. It caught my interest in a way that no other problem had before or has since, as I sensed that it could open a door to a new branch of mathematics. While the conjecture was vaguely related to Poincaré's classic problem, it struck me as more general because if Calabi's hunch were true, it would lead to a large class of mathematical surfaces and spaces that we didn't know anything about—and perhaps a new understanding of spacetime. For me the conjecture was almost inescapable: Just about every road I pursued in my early investigations of curvature led to it.

Before discussing the proof itself, we first need to go over the aforementioned concepts that underlie it. The Calabi conjecture pertains strictly to complex manifolds. These manifolds, as we've said, are surfaces or spaces, but unlike the two-dimensional surfaces we're familiar with, these surfaces can be of any even dimension and are not confined to the usual two. (The restriction to even dimensions pertains only to complex manifolds but, in general, a manifold can be of any dimension, even or odd.) By definition, manifolds resemble Euclidean space on a small or local scale but can be very different on a large or so-called global scale. A circle, for instance, is a one-dimensional manifold (or *one-manifold*), and every point you pick on that circle has a "neighborhood" around it that looks like a line segment. But taken as a whole, a circle looks nothing like a straight line. Moving up a dimension, we live on the surface of a sphere, which is a two-manifold. If you pick a small enough spot on the earth's surface, it will look almost perfectly flat—like a disk or a portion of a plane—even though our planet is curved overall and thus non-Euclidean. If you instead look at a much larger neighborhood around that point, departures from Euclidean behavior will become manifest and we'll have to correct for the curvature.

One important feature of manifolds is their *smoothness*. It's built into the definition, because if every local patch on a surface looks Euclidean, then your surface has to be smooth overall. Geometers will characterize a manifold as smooth even if it has some number of funny points where things are not even locally Euclidean, such as a place where two lines intersect. We call such a point a *topological singularity* because it can never be smoothed over. No matter how small you draw the neighborhood around that point, the cross made by the intersecting lines will always be there.

This sort of thing happens in Riemannian geometry all the time. We may start out with a smooth object that we know how to handle, but as we move toward a certain limit—say, by making a shape pointier and pointier or an edge ever sharper—it develops a singularity nevertheless. We geometers are so liberal, in fact, that a space can have an infinite number of singularities and still qualify in our eyes as a kind of manifold—what we call a singular space or singular manifold, which lies at the limit of a smooth manifold. Rather than two lines intersecting in a point, think instead of two planes intersecting in a line.

So that's roughly what we mean by manifolds. Now for the "complex" part. Complex manifolds are surfaces or spaces that are expressed in terms of complex numbers. A complex number takes the form of $a + ib$, where a and b are real numbers and i is the "imaginary unit," defined as the square root of negative one. Much as two-dimensional numbers of the form (x, y) can be graphed on two axes named x and y, one-dimensional complex numbers of the form $a + ib$ can be graphed on two axes referred to as the real and imaginary axes.

Complex numbers are useful for several reasons—one being that people would like to be able to take the square root of negative numbers as well as positive numbers. Armed with complex numbers, one can solve quadratic equations of the form $ax^2 + bx + c = 0$, regardless of the values of a, b, and c, using the quadratic formula that many first learned in high school: $x = [-b \pm \sqrt{(b^2 - 4ac)}]/2a$. When complex numbers are allowed, you don't have to throw up your arms in despair when $b^2 - 4ac$ is negative; you can still solve the equation.

Complex numbers are important, and sometimes essential, for solving polynomial equations—equations, that is, involving one or more variables and constants. The goal in most cases is to find the *roots* of the equation, the points where the polynomial equals zero. Without complex numbers, some problems of this sort are insoluble. The simplest example of that is the equation $x^2 + 1 = 0$, which has no real solutions. The only time that statement is true, and the equation equals zero, is when $x = i$ or when $x = -i$.

Complex numbers are also important for understanding wave behavior, in particular the phase of a wave. Two waves of the same size (amplitude) and frequency can either be in phase, meaning they line up and add to each other constructively, or be out of phase, meaning they partially or completely cancel each other in what's called destructive interference. When waves are expressed in terms of complex numbers, you can see how the phases and amplitudes add up simply by adding or multiplying those complex numbers. (Doing the calculation without complex numbers is possible but much more difficult, just as it's possible to compute the motions of the planets in the solar system from the perspective of Earth, though the equations become much simpler and more elegant when you put the sun at the center of your frame of reference.) The value of complex numbers for describing waves makes these numbers crucial to physics. In quantum mechanics, every particle in nature can also be described as a wave, and quantum theory itself will be an essential component of any theory of quantum gravity—any attempt to write a so-called theory of everything. For that task, the ability to describe waves with complex numbers is a considerable advantage.

The first well-known calculation involving complex numbers was contained in a book published in 1545 by the Italian mathematician Girolamo Cardano. But it wasn't until about three hundred years later that complex geometry was established as a meaningful discipline. The person who really brought complex geometry to the fore was Georg Friedrich Bernhard Riemann—the architect of the first complex manifolds ever seriously examined, so-called Riemann surfaces. (These would become important in string theory more than a century after Riemann died. When a tiny loop of *string*—the basic unit of string theory—moves through higher-dimensional spacetime, it sweeps out a surface that is none other than a Riemann surface. Such surfaces have proven to be quite useful for doing computations in string theory, making them one of the most studied surfaces in theoretical physics today. And Riemann surface theory itself has benefited from the association as well, with equations drawn from physics helping to reinvigorate the math.)

Riemann surfaces, like ordinary two-dimensional manifolds, are smooth, but being complex surfaces (of one complex dimension), they have some added structure built in. One feature that automatically comes with complex surfaces, though not always with real ones, is that all the neighborhoods of the surface are related to each other in specific ways. If you take a small patch of a curved Riemann surface and project it onto a flat surface and do the same with other small patches around it, you'll end up with a map, similar to what you'd end up with if you tried to represent a three-dimensional globe in a two-dimensional world atlas. When you make a map like this with Riemann surfaces, the distance between things can get distorted but the angles between things are always maintained. The same general idea was used in Mercator projection maps, first introduced in the sixteenth century, which treat Earth's surface as a cylinder rather than a sphere. This angle-preserving characteristic, known as conformal mapping, was important in navigation centuries ago and helped keep ships on course. Conformal mapping can help simplify calculations involving Riemann surfaces, making it possible to prove things about these surfaces that can't be proved for noncomplex ones. Finally, Riemann surfaces, unlike ordinary manifolds, must be orientable, meaning that the way we measure direction—the frame of reference that we choose—remains consistent regardless of where we are in space. (This is not the case on a Möbius strip, a classic example of a nonorientable surface, where directions are reversed—down becomes up, left becomes right, clockwise becomes counterclockwise—as you loop around to your original spot.)

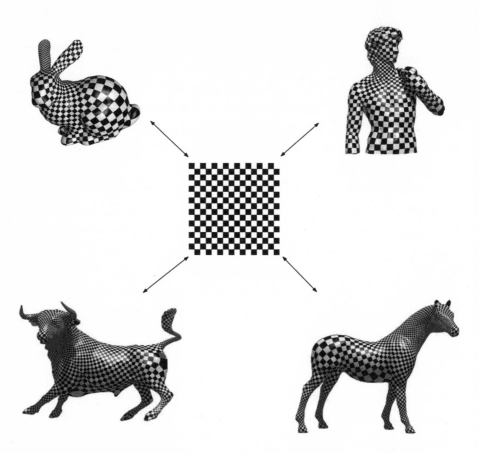

4.2—These two-dimensional surfaces—bull, bunny, David, and horse—are all examples of Riemann surfaces, which are of great importance in both mathematics and string theory. One can place a checkerboard pattern onto these surfaces by picking a point on the checkerboard, feeding its coordinates into a mathematical "function," and generating a point on, say, the bunny instead. But the transposed checkerboard cannot be perfect (unless it is being mapped onto a two-dimensional torus), owing to the presence of singular points—such as might be found on the north and south poles of a sphere—that will inevitably crop up on a surface whose Euler characteristic (described later in this chapter) is nonzero.

Nevertheless, the process is still *conformal*, meaning that angles—including the right angles of the checkerboard—are always preserved in moving from one surface to the other. Although the size of things, such as the checkerboard squares, may get distorted, the corners of the squares are always 90 degrees. This angle-preserving property is one of the special features of a Riemann surface.

As you move from patch to patch on a Riemann surface, you will invariably change coordinate systems, as only a tiny neighborhood around a given point looks Euclidean. But these patches need to be stitched together in just the right way so that movements between them always preserve angles. That's what we mean by calling such a movement, or "transformation," *conformal*. Complex manifolds, of course, come in higher dimensions as well—Riemann surfaces being just the one-dimensional version of them. Regardless of dimensions, the different regions or patches of a manifold have to be properly attached in order for the manifold to qualify as complex. However, on a higher-dimensional complex manifold, angles are not preserved during movements between one patch of the manifold and another and between one coordinate system and another. These transformations are not conformal, strictly speaking, but are rather a generalization of the one-dimensional case.

The spaces Calabi envisioned not only were complex, but also had a special property called *Kähler geometry*. Riemann surfaces automatically qualify as Kähler, so the real meaning of the term only becomes apparent for complex manifolds of two (complex) dimensions or higher. In a Kähler manifold, space looks Euclidean at a single point and stays close to being Euclidean when you move away from that point, while deviating in a specific way. To qualify that last statement, I should say that space looks "complex Euclidean," rather than just plain old Euclidean, meaning that the space is even-dimensional and that some of the coordinates defining a point are expressed in terms of complex numbers. This distinction is important because only complex manifolds can have Kähler geometry. And Kähler geometry, in turn, enables us (among other things) to measure distance using complex numbers. The Kähler condition, named after the German mathematician Erich Kähler, provides an indication of how close a space comes to being Euclidean based on criteria that are not strictly related to curvature.

To quantify the closeness of a given manifold to Euclidean space, we need to know the metric for that manifold. In a flat space, where all the coordinate axes are perpendicular to each other, we can simply use the Pythagorean formula to compute distances. In curved spaces, where coordinate axes are not necessarily perpendicular, things become more complicated, and we have to use a modified version of that formula. Distance calculations then involve metric coefficients—a set of numbers that vary from point to point in space and also depend on how the coordinate axes are oriented. Selecting one orientation for the coordinate axes rather than another will change those numbers. What's important is not the value of the metric coefficients themselves, which is somewhat arbitrary, but rather how they change from place to place in the manifold. For that information tells you where one point lies in relation to every other, which encapsulates everything you need to know about the geometry of the manifold. As we've seen in previous chapters, in four-dimensional space, ten such coefficients are needed. (Actually, there are sixteen numbers in all because the metric tensor in this case is a four-by-four matrix. However, the metric tensor is always symmetrical around a diagonal axis running from the upper left corner of the matrix to the lower right. There are four numbers on the diagonal itself, and two sets of six numbers on either side of the diagonal that are the same. Owing to this symmetry, we often need to concern ourselves with just ten numbers—four along the diagonal and six on either side—rather than all sixteen.)

Still, this doesn't explain how a metric works. Let's take a reasonably simple example in one complex or two real dimensions—the Poincaré metric on the unit disk in a plane centered at the origin $(0, 0)$ of the x-y coordinate system. This is just the set of points (x, y) that satisfies the inequality $x^2 + y^2 < 1$. (Technically this is referred to as an "open" unit disk because it lacks a boundary, the unit circle itself, defined by $x^2 + y^2 = 1$.) Because we are in two dimensions, the metric tensor of the Poincaré metric is a two-by-two matrix. Each spot in the matrix is filled with a coefficient of the form Gij where i refers to the row and j to the column. The array, therefore, looks like this:

$$G_{11} \quad G_{12}$$
$$G_{21} \quad G_{22}$$

Owing to symmetry, as discussed above, G_{12} has to be the same as G_{21}. For the Poincaré metric, these two "off-diagonal" numbers are defined as zero. The other two numbers, G_{11} and G_{22}, do not have to be the same, but in the Poincaré metric, they are: Both, by definition, are equal to $4/(1 - x^2 - y^2)^2$. For any pair of x and y you choose sitting within this disk, the metric tensor tells you the coefficients. For $x = 1/2$ and $y = 1/2$, for instance, G_{11} and G_{22} both equal 16, whereas the other two coefficients are 0, as they are everywhere, for any point (x, y) in the unit disk.

Now that we have these numbers, these coefficients, what do we do with them? And how do they relate to distance? Let's draw a little curve inside the unit disk, but instead of thinking of a curve as a static item, let's think about it as the path traced out by a particle moving in time from a point A to point B. What is the length of that path, according to the Poincaré metric?

To answer that question, we take the curve s and break it up into tiny line segments—the tiniest line segments you can possibly imagine—and add them up. We can approximate the length of each of those segments by using the Pythagorean theorem. First, we define the x, y, and s values parametrically; that is, as functions of time. Thus, $x = X(t)$, $y = Y(t)$, and $s = S(t)$. The derivatives of those time functions, $X'(t)$ and $Y'(t)$, we treat as the legs of a triangle; when used in the Pythagorean theorem, $\sqrt{([X'(t)]^2 + [Y'(t)]^2)}$ gives us the derivative $S'(t)$—that is, the approximate length of one of those minuscule segments we created. Integrating that derivative over A to B gives us the length of the entire curve. Each line segment, in turn, represents a tangent to the curve, known in this context as a *tangent vector*. But, because we are on the Poincaré disk, we must integrate the product of the Pythagorean result times the metric—namely, $\sqrt{([X'(t)]^2 + [Y'(t)]^2)} \times \sqrt{(4/[1 - x^2 - y^2]^2)}$—to correct for the curvature.

To simplify the picture further, let's set $Y(t)$ to 0 and thereby restrict ourselves to the x-axis. We'll start at 0 and move at a constant rate along the x-axis toward 1. If time is moving from 0 to 1 as well, then $X(t) = t$, and if $Y(t) = 0$ (as we're assuming in this instance), then $X'(t) = 1$ because we're taking the derivative of X with respect to time. But since X in this case equals time, we're really taking the derivative of X with respect to itself, and that's always 1. If you think of the derivative in terms of a ratio, this last point becomes fairly obvious: The derivative of X here is the ratio of the change in X to the change in X, and any ratio of that form—with the same quantity in the numerator and denominator— has to be 1.

So the fairly messy product of the above paragraph, which is the thing we have to integrate to get the distance, now reduces to $2/(1-x^2)$. It's fairly easy to see that as x approaches 1, this factor approaches infinity and its integral approaches infinity as well.

One caveat is that just because the metric coefficients—the G_{11} and G_{22} terms here—go to infinity doesn't necessarily mean the distance to the boundary goes there, too. But this does indeed happen in the case of the Poincaré metric and the unit disk. Let's take a closer look at what happens to those numbers as we move outward in space and time. Starting at the origin, where $x = 0$ and $y = 0$, G_{11} and G_{22} equal 4. But when we get near the edge of the disk, where the sum of the squares of x and y is close to 1, the metric coefficients get really big, meaning the tangent vector lengths get big as well. When $x = 0.7$ and $y = 0.7$, for example, G_{11} and G_{22} equal 10,000. For $x = 0.705$ and $y = 0.705$, they're more than 100,000; and for $x = 0.7071$ and $y = 0.7071$, they're more than 10 billion. Moving closer to the disk boundary, these coefficients not only get bigger, but go to infinity—and so do the distances to the edge. If you're a bug on this surface crawling toward the edge, the bad news is that you'll never reach it. The good news is that you're not missing much, since this surface—taken on its own—doesn't really have an edge. When you put an open unit disk inside a plane, it has a boundary—in the form of a unit circle—as part of the plane. But as an object unto itself, this surface has no boundary, and any bug seeking to get there will die unfulfilled. This unfamiliar, and perhaps counterintuitive, fact is a result of the negative curvature of a unit disk as defined by the Poincaré metric.

We've been spending some time talking about the metric in order to get some sense of what a Kähler metric and a Kähler manifold equipped with such a metric are all about. Whether a particular metric is Kähler is a function of how the metric changes as you move from point to point. Kähler manifolds are a subclass of a set of complex manifolds known as Hermitian manifolds. On a Hermitian manifold, you can put the origin of a complex coordinate system at any point, such that the metric will look like a standard Euclidean metric at that point. But as you move away from that point, the metric becomes increasingly non-Euclidean; more specifically, when you move away from the origin by a distance epsilon, ε, the metric coefficients themselves change by a factor on the order of ε. We characterize such manifolds as Euclidean to first order. So if ε were one-thousandth of an inch, when we move by ε, the coefficients in our

Hermitian metric would remain constant to within a thousandth of an inch or thereabouts. Kähler manifolds are Euclidean to second order, meaning that the metric is even more stable; the metric coefficients on a Kähler manifold change by ε^2 as you move from the origin. To continue the previous example, where $\varepsilon = 0.001$ inch, the metric would change by 0.000001 inch.

So what prompted Calabi to single out Kähler manifolds as the ones of greatest interest? To answer that, we need to consider the range of choices available. If you want to be really stringent, you could insist, for example, that the manifolds must be totally flat. But the only compact manifolds that are totally flat happen to be donuts or tori or close relatives thereof, and this holds for any dimension of two or higher. As manifolds go, tori are rather simple and can thus be rather limiting. We'd prefer to see more variety—a broader range of possibilities. Hermitian manifolds, on the other hand, don't constrain things enough: the possibilities are just too great. Kähler manifolds, lying between Hermitian and flat, have the kind of properties that we geometers often seek out. They have enough structure to make them easier to work with, yet not so much structure as to be overly restrictive in the sense that no manifold can match your detailed specifications.

Another reason for focusing on Kähler manifolds is that we can use tools introduced by Riemann—some of which were exploited by Einstein—to study these manifolds. These tools work on Kähler manifolds, which are a restrictive class of Hermitian manifolds, but do not apply to Hermitian manifolds in general. We'd like to be able to use these tools, because they were already quite powerful when Riemann first developed them, and mathematicians have had more than a century in which to enhance them further. That makes Kähler manifolds a particularly appealing choice because we have the technology at our disposal to really probe them.

But that's not all. Calabi was interested in these manifolds because of the kinds of symmetry they possess. Kähler manifolds, like all Hermitian manifolds, have a rotational symmetry when vectors on them are multiplied by the imaginary unit i. In one complex dimension, points are described by a pair of numbers (a, b) taken from the expression $a + bi$. Let's assume that (a, b) defines a tangent vector sticking out from the origin. When we multiply that vector by i, its length is preserved, although it gets rotated by 90 degrees. To see how the rotation works, let's start out at the point (a, b), or $a + ib$. Multiplying by i yields $ia - b$ or, equivalently, $-b + ia$, corresponding to a new point $(-b, a)$ on the complex plane that defines a vector that's orthogonal (or perpendicular) to the original vector but of the same length.

You can see that these vectors really are perpendicular by plotting the points (a, b) and $(-b, a)$ on simple Cartesian coordinates and measuring the angle between the vectors going to these points. The operation we're talking about here—switching the x coordinate to $-y$ and switching the y coordinate to x—is called the J transformation, which happens to be the real-number analogue of multiplying by i. In this case, doing the J operation twice, or J^2, is the same as multiplying by -1. It's best to continue this discussion by talking about J rather than i since any way we might try to picture this—either in our heads or on paper—will likely involve real coordinates rather than the complex plane. Just keep in mind, once again, that the J operation is a way of interpreting complex multiplication by i as a transformation of two-dimensional coordinates.

All Hermitian manifolds have this kind of symmetry: J transformations rotate vectors 90 degrees while keeping their length unchanged. Kähler manifolds (being a more restrictive subset of the Hermitian case) have this symmetry as well. In addition, Kähler manifolds have what's called *internal symmetry*—a subtle kind of symmetry that must remain constant as you move between any two points in the space or it's not Kähler geometry. Many of the symmetries we see in nature relate to rotation groups. A sphere, for example, has *global symmetry*—so called because it applies to every point on the object at once. One such symmetry is rotational invariance, which means you can turn the sphere every which way and it still looks the same. The symmetry of a Kähler manifold, on the other hand, is more local because it only applies to first derivatives of the metric. However, through the techniques of differential geometry, which involve integrating over the whole manifold, we can see that the Kähler condition and its associated symmetry do imply a specific relation between different points. In this way, the symmetry that we initially described as local has, by virtue of calculus, a broader, more global reach that establishes a link between different points on the manifold.

The crux of this symmetry relates to the rotation caused by the J operation and something called *parallel translation* or *transport*. Parallel transport, like the

J operation, is a linear transformation: It's a way of moving vectors along a path on a surface or manifold that keeps the lengths of those vectors constant while keeping the angles between any two vectors constant as well. In instances where parallel transport is not easy to visualize, we can determine the precise way of moving vectors around from the metric by solving differential equations.

On a flat (Euclidean) plane, things are simple: You just keep the direction and length constant for each vector. On curved surfaces and general manifolds, the situation is analogous to maintaining a constant bearing in Euclidean space but much more complicated.

Here's what's special about a Kähler manifold: If we take a vector V at point P and parallel-transport it along a prescribed path to point Q, we'll get a new vector at Q called W_1. Then we'll do the J operation on W_1, rotating it 90 degrees, which gives us a new vector, JW_1, that's been transformed by the J operation. Alternatively, we can start with vector V at point P and J transform it (or rotate it) right there, giving us the new vector JV (still affixed to point P). We then parallel-transport vector JV to point Q, where we'll rename it W_2. If the manifold is Kähler, the vectors JW_1 and W_2 will always be the same, regardless of what path you take in getting from P to Q. One way of putting this is that on a Kähler manifold, the J operation remains invariant under parallel transport. That's not the case for complex manifolds in general. Another way of putting it is that on a Kähler manifold, parallel-transporting a vector and then transforming it by the J operation is the same as transforming the original vector by J and then parallel-transporting it. These two operations commute—the order in which you do them doesn't matter. That's not always true, as Robert Greene memorably describes it, "just as opening the door and then walking out of the house is not the same as walking out of the house (or trying to!) and then opening the door."[2]

The basic notion of parallel transport is illustrated in Figure 4.3 for a surface of two real dimensions, or one complex dimension, because that's what we can draw. But this case is rather trivial, because the possibilities for rotation are so limited. There are only two choices: rotating to the left and to the right.

But in complex dimension two, which is real dimension four, there are an infinite number of vectors of a given length perpendicular to a given vector. This infinitude of vectors defines a tangent space, which you might picture—in two

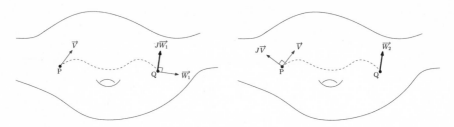

4.3—In the first panel, we parallel-transport vector V from point P to Q, where the vector is given the new name W_1. We then perform the J operation on W_1, which rotates it by 90 degrees. The reoriented vector is called JW_1. In the second panel, we perform the J operation on vector V at point P, which gives us a realigned vector (again rotated 90 degrees), called JV. We then parallel-transport JV to point Q, where it is assigned the new name, W_2. In both cases, the resultant vectors—JW_1 and W_2—are the same. This is one of the hallmarks of a Kähler manifold—namely, that you get the same result, the same vector, regardless of whether you parallel-transport a vector and then transform it by J or transform the vector by J first and then parallel-transport it. The two operations *commute*, meaning that the order in which you do them does not matter.

dimensions rather than four—as a giant piece of plywood balanced on top of a basketball. In this setting, knowing that a vector is perpendicular to the one you have in hand does little to narrow down the field of possibilities—unless the manifold happens to be Kähler. In that case, if you know how the J transformation works at one point, you know what vector you'll get when you do the J transformation at some remote point, because you can carry your result from the first point to the second via parallel transport.

There's another way of showing what this simple operation, the J transformation, has to do with symmetry. It's what we call a *fourfold symmetry* because each time you operate by J, you rotate the vector 90 degrees. When you do it four times, or 360 degrees, you've come full circle and are back where you started. Another way to think of it is that transforming by J twice is the same as multiplying by –1. Transforming by J four times is –1 × –1 = 1. Back to identity, as we say.

Given that this fourfold symmetry applies only to the tangent space of a single point, in order for it to be useful, it has to behave consistently as you move around in the space. This consistency is an important feature of internal symmetry. You might compare it to a compass that has twofold symmetry in the sense that it can only point in two directions: north and south. If you moved

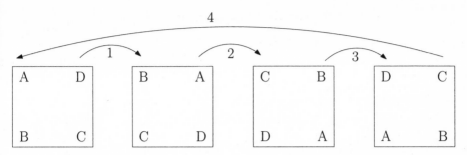

4.4—A demonstration of the simple and rather obvious fact that a square has fourfold symmetry about its center, which is another way of saying that if you rotate it by 90 degrees, four times consecutively, you'll end up where you started. Since the J transformation also involves a 90-degree rotation, this operation has fourfold symmetry as well, with four rotations bringing you back to the starting point. (Technically speaking, the J operation acts on tangent vectors, so it's only roughly analogous to the rotation of an object like a square.) The J transformation, as discussed in the text, is the real-number analogue of the *i* multiplication. Multiplying by *i* four times is the same as multiplying by 1, which—like doing the J operation four times—will inevitably bring you back to where you started.

this compass around in space and it randomly pointed north or south, with no rhyme or reason to it, you might conclude that your space was not particularly symmetrical, nor did it have a discernible magnetic field (or that it was time to buy a new compass). Similarly, if your J operation gave you different results depending on where and how you moved in your manifold, it would lack the order and predictability that symmetry normally confers. Not only that, you'd know that your manifold was not Kähler.

This internal symmetry, which in many ways defines Kähler manifolds, is restricted to the space tangent to the manifolds. That can be advantageous because, on a tangent space, the results of an operation do not depend on your choice of coordinates. In general, that's what we look for in both geometry and physics—that the results of an operation are independent of the coordinate system that was selected. Simply put, we don't want the answers we get to be a consequence of an arbitrary decision regarding how the coordinate axes were oriented or which point was picked as the origin.

This symmetry imposes another set of constraints on the mathematical world Calabi was imagining, drastically simplifying it and making the problem of its proof seem potentially soluble. But it had other consequences that he did not foresee; indeed, this internal symmetry that he was proposing is a special kind of supersymmetry—an idea that would prove to be of vital importance to string theory.

The last two pieces in our puzzle—Chern class and Ricci curvature—are related and grow out of attempts made by geometers to generalize Riemann surfaces from one dimension into many, and then to characterize the differences between those generalizations mathematically. This leads us to an important theorem that applies to compact Riemann surfaces—as well as to any other compact surface without a boundary. (The definition of a *boundary* in topology is rather intuitive: A disk has a boundary or clearly defined edge, whereas a sphere has none. You can move in any direction you want on the surface of a sphere, and go as far as you want, without ever reaching, or approaching, an edge.) The theorem was formulated in the nineteenth century by Carl Friedrich Gauss and the French mathematician Pierre Bonnet, connecting the geometry of a surface to its topology.

The Gauss-Bonnet formula holds that the total Gauss curvature of such surfaces equals 2π times the Euler characteristic of that surface. The Euler characteristic—known as χ ("chi")—in turn, is equal to $2 - 2g$, where g is the genus or number of "holes" or "handles" that a surface has. The Euler characteristic of a two-dimensional sphere, for example, which has zero holes, is 2. Euler had previously devised a separate formula for finding the Euler characteristic of any polyhedron: $\chi = V - E + F$, where V is the number of vertices, E the number of edges, and F the number of faces. For a tetrahedron, $\chi = 4 - 6 + 4 = 2$, the same value we derived for the sphere. For a cube with 8 vertices, 12 edges, and 6 faces, $\chi = 8 - 12 + 6 = 2$, again the same as the sphere. It makes sense for those topologically identical (though geometrically distinct) objects to have the same value of χ, given that the Euler characteristic relates to an object's topology rather than its geometry. The Euler characteristic, χ, was the first major *topological invariant of a space:* a property that remains constant, or invariant, for spaces that may look dramatically different—like a sphere, tetrahedron, and cube—yet are topologically equivalent.

Going back to the Gauss-Bonnet formula, the total Gauss curvature of a two-dimensional sphere is thus 2π times 2, or 4π. For a torus, it's 0, as χ for a two-

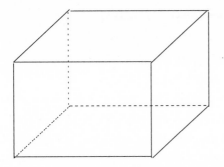

$$V = 8$$
$$E = 12$$
$$F = 6$$
$$\chi = V - E + F = 2$$

4.5—An orientable (or two-sided) surface is described topologically by its Euler characteristic, or Euler number. There's a simple formula for computing the Euler characteristic of a polyhedron, which, loosely speaking, is a geometric object with flat faces and straight edges. The Euler characteristic, represented by the Greek letter χ (chi), equals the number of vertices minus the number of edges plus the number of faces. For the rectangular prism (or "box") in this example, that number turns out to be 2. It's also 2 for a tetrahedron $(4 - 6 + 4)$ and for a square pyramid $(5 - 8 + 5)$. The fact that these spaces have the same Euler characteristic (2) isn't surprising, since all these objects are topologically equivalent.

dimensional torus with one hole is 0 $(2 - 2g = 2{-}2 = 0)$. Attempting to generalize the Gauss-Bonnet principle to higher dimensions leads us to Chern classes.

Chern classes were developed by my mentor and adviser S. S. Chern as a crude method for mathematically characterizing the differences between two manifolds. Simply put, if two manifolds have different Chern classes, they cannot be the same, though the converse is not always true: Two manifolds can have the same Chern classes and still be different.

For Riemann surfaces of one complex dimension, there's just one Chern class, the first Chern class, which in this case equals the Euler characteristic. The number of Chern classes that a manifold can be assigned depends on the number of dimensions. A manifold of two complex dimensions, for instance, has a first and a second Chern class. For manifolds of keen interest in string theory— those with three complex dimensions (or six real ones)—there are three Chern classes. The first Chern class assigns integer coefficients to subspaces, or submanifolds, of two real dimensions sitting inside the six-dimensional space (much as two-dimensional surfaces such as paper can be fit into your three-dimensional office). The second Chern class assigns numbers to sub-manifolds of four real dimensions inside the six-dimensional space. The third Chern class assigns a single number—χ, the Euler characteristic—to the manifold itself, which has three complex dimensions and six real dimensions. For a manifold of n complex dimensions, the last Chern class—or nth Chern class—is always equal to the Euler characteristic.

But what does the Chern class really tell us? In other words, what's the purpose of assigning all those numbers? It turns out the numbers don't tell us anything of great value about the submanifolds themselves, but they can tell us a lot about the bigger manifold. (That's a fairly common practice in topology: We try to learn about the structure of complex, higher-dimensional objects by looking at the number and types of subobjects they can hold.)

Suppose, for example, you've assigned a different number to every person in the United States. The number given to an individual doesn't really tell you anything about him or her. But all those numbers taken together can tell you something interesting about the bigger "object"—the United States itself—such as the size of its population, or the rate of population growth.

Here's another example that can provide a picture to go along with this rather abstract concept. As we often do, we'll start by looking at a fairly simple object, a sphere—a surface of one complex dimension or two real dimensions. A sphere has only one Chern class, which in this case equals the Euler characteristic. In Chapter 2, as you may recall, we discussed some of the meteorological (and fluid dynamical) implications of living on a spherically shaped planet. Let's suppose the wind blows from west to east at every spot on Earth's surface. Well, at almost every spot. You can imagine the wind blowing in the easterly direction on the equator and at every possible latitudinal line north or south of the equator. At two points, however, lying at the absolute centers of the north and south poles (which can be regarded as singularities), the wind does not blow at all—an inevitable consequence of spherical geometry. For a surface like that, which has those special points that stick out like a sore thumb, the first Chern class is not equal to zero. In other words, it is not vanishing.

Now let's consider the donut. Winds could blow on the surface in any direction you want—in long rings around the hole, in short rings through the hole, or in more complicated spiral patterns—without ever hitting a singularity, a place where the flow stops. You can keep going around and around and never

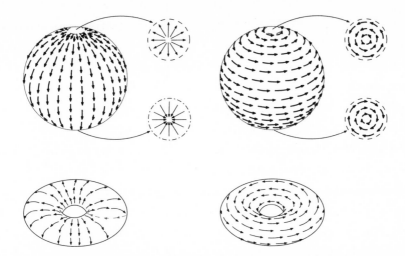

4.6—The *first Chern class* (which has the same value as the Euler characteristic for two-dimensional surfaces such as these) relates to places where the flow in a vector field totally shuts down. One can see two such spots on the surface of a sphere such as the familiar globe. If, for example, everything flows from the north pole to the south pole (as in the left-hand sphere), there will be zero net flow at each of the poles, because all the vectors representing flows will cancel each other out. Similarly, if everything flows from west to east (as in the right-hand sphere), there will be exactly two dead spots—again, one on the north pole and one on the south pole—where nothing flows at all, because at these spots, there is no east or west.

This is not the case on the surface of a donut, however, where things can flow vertically (as in the left-hand donut) or horizontally (as in the right-hand donut) without ever hitting a dead spot. That's why the first Chern class is zero for a donut, which lacks these singular spots, but not for a sphere.

4.7—Determining the first Chern class of an object comes down to finding places where flows in a vector field drop to zero. Places like that can be found in the center of a vortex, such as the eye of a hurricane—a circular region of calm weather, anywhere from 2 to 200 miles in diameter, surrounded by some of the stormiest conditions found on Earth. This photo was taken of Hurricane Fran in 1996, just before the storm ravaged the eastern United States, causing billions of dollars' worth of damage. (Photo by Hasler, Chesters, Griswold, Pierce, Palaniappan, Manyin, Summey, Starr, Kenitzer, and de La Beaujardière, Laboratory for Atmospheres, NASA Goddard Space Flight Center)

hit a snag. For a surface like that, the first Chern class is equal to zero or vanishing, as we say.

To pick another example, so-called K3 surfaces of complex dimension two and real dimension four have a first Chern class of zero (Chapter 6 discusses K3 surfaces in more detail). According to Calabi's conjecture, that would enable them to support a Ricci-flat metric, much as a torus could. But, unlike a two-dimensional torus (whose Euler characteristic is zero), the value of χ for a K3 surface is 24. The point is that the Euler characteristic and first Chern class can be quite different in higher dimensions, even though they are the same in the case of one complex dimension.

The next item on our list is Ricci curvature, which is a concept essential to understanding what the Calabi conjecture is all about. Ricci curvature is a kind of average of a more detailed type of curvature known as sectional curvature. To see how it works, let's start with a simple picture: a sphere and the space (a plane) tangent to its north pole. The plane—which is perpendicular to the line containing the sphere's center and the selected point on the sphere—contains all the tangent vectors to the sphere at that point. (Similarly, a three-dimensional surface has a three-dimensional tangent space consisting of all vectors tangent to the point, and so on for higher dimensions.) Every vector in that tangent plane is also tangent to a great circle on the globe that runs through the north and south poles. If we take all the great circles that are tangent to vectors in the plane and put those circles together, we can assemble a new two-dimensional surface. (In this case, that surface will just be the original sphere, but in higher dimensions, the surface so constructed will be a two-dimensional submanifold sitting within a larger space.) The sectional curvature of the tangent plane is simply the Gauss curvature of the newly formed surface associated with it.

To find the Ricci curvature, pick a point on the manifold and find a vector tangent to that point. You then look at all the two-dimensional tangent planes that contain that vector; each of those planes has a sectional curvature attached to it (which, as we've just said, is the Gauss curvature of the surface associated with that plane). The Ricci curvature is the average of those sectional curvatures. A Ricci-flat manifold means that for each vector you pick, the average sectional curvature of all the tangent planes containing that vector equals zero, even though the sectional curvature of an individual plane may not be zero.

As you may have surmised, this means our example of a two-dimensional sphere, where one might take a vector tangent to the north pole, is rather uninteresting because there's only one tangent plane containing that vector. In this case, the Ricci curvature is just the sectional curvature of that plane, which, in turn, is just the Gauss curvature of a sphere (which is 1 for a sphere of unit radius). But when you go to higher dimensions—anything beyond one complex or two real dimensions—there are lots of tangent planes to choose from, and as a consequence, a manifold can be Ricci flat without being flat overall—that is, without having zero sectional curvature and zero Gauss curvature.

The sectional curvature completely determines the Riemann curvature, which in turn encodes all the curvature information you could possibly want about a surface. In four dimensions, this takes twenty numbers (and more for higher dimensions). The Riemann curvature tensor can itself be split into two terms, the Ricci tensor and something called the Weyl tensor, which we won't go into here. The main point is that of the twenty numbers or components needed to describe four-dimensional Riemann curvature, ten describe Ricci curvature while ten describe Weyl curvature.

The Ricci curvature tensor, a key term in the famous Einstein equation, shows how matter and energy affect the geometry of spacetime. In fact, the left-hand side of this equation consists of what's called the modified Ricci tensor, whereas the right-hand side of the equation consists of the *stress energy tensor*, which describes the density and flow of matter in spacetime. Einstein's formulation, in other words, equates the flow of matter density and momentum at a given point in spacetime to the Ricci tensor. Since the Ricci tensor is just part of the total curvature tensor, as discussed above, we cannot rely on it alone to determine the curvature completely. But if we draw on our knowledge of the global topology, we may have hopes of deducing the curvature of spacetime.

In the special case where the mass and energy are zero, the equation reduces to this: The modified Ricci tensor = 0. That's the vacuum Einstein equation, and although it may look simple, you have to remember it's a nonlinear partial differential equation, which is almost never easy to solve. Moreover, the vacuum Einstein equation is actually ten different nonlinear partial differential equations bundled together, because the tensor itself consists of ten independent terms. This equation is similar, in fact, to the Calabi conjecture, which sets the Ricci curvature to zero. It's not too surprising that one can find a trivial solution to the vacuum Einstein equation—a trivial solution being the least interesting one in which you have a spacetime where nothing happens: no matter, no gravity, and nothing much going on. But there's a more intriguing possibility, and this is precisely what the Calabi conjecture gets at: Can the vacuum Einstein equation have a nontrivial solution, too? The answer to that question is yes, as we shall see in due time.

Soon after Chern came up with the idea of Chern classes in the mid-1940s, he showed that if you have a manifold with zero Ricci curvature—with a certain geometry, that is—then its first Chern class must also be zero. Calabi flipped that over, asking whether certain topological conditions are sufficient in themselves to dictate the geometry—or, more precisely, to *allow* that particular geometry to be dictated. Reversals of this sort are not always true. For example, we know that a smooth surface (without edges) whose Gauss curvature is greater than one must be bounded or compact. It cannot wander off to infinity. But as a general matter, compact, smooth surfaces need not have a metric with a Gauss curvature greater than one. A donut, for instance, is perfectly smooth and compact, yet its Gauss curvature cannot always be positive, let alone greater than one. Indeed, a metric of zero Gauss curvature, as discussed previously, is entirely possible, whereas a metric with positive curvature everywhere is not.

So Calabi's conjecture faced two big challenges: Just because it was the converse of a well-established proposition was not sufficient to make it true. And even if it were true, proving the existence of a metric that meets the desired requirements would be extremely difficult. Like the Poincaré conjecture that came before it, Calabi's conjecture—or, I should say, an important case of this conjecture—can be summed up in a single sentence: A compact Kähler manifold with a vanishing first Chern class will admit a metric that is Ricci flat. Yet it took more than two decades to prove the contents of that sentence. And several decades after the proof, we're still exploring the full range of its implications.

As Calabi recalls, "I was studying Kähler geometry and realized that a space that admits a Kähler metric admits other Kähler metrics as well. If you can see one, you can easily get all the others. I was trying to find out if there is one metric that is better than the others—a 'rounder' one, you might say—one that gives you the most information and smoothes out the wrinkles." The Calabi conjecture, he says, is about trying to find the "best" metric.[3]

Or as Robert Greene puts it, "you're trying to find the one metric given to you by God."[4]

The best metric for the purposes of geometry sometimes means "homogenous," which implies a certain uniformity. When you know one part of a surface, you pretty much know it all. Owing to its constant curvature—and by that we also mean constant sectional curvature—a sphere is like that. The pinnacle of regularity, it looks the same everywhere (unlike, say, a football, whose sharp ends stand out, literally and physically, from other parts of the surface). While spheres are attractive for this reason, a Calabi-Yau manifold of complex dimension greater than one cannot have constant sectional curvature, unless it is completely flat (in which case its sectional curvature is zero everywhere). If this property is ruled out in the manifolds we seek, which are not totally flat and uninteresting, "the next best thing is to try to make the curvature as nearly constant as possible," Calabi says.[5] And the next best thing happens to be constant Ricci curvature—or, more specifically, zero Ricci curvature.

The full Calabi conjecture is more general than just setting the Ricci curvature to zero. The constant Ricci curvature case is also very important, especially the negative curvature case, which I eventually drew on to solve some noted problems in algebraic geometry (as will be discussed in Chapter 6). Nevertheless, the zero Ricci curvature case is special because we're saying that the curvature is not only constant, but it's actually zero. And this posed a special challenge: finding the metric for a manifold, or class of manifolds, that comes close to perfection without being totally boring.

There was a catch, however. Two decades after Calabi issued his proposition, very few mathematicians—save for the author of the conjecture himself—believed it to be true. In fact, most people thought it was too good to be true. I was among them, but was not content to stay on the sidelines, quietly harboring my doubts. On the contrary, I was dead-set on proving this conjecture wrong.

Five

PROVING CALABI

A mathematical proof is a bit like climbing a mountain. The first stage, of course, is discovering a mountain worth climbing. Imagine a remote wilderness area yet to be explored. It takes some wit just to find such an area, let alone to know whether something worthwhile might be found there. The mountaineer then devises a strategy for getting to the top—a plan that appears flawless, at least on paper. After acquiring the necessary tools and equipment, as well as mastering the necessary skills, the adventurer mounts an ascent, only to be stopped by unexpected difficulties. But others follow in their predecessor's footsteps, using the successful strategies, while also pursuing different avenues—thereby reaching new heights in the process. Finally someone comes along who not only has a good plan of attack that avoids the pitfalls of the past but also has the fortitude and determination to reach the summit, perhaps planting a flag there to mark his or her presence. The risks to life and limb are not so great in math, and the adventure may not be so apparent to the outsider. And at the end of a long proof, the scholar does not plant a flag. He or she types in a period. Or a footnote. Or a technical appendix. Nevertheless, in our field there are thrills as well as perils to be had in the pursuit, and success still rewards those of us who've gained new views into nature's hidden recesses.

Eugenio Calabi found his mountain some decades earlier, but in the early 1970s, I (among many others) still needed some convincing that it was more than a molehill. I didn't buy the provocative statement he'd put before us. As I saw it, there were a number of reasons to be skeptical. For starters, people were doubtful that a nontrivial Ricci-flat metric—one that excludes the flat torus—could exist on a compact manifold without a boundary. We didn't know of a single example, yet here was this guy Calabi saying it was true for a large, and possibly infinite, class of manifolds.

Then there was the fact that Calabi was, as Robert Greene puts it, taking a very general topological condition and using it to find a very specific geometric corollary that was uniform over the whole space. That doesn't happen in real manifolds that are lacking in complex structure, but the conjecture suggested that with complex manifolds, it does.[1] To elaborate on Greene's point, the Calabi conjecture basically says, starting with the case of one complex dimension (and two real dimensions), that if you have a general topology or shape where the average curvature is zero, then you can find a metric, or geometry, where the curvature is zero everywhere. In higher dimensions, the conjecture specifically refers to Ricci curvature (which is the same as Gauss curvature in real dimension two but is different in dimensions above two), and the condition of the average Ricci curvature being zero is replaced by the condition of the first Chern class being zero. Calabi asserts that if this topological condition of a zero first Chern class is met, then there exists a Kähler metric with zero Ricci curvature. You've thus replaced a rather broad, nonspecific statement with a much stronger, more restrictive statement, which is why Greene (and practically everyone else) regarded the proposition as surprising.

I was also wary for some additional technical reasons. It was widely held that no one could ever write down a precise solution to the Calabi conjecture, except perhaps in a small number of special cases. If that supposition were correct—and it was eventually proven to be so—the situation thus seemed hopeless, which is another reason the whole proposition was deemed too good to be true.

There's a comparison to number theory we can make. While there are many numbers we can write down in a straightforward way, there is a much larger class of numbers that we can never write down explicitly. These numbers, called *transcendental*, include e (2.718 . . .) and π (3.1415 . . .), which we can write out to a trillion digits or more but can never write out in full. In technical terms, this is because the numbers cannot be constructed by algebraic manipulation, nor can they be the solution to a polynomial equation whose coefficients are rational numbers. They can only be defined by certain rules, which means they can be narrowed down to a large degree without ever being spelled out verbatim.

Nonlinear equations, such as those pertaining to the Calabi conjecture, are similar. The solution to a nonlinear equation is itself a function. We don't expect to be able to solve them in a clean, explicit manner—as in writing down an exact formula for the solution—because that's simply not possible in most cases. We try to approximate them with the functions we know very well: some polynomial functions, trigonometric functions (such as sine, cosine, and tangent curves), and a few others. If we cannot approximate them with functions we know how to handle, we have trouble.

With that as a backdrop, I tried to find counterexamples to the Calabi conjecture in my spare time. There were moments of excitement: I'd find a line of attack that seemed likely to disprove the conjecture, only to later discover a flaw in my ostensibly impeccable constructions. This happened repeatedly. In 1973, inspiration seized me. This time, I felt I really had something. The approach I took, which we call proof by contradiction, is similar to the approach Richard Schoen and I used for the positive mass conjecture proof. As far as I could tell, my argument was airtight.

Coincidentally, the idea came to me during an international geometry conference held at Stanford in 1973—the same conference at which Geroch spoke about the positive mass conjecture. As a general matter, conferences are a great way of keeping abreast of developments in your field and in fields outside your own specialty, and this one was no exception. They're a terrific venue for exchanging ideas with colleagues you don't see every day. Nevertheless, it's rare for a single conference to change the course of your career. Twice.

During the meeting, I casually mentioned to some colleagues that I might have found a way to disprove Calabi, once and for all. After some prompting, I agreed to discuss my idea informally one night after dinner, even though I was already scheduled to give several other formal talks. Twenty or so people showed up for my presentation, and the atmosphere was charged. After I made my case, everyone seemed to agree that the reasoning was sound. Calabi was there, and he raised no objections, either. I was told that by virtue of this work,

I'd made a big contribution to the conference, and afterward, I felt quite proud of myself.

Calabi contacted me a few months later, asking me to write down the argument, as he was puzzled over certain aspects of it. I then set out to prove, in a more rigorous way, that the conjecture was false. Upon receiving Calabi's note, I felt that the pressure was on me to back up my bold assertion. I worked very hard and barely slept for two weeks, pushing myself to the brink of exhaustion. Each time I thought I'd nailed the proof, my argument broke down at the last second, always in an exceedingly frustrating manner. After those two weeks of agony, I decided there must be something wrong with my reasoning. My only recourse was to give up and try working in the opposite direction. I had concluded, in other words, that the Calabi conjecture must be right, which put me in a curious position: After trying so hard to prove that the conjecture was false, I then had to prove that it was true. And if the conjecture were true, all the stuff that went with it—all the stuff that was supposedly too good to be true—must also be true.

Proving the Calabi conjecture meant proving the existence of a Ricci-flat metric, which meant solving partial differential equations. Not just any partial differential equations, but highly nonlinear ones of a certain type: complex Monge-Ampère equations.

Monge-Ampère equations are named after the French mathematician Gaspard Monge, who started studying equations of this sort around the time of the French Revolution, and the French physicist and mathematician André-Marie Ampère, who continued this work a few decades later. These are not easy equations to work with.

Perhaps the simplest example of one that we can draw from the real world, Calabi suggests, concerns a flat plastic sheet affixed to an immovable rim. Suppose the surface either stretches or shrinks. The question is, when that happens, how does the surface bend or otherwise change from being totally flat? If the middle expands, it will create a bulge extending upward with positive curvature, and the solution to the Monge-Ampère equation will be of an *elliptic* form. Conversely, if the interior shrinks instead, and the surface becomes a saddle with negative curvature everywhere, the solution will be of a *hyperbolic* form. Finally, if the curvature is zero everywhere, the solution will be of a *parabolic* form. In each case, the original Monge-Ampère equation you want to solve is the same, but, as Calabi explains, "they must be solved by completely different techniques."[2]

Of the three forms of differential equations, we have the best techniques for analyzing the elliptic type. Elliptic equations pertain to simpler, stationary situations where things do not move around in time or space. These equations model physical systems that are no longer changing with respect to time, such as a drum that has stopped vibrating and returned to equilibrium. The solutions to elliptic equations, moreover, are the easiest of the three to understand because when you graph them as functions, they look smooth, and you rarely encounter troublesome singularities, though singularities can crop up in the solutions to some nonlinear elliptic equations.

Hyperbolic differential equations model things like waves or vibrations, which may never reach equilibrium. The solutions to these equations, unlike the elliptic ones, generally have singularities and are therefore much harder to work with. While we can handle linear versions of hyperbolic equations reasonably well—in cases where changing one variable leads to a proportional change in another—we don't really have effective tools for handling nonlinear hyperbolic equation's in a way that controls the singularities.

Parabolic equations lie somewhere in between. They model a stable physical system—such as a vibrating drum—that will eventually reach equilibrium but may not have gotten there yet, which brings time evolution or change into the picture. These equations are less prone to singularities than hyperbolic ones, and any singularities there are more readily smoothed out, which again places them somewhere between elliptic and hyperbolic in terms of difficulty.

And there are even tougher mathematical challenges to be had. Although the simplest Monge-Ampère equations have only two variables, many have more. These equations are beyond hyperbolic (and are sometimes called ultrahyperbolic), the solutions of which we know even less about. As Calabi puts it, "we have no clue about these other solutions that go beyond the familiar three because we have absolutely no physical picture to draw on."[3] As a result of the varying difficulty of the three types of equations, most of the contributions from geometric analysis to date have involved either elliptic or parabolic equations.

We're interested in all these equations, of course, and there are plenty of exciting problems that involve hyperbolic equations (such as the full Einstein equations) that we'd like to address, if only we had the wherewithal.

The equations of the Calabi conjecture, fortunately, were of the nonlinear, elliptic variety. That's because, even though it is related to the Einstein equation, which is itself hyperbolic, the Calabi conjecture is based on a somewhat different geometric framework. In this case, we assume that time is frozen, like the scene in *Sleeping Beauty* where nobody moves for a hundred years. That placed the conjecture in the somewhat simpler elliptic category, with time taken out of the equation. For this reason, I felt hopeful that the tools of geometric analysis—including some of those we've touched on so far—might be successfully brought to bear on the problem.

Even with those tools at my disposal, there was still a good deal of preparatory work to be done. Part of the challenge was that no one had ever solved a complex Monge-Ampère equation before except in one dimension. Just as a mountain climber constantly strives to reach higher elevations, I was going to need to push to higher dimensions. To gird myself for higher-dimensional Monge-Ampère equations (whose nonlinearity goes without saying), I set out with my friend S. Y. Cheng to tackle some higher-dimensional cases, starting with problems expressed in real numbers before getting to the more difficult, complex equations.

First we took on a famous problem posed by Hermann Minkowski around the turn of the twentieth century. The Minkowski problem involves taking a prescribed set of information and then determining whether a structure that meets those criteria actually exists. Take a simple polyhedron. Upon examining such a structure, you could characterize it by counting the numbers of faces and edges and measuring their dimensions. The Minkowski problem is like the flip side of that: If you are told the shape, area, number, and orientation of the faces, can you determine whether a polyhedron meeting those criteria actually exists and, if so, whether such an object is unique?

The problem is actually more general than that, because it could apply to an arbitrary convex surface rather than just a polyhedron. Instead of talking about which way the faces are pointing, you could specify the curvature in terms of perpendicular (or "normal") vectors at each spot on the surface, which is equivalent to describing which way the surface is pointing. You could then ask

5.1—The mathematician S. Y. Cheng (Photo by George M. Bergman)

whether an object with the specified curvature could exist. The useful thing is that this problem needn't only be presented in purely geometric terms. It can also be written as a partial differential equation. "If you can solve the geometric problem," says Erwin Lutwak of the Polytechnic Institute of New York University (NYU), "there is this huge bonus: You're also solving this horrific partial differential equation. That interplay between geometry and partial differential equations is one of the things that makes this problem so important."[4]

Cheng and I found a way to solve the problem, and our paper on the subject was published in 1976. (As it turned out, another independent solution had been presented several years earlier in a 1971 paper by the Russian mathematician Aleksei Pogorelov. Cheng and I had not seen that paper, because it was published in Pogorelov's native tongue.) In the end, it came down to solving a complex nonlinear partial differential equation of the sort that had never been solved before.

Even though no one had previously managed to solve a problem of this exact type (except for Pogorelov, whose work we weren't aware of), there was a well-prescribed procedure for attacking nonlinear partial differential equations. The approach, called the continuity method, is based on making a series of estimates. Although the general approach was by no means new, the trick, as always, lies in coming up with a strategy specifically tailored to the problem at hand. The basic idea is to try to successively approximate a solution through a process that keeps yielding better and better results. The essence of the proof lies in showing that this process will, with a sufficient number of iterations, eventually converge on a good solution. If everything works out, what you're left with in the end is not the solution to an equation that you can write down as an exact formula but merely a proof that a solution to this equation exists. In the case of the Calabi conjecture and other problems of this sort, showing that a solution can be found to a partial differential equation is equivalent to an existence proof in geometry—showing that for a given "topological" condition, a specified geometry is indeed viable. This is not to say that you'd know nothing about the solution whose existence you've just proved. That's because the scheme you used to prove the existence of a solution can often be turned into a numerical technique for approximating the solution on a computer. (We'll talk about numerical techniques in Chapter 9.)

The continuity method is so named because it involves taking the solution to an equation you know how to solve and continuously deforming it until you arrive at a solution to the equation you want to solve. This procedure, which was used in the Calabi conjecture proof, is typically broken down into two parts, one of which only works in the immediate vicinity of the known solution.

One of these parts we'll call Newton's method, because it's roughly based on a technique Isaac Newton invented more than three hundred years ago. To see how Newton's method works, let's take a function, $y = x^3 - 3x + 1$, which describes a curve that intersects the x-axis at three points (or *roots*). Newton's approach enables us to find where those three roots lie, which is not obvious just by looking at the equation. Let's assume we can't solve the equation outright, so we guess that one root can be found at a point called x_1. If we take a tangent to the curve at that point, the tangent line will intersect the x-axis at another point, which we'll call x_2, that is even closer to the root. If we then take a tangent to the curve at x_2, it will intersect the x-axis closer still to the root at x_3. The process will converge quickly on the root itself if the initial guess is not too far off the mark.

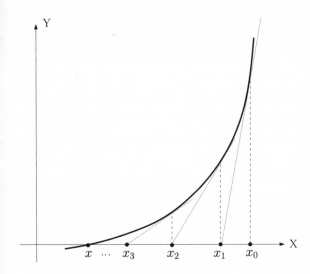

5.2—An introduction to *Newton's method*. To find out where a particular curve, or function, crosses the X-axis, we start off with our best guess—a point we'll call x_0. We then take the tangent of the curve at x_0 and see where the tangent line crosses the X-axis (at a point we'll call x_1). We continue this process at x, and so forth. Assuming our initial guess was not too far off, we'll come closer and closer to the true answer, point x.

As another example, let's suppose we have a whole series of equations, E_t, only one of which, E_0 (pertaining to the case when $t = 0$), we know how to solve. But the equation we really want to solve is E_1, which concerns the case when $t = 1$. We could use Newton's method when we're very close to 0, where we know the solution, but this approach may not carry us all the way to 1. In that event, we'll need to bring in another, more broadly applicable estimation technique.

So how do we do that? Imagine somebody shot a missile over the Pacific Ocean that landed within a hundred-mile radius of the Bikini Atoll. That gives us some idea of where the missile could be—its general position, in other words—but we'd like to know more, such as its velocity or its acceleration over the course of its flight or how its acceleration changed. We'd do this with calculus, taking the first, second, and third derivatives of an equation describing the rocket's location. (We could take even more derivatives, but for second-order elliptic equations of the type I was interested in, the third derivative is usually high enough.)

Just getting these derivatives is not enough, even though this alone may be extremely challenging. We also have to be able to "control" them. And by that I mean putting bounds on them—making sure they cannot get too big or too small. Making sure, in other words, that the solutions we get are "stable" and do not blow up on us, in which case they'd disqualify themselves as solutions, dashing our dreams in the process. So we start with the zeroth derivative, the location of the missile, and see if we can set some upper and lower bounds—make estimates, in other words, that suggest that an answer is at least possible. We'd then do the same for each higher-order derivative, making sure that they are not too big or too small and that the functions describing them do not fluctuate too wildly. This involves making a priori estimates for the velocity, acceleration, change in acceleration, and so forth. If we can establish control in this way, from the zeroth through the third derivative, normally we have good control over the equation as a whole and have a decent chance of solving it. This process of estimation, and proving that the estimates themselves can be controlled, is typically the hardest part of the whole process.

So in the end, it all boils down to estimates. There was something ironic in my recognizing their relevance to the problem I faced. When I first entered graduate school, I remember encountering two Italian postdocs in the halls of the Berkeley math department. They were jumping up and down, screaming jubilantly. I asked them what had happened and was told they'd just obtained an estimate. When I asked them what an estimate was, they acted as if I were an ignoramus who had no business setting foot in the building. That's when I decided I might want to learn about a priori estimates. Calabi received a similar dressing-down some decades earlier from his friend and collaborator Louis Nirenberg: "Repeat after me," Nirenberg said, "you cannot solve partial differential equations without a priori estimates!"[5] And in the early 1950s, Calabi had written to André Weil about his conjecture. Weil, who thought the "technology" of that era was simply not mature enough to yield a solution, asked him, "How are you going to get the estimates?"[6]

Two decades later, which is when I entered the game, the problem itself had not changed. It was still incredibly difficult, yet the tools had advanced to the point that a solution was within the realm of possibility. It was a question of figuring out a line of attack or at least establishing some sort of foothold. So I picked an easier equation and then sought to show that the solution to it could eventually be "deformed" to the solution of the harder equation.

Suppose you want to solve the equation $f(x) = x^2 - x$ when $f(x) = 0$. For starters, we could try $x = 2$, which doesn't quite work; $f(2) = 2$, not 0. Nevertheless, I now have a solution, if not to the original equation then to something similar. So next I rewrite the equation as $f(x) = 2t$. I know how to solve it when $t = 1$, but I'd really like to solve it for $t = 0$, which is the equation we started with in the first place. So what do I do? I look at the parameter t. I know how to solve the equation for $t = 1$, but what happens when I move t a bit so that it's not exactly 1 but is still close? I then guess that when t is close to 1, there exists a solution to $f(t)$ that is close to 2. That assumption turns out to be true in most cases, which means that when t is close to 1, I can solve the equation.

Now I want to make t smaller and smaller and approach 0 so that we can get to our original equation. I keep on making t smaller and smaller, and for each value I pick, I can find a solution. So I have a sequence of points where I can solve the equation, and each of these points corresponds to a different value of x, which I'll collectively call x_i. The whole point of this exercise is to prove that the sequence xi converges on a particular value. To do that, we must show that xi is bounded and cannot move to infinity, because for any bounded sequence, at least some part of it must converge. Showing that x_i converges is equivalent to showing that we can reduce the value of t all the way down to 0 without running into any insurmountable difficulties. And if we can do that, we'll have solved the equation by showing that $t = 0$ can be solved. In other words, we've shown that a solution to our original function, $x^2 - x = 0$, must exist.

This is exactly the kind of argument I used in the Calabi proof. A key part of that proof, in fact, was showing that xi is a convergent sequence. Of course, in tackling the Calabi conjecture, the starting equation is obviously more complicated than $x^2 - x = 0$. In the Calabi case, x is a function rather than a number, which increases the complexity immensely because the convergence of a function is normally a very tough thing to prove.

Again, we break the big problem into smaller pieces. The equation of the Calabi conjecture is a second-order elliptic equation, and to solve equations like this, we need to work out the zeroth-, first-, second-, and third-order estimates. Once those estimates are completed, and you can prove they converge to the desired solution, you've proved the entire conjecture. That's easier said than done, of course, because solving those four estimates is not necessarily easy at all. I guess that's why they call it work.

But there were still more preliminaries to be done before Cheng and I could attack complex Monge-Ampère equations. We started working on the Dirichlet problem, named after the German mathematician Lejeune Dirichlet. It's what we call a boundary value problem, and tackling such problems is usually the first thing we do when trying to solve an elliptic differential equation. We discussed an example of a boundary value problem in Chapter 3, the Plateau problem, which is often visualized in terms of soap films and which stated that for an arbitrary closed curve, one could find a minimal surface stretching across that same boundary. Every point on that surface is also a solution to a particular differential equation. The question boils down to this: If you know the boundary of the solution to such an equation, can you find the interior surface that not only connects to the boundary but solves the equation as well? While the Calabi conjecture is not a boundary value problem, Cheng and I needed to test the methods, which we could then bring to bear on complex Monge-Ampère equations like Calabi's. For practice, we were trying to solve the Dirichlet problem in certain domains in complex Euclidean spaces.

To solve the Dirichlet problem, one has to go through the same steps we've just outlined, finding the zeroth-, first-, second-, and third-order estimates for the boundary. But we also need to make the same set of estimates for the interior of the curve, as there could be discontinuities, singularities, or other departures from smoothness in the "soap bubble" under consideration. That means eight estimates in all.

By early 1974, Calabi and Nirenberg, who were also working on this Dirichlet problem, had gotten the second-order estimate, as had Cheng and I. The zeroth-order estimate turned out to be relatively easy. And the first-order estimate could be derived from the zeroth and second. That left the third order, wherein lay the key to the whole Dirichlet problem.

The tools required for finding the solution first emerged in the late 1950s, while I was still in grade school, when Calabi solved a major problem in geometry that later proved critical in understanding how to obtain the third-order interior estimate for the real Monge-Ampère equation. Calabi's contributions to this estimate stemmed, in part, from a coincidence. He had been working on a seemingly unrelated problem in affine geometry (a generalization of Euclidean geometry that, being somewhat far afield, I won't take the time to explain here), while Nirenberg and Charles Loewner of Stanford were working on the Dirichlet problem of a Monge-Ampère equation with a singular boundary (like the cresting edge of a wave) rather than a smooth boundary. After seeing the equation that Nirenberg and Loewner had studied, Calabi realized it was related to his work on affine geometry. Calabi and Nirenberg were able to figure out how Calabi's efforts from the 1950s could be applied to the third-order interior estimate problem we were facing in the 1970s. "A lot of mathematical discoveries occur through lucky accidents like that," Calabi notes. "It's often a matter of connecting up ideas that might seem un-related and then exploiting this newfound connection."[7]

Later in 1974, Calabi and Nirenberg an-nounced the solution to the boundary value problem for the complex Monge-Ampère equation. But it turns out they had made a mistake: The third-order boundary estimate was still missing.

5.3—The mathematician Louis Nirenberg

Cheng and I soon came up with our own solution for the third-order boundary estimate. Chern had invited us to join him and Nirenberg for dinner, during which we'd present our proof. Nirenberg was a big shot at the time, while we were barely out of graduate school, so we checked our proof carefully the night before the dinner and, to our dismay, found mistakes. We stayed up the entire night fixing those mistakes and thereby fixing the proof as well. We showed it to Nirenberg the next night before dinner. He thought everything looked fine; we thought it looked fine, too, and everyone enjoyed the meal. But after dinner, Cheng and I went through the proof again and found additional mistakes. It wasn't until late in 1974, about six months later, that Cheng and I solved the boundary value problem. We did it by studying an equation similar to what Loewner and Nirenberg had looked at, only in higher dimensions. The method we used bypassed the third-order boundary estimate, showing why it was unnecessary.

That work done, I was ready to take on the complex case of Calabi itself—a problem that was set on a complex manifold, in contrast to the Dirichlet problem, which was set in complex Euclidean space. So eager was I, in fact, that we didn't get around to publishing the paper on the Dirichlet problem until 1979—about five years later.

With the Dirichlet problem put to rest, much of the work that lay ahead involved generalizing or translating the estimates for the real Monge-Ampère equation to the complex case. But I was to travel alone from this point on, as Cheng's interests lay in other directions.

Sometime in 1974, Calabi and Nirenberg, along with J. J. Kohn of Princeton, started to work on the complex version of the Dirichlet problem in Euclidean space. They made some headway on the third-order estimate, and I was able to extend their result for curved space. Later that year, I got some ideas for the second-order estimate of the Calabi conjecture as well, drawing on some work I'd done in 1972 on something called the Schwarz lemma. The lemma, or mini-theorem, dates back to the nineteenth century and originally had nothing to do with geometry until it was reinterpreted by Lars Ahlfors of Harvard in the first half of the twentieth century. Ahlfors's theorem pertained just to Riemann surfaces (of one complex dimension, by definition), but I generalized his theorem to any complex dimension.

I finished the preliminaries on the second-order estimate for the Calabi conjecture during the summer of 1975. (A year later, I learned that the French mathematician Thierry Aubin had independently arrived at an estimate for the second-order problem.) In addition to determining the second-order estimate, I showed how it depended on the zeroth-order estimate and how one could go from the zeroth to the second. After I'd finished the work on the second-order estimate, I knew that the entire proof now rested on a single question, the zeroth-order problem. Once that was knocked off, I could not only get the second, but the first as well, which was a kind of freebie. For when we had the zeroth- and second-order estimates, we'd know exactly how to do the first-order estimate. It was one of those lucky breaks, a matter of how the cookie crumbles, and this time, it crumbled the right way. Not only that, but the third-order estimate also depended on the zeroth and second, so in the end, it all came down to the zeroth-order estimate. With that in hand, everything else would fall into place. Without it, I would have nothing.

I had done this latest work at NYU's Courant Institute, securing a visiting position through the help of Nirenberg. Before long, my fiancé Yu-Yun, who had been working at Princeton, had been offered a job in Los Angeles. I took another visiting faculty position at UCLA to be with her. In 1976, we drove across country together, planning to get married when we got to California. (And we did, in fact, get married in California.) It was a memorable trip: We were in love. The scenery was beautiful. And we shared our visions of a future life together. But I must confess to being more than a bit distracted: I still had the Calabi conjecture on my mind and the zeroth-order estimate in particular, which was proving to be most obstinate. I spent the entire year working on it. Finally, in September 1976, right after our wedding, I found the zeroth-order estimate, and the rest of the proof fell into place with it. Married life appears to have been just what I needed.

The problem of getting the zeroth-order estimate was similar to getting the other estimates: You have an equation or a function that you'd like to put some constraints on—both from above and from below. In other words, you want to put the function in a metaphorical box and show that the box doesn't have to be too big for the function to "fit in." If you can do that, you've bounded the function from above. At the other end, you need to show that the function cannot be so small that it will somehow "leak out" of the box, thereby bounding it from below.

One way to approach a problem of this sort is to take the *absolute value* of the function, which tells you the function's overall magnitude—how big it gets in either the negative or the positive direction. In order to control a function, u, we just have to show that its absolute value at any point in space is less than or equal to a constant, c. Since c is a well-defined number, we've shown that u cannot be arbitrarily big or small. What we want to prove, in other words, is a simple inequality, that the absolute value of u was less than or equal to c: $|u| \le c$. While that may not look so hard, it can be challenging when u happens to be a particularly messy object.

I won't go through the details of the proof, but I will say that it drew on the second-order Monge-Ampère problem, which I'd already solved. I also used a famous inequality by Poincaré as well as an inequality by the Russian mathematician Sergei Sobolev. Both inequalities involved integrals and derivatives (of various orders) of the absolute value of u taken to a certain power or exponent. The last part, involving exponents, is crucial for making estimates, because if you can show that u, in various forms, taken to the pth power is neither too big nor too small—even when p is a very large number—you've done your work. You've stabilized the function. In the end—by using these inequalities and various theorems, as well as some lemmas I'd developed along the way—that's what I was able to do. With the zeroth-order estimate in hand, the case was closed, or so it appeared.

Of course, people often say the proof of the pudding is in the eating, which presumably means that even when something looks good, you don't know for sure until you put it to a test. And this time, with regard to this proof, I wasn't taking any chances. I had already embarrassed myself once in 1973, when I'd claimed at Stanford, in a highly public forum, that I knew how to disprove the Calabi conjecture. My alleged proof of the negative case had been flawed, and if my proof of the affirmative case were similarly flawed, I would not be doing much to bolster my reputation. In fact, I was convinced that at this stage of my career, still in my twenties, I couldn't afford to be wrong again—or at least not wrong on something so important and high-profile.

So I checked it and rechecked it, going through the proof four times in four different ways. I checked it so many times that I vowed that if I was wrong about this, I'd give up mathematics altogether. But try as I might, I couldn't find anything wrong with my argument. As far as I could tell, it all added up. Since in those days, there was no World Wide Web where I could just post a draft of a paper and solicit comments, I did it the old-fashioned way: I mailed a copy of my proof to Calabi and then went to Philadelphia to discuss it with him and other geometers at the University of Pennsylvania math department, including Jerry Kazdan.

Calabi thought the proof appeared solid, but we decided to meet with Nirenberg and go through it together, step by step. As it was hard to find a time when all three of us were free, we met on Christmas Day 1976, when none of us had any other pressing business engagements. No flaws were uncovered during that meeting, but it would take more time to be sure. "Tentatively, it looked good," Calabi recalls. "Still it was a difficult proof that would take about a month to check thoroughly."[8]

By the end of that review period, both Calabi and Nirenberg had signed off. It appeared that the Calabi conjecture had been proved, and in the thirty-plus years that have transpired since, nothing has come up to alter that judgment. By now, the proof has been cross-examined so many times by so many people it's virtually impossible to imagine that a significant defect will ever be exposed.

So what exactly had I accomplished? In proving the conjecture, I had validated my conviction that important mathematical problems could be solved by combining nonlinear partial differential equations with geometry. More specifically, I had proved that a Ricci-flat metric can be found for compact Kähler spaces with a vanishing first Chern class, even though I could not produce a precise formula for the metric itself. All I could say was that the metric was there without being able to say exactly what it was.

Although that might not sound like much, the metric I proved to be "there" turned out to be pretty magical. For as a consequence of the proof, I had confirmed the existence of many fantastic, multidimensional shapes (now called Calabi-Yau spaces) that satisfy the Einstein equation in the case where matter is absent. I had produced not just *a* solution to the Einstein equation, but also the largest class of solutions to that equation that we know of.

I also showed that by changing the topology continuously, one could produce an infinite class of solutions to the key equation of the Calabi conjecture (now called the Calabi-Yau equation), which is itself a special case of the Einstein equation. The solutions to this equation were themselves topological spaces, and the power of the proof lay in the fact that it was completely general. In other words, I had proved the existence of not just one example of such spaces, or a special case, but a broad class of examples. I showed, moreover, that when you fix certain topological data—such as some complex submanifolds within the bigger manifold—there is only one possible solution.

Before my proof, the only known compact spaces that satisfied the requirements set down by Einstein were locally homogeneous, meaning that any two points near each other will look the same. But the spaces I identified were both inhomogeneous and asymmetrical—or at least lacking in a sweeping global symmetry, though they were endowed with the less visible internal symmetry we discussed in the last chapter. To me, that was like overcoming a great barrier, because once you go beyond the strictures of global symmetry, you open up all kinds of possibilities, making the world more interesting and more confusing at the same time.

At first, I just reveled in the beauty of these intricate spaces and the beauty of the curvature itself, without thinking about possible applications. But before long, there would be many applications, both inside and outside mathematics. While we once believed that Calabi's proposition was "too good to be true," we'd just found out that it was even better.

Six

THE DNA OF STRING THEORY

When searching for diamonds, if you're lucky you may find other precious gems as well. When I announced the proof of the Calabi conjecture in a two-page 1977 paper, which was followed by a seventy-three-page 1978 elaboration, I also announced the proof of five other related theorems. The windfall was, in many ways, a consequence of the unusual circumstances under which I approached the Calabi conjecture—first trying to prove it wrong before shifting gears and trying to prove it right. Luckily, it turns out that none of that effort was wasted: I could use every apparent misstep, every ostensible dead end. For my alleged counter-examples—ideas logically implied by the conjecture that I'd hoped to prove false—turned out to be true, as was the conjecture itself. These failed counter-examples were, in fact, real examples and soon became mathematical theorems in themselves—a few of some note.

The most important of these theorems led to the proof of the Severi conjecture—a complex version of the Poincaré conjecture that had remained unsolved for more than two decades. But before getting to that, I had to solve an important inequality for classifying the topology of surfaces, which I became interested in partly as a result of a talk by Harvard mathematician David Mumford, who was passing through California at that time. The problem, first posed by Antonius van de Ven of Leiden University, concerned an inequality among Chern classes for Kähler manifolds. Van de Ven had initially proved that the second Chern class of a manifold multiplied by 8 must be greater than or equal to the square of the first Chern class of that same manifold. Many people believed this inequality could be made much stronger by inserting the constant 3 into the expression in place of 8. Indeed, 3 was considered the optimal value. The question Mumford posed was whether this more stringent inequality could be proved. (Mumford's version was more stringent because it claimed that a certain quantity, the second Chern class, was greater than something else—hence the inequality—not only when multiplied by 8 but even when multiplied by the smaller number 3.)

He asked this question in September 1976 during a lecture at the University of California, Irvine; I was attending the lecture not long after finishing my proof of the Calabi conjecture. In the middle of the talk, I became almost certain that I'd encountered this problem before. In a discussion after the lecture, I told Mumford I thought I could prove this more difficult case. When I got home, I checked my calculations and found out that just as I'd suspected, I had tried to use this kind of inequality in 1973 to disprove the Calabi conjecture; now I could use the Calabi-Yau theorem to prove that this inequality held. What's more, I was able to use a special case of that statement—namely, the equality case (3 times the second Chern class *equals* the square of the first Chern class)—to solve the Severi conjecture, as well.

These two theorems, which proved the Severi conjecture and the more general inequality (sometimes called the Bogomolov-Miyaoka-Yau inequality to acknowledge two other contributors to this problem), were the first major spin-offs of the Calabi proof, but many others were to follow. Calabi's conjecture was, in fact, broader than I've indicated so far. It applied not only to the case of zero Ricci curvature, but to the cases of constant negative and constant positive Ricci curvature as well. No one has yet proved the positive curvature case in its full generality, where Calabi's original conjecture is known to be false. (I formulated a new conjecture that specified a condition under which metrics with positive Ricci curvature could exist. Over the past two decades, many mathematicians, including Simon Donaldson, have made serious contributions toward this conjecture, but it has yet to be proven.) Nevertheless, I did prove the negative curvature case as part of my overall argument, and this was independently proved by the French mathematician Thierry Aubin as well. The solution to the negative curvature case established the existence of a broad class of objects called Kähler-Einstein manifolds, thereby establishing a new field of geometry that has turned out to be surprisingly fruitful.

It's fair to say I had a good time pursuing the immediate applications of the Calabi conjecture, knocking off a half dozen or so proofs in short order. It turns out that once you know a metric exists, there are all kinds of consequences. You can use that knowledge to work backward and deduce things about manifold topology, without knowing the exact metric. We can use those properties, in turn, to identify unique features of a manifold in the same way we can identify a galaxy without knowing everything about it, or much as we don't need a full deck of cards to be able to deduce a great deal about the deck (including the total number of cards and the markings on each one). To me, that's kind of magical and says even more about the power of mathematics than a situation in which we have every last detail in front of us.

It was gratifying to reap some of the rewards of my arduous labors and to see others pursue avenues that hadn't occurred to me. Despite that good fortune, something was gnawing at me. Deep down, I felt certain that this work would have implications for physics, in addition to mathematics, though I did not know exactly how. In a way, this sentiment was rather obvious since the differential equations you are trying to solve in the Calabi conjecture—in the zero Ricci curvature case—are literally Einstein's equations of empty space, corresponding to a universe with no background energy, or a cosmological constant of no value. (These days, the cosmological constant is generally thought to be positive and synonymous with the dark energy that is pushing the universe to expand). Likewise, Calabi-Yau manifolds are regarded as solutions to Einstein's differential equations, just as a unit circle is the solution to the equation $x^2 + y^2 = 1$.

Of course, you need many more equations to describe Calabi-Yau spaces than it takes to describe a circle, and the equations themselves are much more complicated, but the basic idea is the same. The Calabi-Yau equations not only satisfy the Einstein equations, they do so in a particularly elegant way that I, at least, found arresting. That made me think they had to fit into physics somewhere. I just didn't know where.

Nor was there much I could do about it other than telling my physics postdocs and physicist friends why I thought the Calabi conjecture, and the so-called Yau's theorem that emerged from it, might be important for quantum gravity. The main problem was that I didn't understand quantum gravity well enough then to follow up on my intuition. I brought up the idea from time to time, but mainly had to sit back and see what came of it.

The years passed, and as other mathematicians and I continued to follow up on the Calabi conjecture in the course of pushing through our broader agenda in geometric analysis, there were some stirrings in the physics world going on behind the scenes and, for a little while at least, unbeknownst to me. It started in 1984, which turned out to be a landmark year in which string theory took great strides from being a somewhat general idea toward becoming an actual, flesh-and-blood theory.

Before getting to those exciting developments, let's say a bit more about this theory that brashly attempts to bridge the gap between general relativity and quantum mechanics. At its core is the notion that the smallest bits of matter and energy are not pointlike particles but are instead tiny, vibrating pieces of string, which assume the form of either loops or open strands. Just as a guitar is capable of playing many different notes, these fundamental strings can vibrate in many different ways. String theory posits that strings vibrating in different ways correspond to the different particles in nature as well as the different forces. And, assuming the theory works—a matter yet to be settled—that's how unification is achieved: These particles and forces share a common bond because they're all manifestations, and excitations, of the same basic string. You could say that's what the universe is made of: When you get down to it, at the most elementary level, it's all strings.

String theory borrows from Kaluza-Klein the general notion that extra dimensions are required for this grand synthesis to be realized. Part of the argument is the same: There is simply not enough room for all the forces—gravity, electromagnetism, weak, and strong—to fit into a single four-dimensional theory. If one were to follow a Kaluza-Klein approach and ask how many dimensions are needed to combine all four forces within a single framework—with five covering gravity and electromagnetism, a couple more for the weak force, and a few more for the strong—you'd need a minimum of eleven dimensions. But it turns out that this approach doesn't quite work—one of the many aspects of the theory we've learned from the physicist Edward Witten.

Fortunately, string theory doesn't go about things in such an ad hoc fashion, picking an arbitrary number of dimensions, scaling up the size of the matrix or Riemann metric tensor, and seeing what forces you can or cannot accommodate. Instead, the theory tells you exactly how many dimensions are needed for the job, and that number is ten—the four dimensions of the "conventional" spacetime we probe with our telescopes, plus six extra dimensions.

The reason string theory demands ten dimensions is fairly complex and stems from the need to preserve symmetry—essential for any purported theory of nature—and to make the theory consistent with quantum mechanics, which surely has to be one of the key ingredients. But the argument essentially boils down to this: The more dimensions you have, the more possible ways a string can vibrate. To reproduce the full range of possibilities in our universe, string theory requires not only a very large number of potential vibrations but a specific number that you only get in ten-dimensional spacetime. (Later in this chapter, we'll discuss a variation, or "generalization," of string theory called M-theory that involves eleven dimensions, but we'll leave off that for now.)

A string confined to one dimension can only vibrate *longitudinally*, by stretching and compressing. In two dimensions, a string can vibrate that way and in the perpendicular, or *transverse*, direction as well. In dimensions three and higher, the number of independent vibrational patterns continues to grow until you reach dimension ten (nine spatial dimensions and one of time), where the mathematical requirements of string theory are satisfied. That's why string theory requires at least ten dimensions. The reason string theory requires exactly ten dimensions to be consistent, no more and no less, relates to something called *anomaly cancellation*, which brings us back to our narrative and the year 1984.

Most string theories developed up to that point had been plagued by anomalies or inconsistencies that rendered any predictions they made nonsensical. The theories, for example, tended to have the wrong kind of left-right symmetry— one that was incompatible with quantum theory. A key breakthrough was made by Michael Green, then at Queen Mary College of London, and John Schwarz of the California Institute of Technology. The main problem Green and Schwarz overcame related to *parity violation*—the idea that the fundamental laws of nature distinguish between left and right and are not symmetric in this respect. Green and Schwarz found ways of formulating string theory such that parity violation was upheld. The quantum effects that had riddled the theory with inconsistencies miraculously canceled themselves out in ten spacetime dimensions, raising hopes that the theory might actually describe nature. Green and Schwarz's achievement marked the start of what was called the first string revolution. And with those anomalies suddenly dispensed with, it was time to see whether the theory could lead to some realistic physics.

Part of the challenge is to see whether string theory can explain why the universe looks the way it does. That explanation must account for the fact that we inhabit a spacetime that looks four-dimensional, while the theory insists it's actually ten-dimensional. The answer to that apparent discrepancy, according to string theory, lies in *compactification*. The notion is not entirely new, as Kaluza and Klein (and Klein, in particular) had already suggested that the extra dimension in their five-dimensional theory was indeed compactified—shrunk down so small we couldn't see it. String theorists have made a similar case, only they have six dimensions to dispose of rather than just one.

Actually, this last remark is somewhat misleading, as we're not really trying to get rid of dimensions. The trick, instead, is to wrap them up in a very exacting way—coiled up within a geometry whose precise contours are critical to the magic act that string theory is trying to pull off. There are many geometries to choose from, each of which leads to a different possible compactification.

The whole idea, according to Harvard physicist Cumrun Vafa, can be summed up in a simple equation that everyone can understand: $4 + 6 = 10$.[1] That's all there is to it, though you might want to rephrase it as $10 - 6 = 4$, indicating that once the six dimensions are concealed (or subtracted away), what is in actuality a ten-dimensional universe appears to have just four. Compactification can, equivalently, be thought of as a funny kind of multiplication known as a Cartesian product—a product in which the number of dimensions are added together rather than multiplied. The relevant equation that describes the *product manifold* in which the four and six dimensions combine suggests that our ten-dimensional spacetime has a substructure; it is literally the product of four- and six-dimensional spacetime, just as a plane is the product of two lines, and a cylinder is the product of a line and a circle. Moreover, the cylinder, as we've seen, is how the Kaluza-Klein concept is often illustrated. If you start by depicting our four-dimensional spacetime as an infinite line that stretches forever in both directions and then snip the line and magnify one of the ends, you'll see that the line actually has some breadth and is more accurately described as a cylinder, albeit one of minuscule radius. And it is within this circle of tiny radius that the fifth dimension of Kaluza-Klein theory is hidden. String theory takes that idea several steps further, arguing in effect that when you look at the cross-section of this slender cylinder with an even more powerful microscope, you'll see six dimensions lurking inside instead of just one. No matter where you are in four-dimensional spacetime, or where you are on the surface of this infinitely long cylinder, attached to each point is a tiny, six-dimensional space. And no matter where you stand in this infinite space, the compact six-dimensional space that's hiding "next door" is exactly the same.

That, of course, is merely a crude, schematic picture that tells us nothing about the actual geometry of this shrunken, six-dimensional world. Imagine, for example, if we took an ordinary sphere—a two-dimensional surface—and reduced it to a point, a zero-dimensional object. In that way, we've compactified two dimensions down to none. We could try to reduce ten to four by shrinking a six-dimensional sphere $(a^2 + b^2 + c^2 + d^2 + e^2 + f^2 = 1)$, but that would not work as the geometry for the extra dimensions; the equations of string theory demand that the six-dimensional space must have a very particular structure that a simple sphere does not possess.

A more complicated shape was clearly needed, and after Green and Schwarz's success with parity violation, the task of finding that shape became quite urgent. Once physicists had the proper manifold in which to curl up the extra six dimensions, they could finally try to do some real physics.

As an initial follow-up step in 1984, Green, Schwarz, and Peter West of King's College decided to look at K3 surfaces—a broad class of complex manifolds that had been studied by mathematicians for more than a century, though K3's attracted the more recent attention of physicists when my proof of the Calabi conjecture showed that these surfaces could support a Ricci-flat metric. "What I understood was that the compact space had to be Ricci flat to insure that the lower-dimensional space we inhabit would not have a positive cosmological constant, which was considered to be a fact of our universe at the time," recalls Schwarz.[2] (In light of the subsequent discovery of dark energy, implying an extremely small but positive cosmological constant, string theorists have devised more complicated ways of producing a tiny cosmological constant in our four-dimensional world from compact, Ricci-flat spaces—a subject to be addressed in Chapter 10.)

A K3 surface—a name that alludes both to the K2 mountain peak and to three mathematicians who explored the geometry of these spaces, Ernst Kummer, the aforementioned Erich Kähler, and Kunihiko Kodaira—was selected for this preliminary inquiry despite the fact that it's a manifold of just four real (and two complex) dimensions rather than the requisite six, in part because Green, Schwarz, and West had been told by a colleague that there were no higher-dimensional analogues of that manifold. Nevertheless, says Green, "I am by no means convinced we would have been able to figure things out . . . even if we had been given the correct information [about the existence of six-dimensional analogues of Ricci-flat K3's] at that time."[3] Going with the tried-and-true K3, Schwarz adds, "wasn't an attempt to do a realistic compactification. We just wanted to play around, see what we got, and see how that meshed with anomaly cancellation."[4] Since then, K3 surfaces have been invaluable to string theorists in this regard as oft-used "toy models" for compactifications. (They are also essential models for exploring the dualities of string theory, which will be discussed in the next chapter.)

At roughly the same time, also in 1984, the physicist Andrew Strominger, now at Harvard but then at the Institute for Advanced Study (IAS) in Princeton, teamed up with Philip Candelas, a mathematically inclined physicist now at Oxford who was then at the University of Texas, to figure out what class of six-dimensional shapes might meet the exacting conditions set by string theory. They knew that the internal space of those shapes had to be compact (to get down from ten dimensions to four) and that the curvature had to satisfy both the Einstein gravity equations and the symmetry requirements of string theory. Their explorations ultimately took them and two other colleagues—Gary Horowitz of the University of California, Santa Barbara, and Witten—to the spaces whose existence I had confirmed in the Calabi conjecture proof (though Witten had followed his own path to those geometric forms). "One of the beautiful things about developments in modern science is that physicists and mathematicians are often led to the same structures for different reasons," Strominger observes. "Sometimes, the physicists are ahead of the mathematicians; sometimes, the mathematicians are ahead of the physicists. This is a case where the mathematicians were ahead. They understood the significance before we did."[5]

Although what Strominger says is true, to an extent, it's also true that mathematicians like me originally had no idea how Calabi-Yau spaces tied into physics. I investigated them because I thought they were beautiful; it was because of their great beauty that I felt physicists ought to be able to do something with them—that they harbored some mysteries worth uncovering. Ultimately, it was up to the physicists themselves to make that connection, bridging the gap between geometry and physics and, in so doing, initiating a long and fertile collaboration between the two fields—a collaboration that flourishes to this day.

How that connection was made turns out to be an interesting tale in itself. Strominger sums it up this way: "Supersymmetry was the bridge to holonomy, and holonomy was the bridge to Calabi-Yau."[6]

As you may recall, we briefly discussed supersymmetry in Chapter 4, in the context of a kind of limited, internal symmetry—as opposed to the more sweeping, global symmetry of an object like a sphere—that Calabi-Yau manifolds (being a class of Kähler manifolds) must possess. This internal symmetry is part of what we mean by supersymmetry, but before we try to paint a clearer picture of that, let's say a few words about holonomy first.

Loosely speaking, *holonomy* is a measure of how tangent vectors on a particular surface get twisted up as you attempt to parallel-transport them on a loop around that surface. Imagine, for example, you're standing on the north pole, holding a spear that is tangent to the earth's surface. For starters, you'll walk directly to the equator, following the direction in which the spear is initially pointing—all the while keeping the spear pointing in the same direction, from your perspective, as you proceed.

When you reach the equator, the spear will be pointing down. Now you'll follow the equator halfway around the earth, keeping the spear pointing down as you do so. Once you've gone halfway around, you'll head to the north pole, with the spear still pointing in the same direction the whole time. When you arrive at the north pole, you'll find that your spear has rotated 180 degrees from its initial direction, despite your best efforts to maintain a fixed bearing.

We could repeat this process any number of times, making shorter or longer trips along the equator, only to discover that our spear has rotated by different

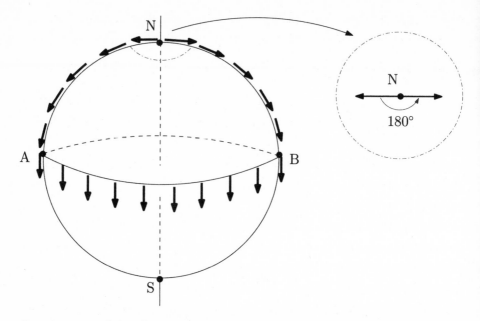

6.1—One way of classifying a space or surface is through *holonomy*, which tells you what happens to a tangent vector as you parallel-transport it—that is, try to keep it pointing in the same direction as you move it along a path that may itself be curved or twisty. In this example, we'll start at the north pole with a tangent vector that's pointing due west and walk to the equator. When we arrive at the equator, our vector is now pointing due south. We'll keep it pointing south as we walk along the equator from A to B, which will take us halfway around the globe. Now we'll walk up to the north pole again, keeping the vector fixed as we do so. When we arrive at the north pole, we'll find that our vector has rotated 180 degrees, despite our best efforts to keep it pointing the same way at all times.

Depending on what path you take on the surface of a globe or sphere, you can end up with any conceivable rotation angle. Once you know the set of all possible angles, you can classify a surface by its holonomy group—a two-dimensional sphere belonging to the special orthogonal group 2, or SO(2).

angles, sometimes less than 180 degrees and sometimes more, depending on the length of our trip along the equator. To determine the holonomy of our planet, a two-dimensional sphere, we need to consider all possible paths—all possible loops—you can make on the surface. It turns out that on the surface of a sphere, you can get any angle of rotation you want, from 0 to 360 degrees, by making your loop larger or smaller. (And you can even go beyond 360 degrees, too, by retracing your steps a second time or more.) We say the two-dimensional sphere belongs to the holonomy group SO(2), or special orthogonal group 2, which is a group that contains all possible angles. (Higher-dimensional spheres belong to $SO[n]$, which are the groups that include all possible orientation-preserving rotations, where n refers to the number of dimensions.)

A Calabi-Yau manifold, on the other hand, belongs to the much more re-strictive $SU(n)$ holonomy group, which stands for the special unitary group of n complex dimensions. The Calabi-Yau manifolds of primary interest to string theory have three complex dimensions, which places them in the $SU(3)$ holonomy group. Calabi-Yau manifolds are a lot more complicated than spheres, and $SU(3)$ holonomy is a lot more complicated than our foregoing example of a vector being rotated, despite our best efforts to keep it pointing in the same direction, as we move it on the surface of a sphere. What's more, because a Calabi-Yau manifold (unlike a sphere) lacks global symmetry, there is no axis around which we can rotate the manifold and leave it unchanged. But it does have a more limited kind of symmetry, as we've discussed, which relates both to holonomy and to supersymmetry. For a manifold to have supersymme-try, it must have what is called a *covariantly constant spinor*. Spinors, though difficult to describe, are analogues of tangent vectors. On a Kähler manifold, there exists one spinor that remains invariant under parallel transport on any closed loop. On Calabi-Yau manifolds—and in the $SU(3)$ group to which they belong—there is an additional spinor that also remains invariant under parallel transport on any closed loop on the manifold.

The presence of these spinors helps ensure the supersymmetry of the man-ifolds in question, and the demand for supersymmetry of the right sort is what pointed Strominger and Candelas to $SU(3)$ holonomy in the first place. $SU(3)$, in turn, is the holonomy group associated with compact, Kähler manifolds with a vanishing first Chern class and zero Ricci curvature. $SU(3)$ holonomy, in other words, implies a Calabi-Yau manifold. Putting this in equivalent terms, if you want to satisfy the Einstein equations as well as the supersymmetry equations— and if you want to keep the extra dimensions hidden, while preserving super-symmetry in the observable world—Calabi-Yau manifolds are the unique solution. They are, as Johns Hopkins physicist Raman Sundrum puts it, "the beautiful mathematical answer."[7]

"I didn't know much mathematics at the time, but I made the connection with Calabi-Yau manifolds through the holonomy group that characterizes man-ifolds," Strominger explains. "I found Yau's paper in the library and couldn't make much sense of it, but from the little I did understand, I realized these manifolds were just what the doctor ordered."[8] While reading my papers is not always a memorable experience, Strominger did tell the *New York Times*—almost twenty years after the fact—of the excitement he felt upon first stumbling across the proof of the Calabi conjecture.[9] Before getting too carried away, though, he called me up first to make sure he'd understood my paper correctly. I told him he had. And it finally dawned on me that after an eight-year wait, physics had finally found Calabi-Yau's.

So supersymmetry was what brought physicists to this arcane bit of mathematics, but I still haven't explained why they considered supersymmetry so all-fired important, apart from the general statement that symmetry is essential for understanding any kind of manifold. As Princeton physicist Juan Maldacena explains, "Supersymmetry not only makes calculations easier, it makes them possible. Why? Because it's easier to describe the motion of a sphere rolling down a hill than the more complicated motion of a football wobbling down a hill."[10]

Symmetry makes all sorts of problems easier to solve. Suppose you want to find all the solutions to the equation $xy = 4$. It would take a while, as there are an infinite number of solutions. If, however, you impose the symmetry condition, $x = y$, then there are just two solutions: 2 and -2. Similarly, if you know that the points on an x-y plane define a circle around the origin, instead of having two variables to worry about—the values of x and y—you only need to worry about one variable, the radius of the circle, to have all the information you need to reproduce the curve exactly. Supersymmetry, in the same way, cuts down the number of variables, thereby simplifying most problems you might want to solve, because it imposes a constraint on the geometric form that the internalized six dimensions can take. That requirement, notes University of Texas mathematician Dan Freed, "gives you Calabi-Yau's."[11]

Of course, we can't insist on the existence of supersymmetry in the universe simply to make our calculations easier. There's got to be more to it than sheer convenience. And there is. One virtue of supersymmetric theories is that they automatically stabilize the vacuum, the ground state of general relativity, so that our universe will not keep descending to lower and lower energy depths. This idea relates to the positive mass conjecture discussed in Chapter 3. In fact, supersymmetry was one of the tools Edward Witten drew upon in his physics-based proof of that conjecture, though supersymmetry did not come up in the more mathematical (and nonlinear) approach pursued by Richard Schoen and me.

But most physicists are interested in supersymmetry for another reason, which is, in fact, how the whole concept originated. To physicists, the most salient aspect of this idea is that of a symmetry that links elementary matter particles like quarks or electrons, which are called *fermions*, and force particles like photons or gluons, which are called *bosons*. Supersymmetry establishes a kinship, a kind of mathematical equivalence, between forces and matter and between these two classes of particles. It claims, in fact, that each fermion has an associated boson partner, called a *superpartner*, and that the reverse holds for every boson. The theory thus predicts a whole new class of particles—with funny names like squarks, selectrons, photinos, and gluinos—that are heavier than their known companions and whose "spin" differs by half an integer. These superpartners have never been seen before, though investigators are looking for them right now in the world's highest-energy particle accelerators (see Chapter 12).

The world we live in, which physicists characterize as "low-energy," is clearly not supersymmetric. Instead, current thinking holds that supersymmetry prevails at higher energies, and in that realm, particles and superpartners would appear to be identical. Supersymmetry "breaks," however, below a certain energy scale, and we happen to live in the realm of broken supersymmetry, where particles and superpartners are distinct in terms of mass and other properties. (After being broken, a symmetry doesn't disappear altogether but instead goes into hiding.) One way to understand the mass difference, says Howard University physicist Tristan Hubsch (a former postdoc of mine), is to think of supersymmetry as the rotational symmetry around, say, a plastic pen that's held upright. Imagine you've fixed the ends of the pen, and are pushing it from two different angles perpendicular to the length of the pen. Each of these movements takes the same amount of energy, no matter which angle you push on, as long as it's perpendicular to the pen, explains Hubsch. "And because these little movements are related by rotational symmetry, you can exchange one for the other."

We can use these pushes to set the pen vibrating with two rotationally symmetric waves. The vibrations would be equivalent, in turn, to two different particles, and the energy of the vibrations would determine the particles' masses.

Because rotational symmetry (or supersymmetry, in the case of string theory) is upheld, the two particles—the particle and superpartner—would have the same mass and would be otherwise indistinguishable.

We can break the rotational symmetry—our stand-in for supersymmetry—by pressing hard on the ends of the pen until it bows. The harder we press, the more severely it bows and the more badly the symmetry is broken. "After symmetry is broken, there are still two kinds of motions, but they are no longer related to each other by rotational symmetry," Hubsch says. One kind of motion—pushing the pen along the direction it is bowed—still takes energy, and the greater it's bowed, the more energy it takes. But if you push on the pen in a direction that is not only perpendicular to the pen itself but also perpendicular to the bowing, the pen will swivel around without requiring any energy (assuming there's no friction between the swiveling pen and the thing holding it in place). In other words, there's an energy difference or gap between these two motions—one requiring the input of energy and one not—which corresponds to the energy (or mass) gap between a massless particle and its massive superpartner in the broken supersymmetry case.[12] Physicists are trying to find signs of this energy gap and thereby establish the existence of massive supersymmetry partners, in high-energy physics experiments now under way at the Large Hadron Collider.

While supersymmetry, in principle, ties forces and matter together in a mathematically beautiful way, string theorists consider it desirable for another reason that goes well beyond the aesthetically pleasing aspects of symmetry. For without supersymmetry, string theory makes little sense. It predicts impossible particles like tachyons, which travel faster than the speed of light and which have a negative mass-squared—that is, their mass is expressed in terms of the imaginary unit i. Our theories of physics can't readily accommodate bizarre entities like that. Supersymmetry may not need string theory—though it owes much of its development to that theory—but string theory certainly benefits from having supersymmetry around. And supersymmetry, as stated above, was the very thing that brought physicists to the doorstep of Calabi-Yau.

Once Strominger and Candelas had their hands on the Calabi-Yau manifold, they were eager to take the next step—to see if this really was the right manifold, the one responsible for the physics we see. They came to Santa Barbara in 1984 eager to pursue this project and soon got in touch with Horowitz, who had moved from IAS to Santa Barbara a year earlier. They knew, moreover, that Horowitz had been my postdoc and, as a result of that association, had probably heard more than he cared to about the Calabi conjecture. When Horowitz found out what Strominger and Candelas were up to—trying to determine the mathematical requirements of string theory's internal space—he too saw that the requisite conditions matched those of Calabi-Yau's. Having more familiarity with this brand of math than the others did, Horowitz became a welcome addition to the team.

Shortly thereafter, Strominger visited Witten back in Princeton and filled him in on what they'd learned so far. It turned out that Witten had independently arrived at pretty much the same place, though he'd taken a different route in getting there. Candelas and Strominger started from the notion that there are ten dimensions in string theory and that these ten must be compactified in some six-dimensional manifold. The physicists then tried to figure out what kind of six-dimensional space would yield the right sort of supersymmetry, among other requirements. Witten, on the other hand, came at the problem from the perspective of a closed string propagating in spacetime, sweeping out a surface of one complex dimension and two real dimensions, otherwise known as a Riemann surface. Like an ordinary surface in differential geometry, a Riemann surface comes equipped with a metric that tells you the distance between two points, as well as a mechanism for telling you the angle between two points. What makes Riemann surfaces stand out is that, with relatively few exceptions, one can find a unique metric with negative (-1) curvature everywhere.

Witten's calculations—in a two-dimensional version of quantum theory called conformal field theory—were quite different from those of his coauthors, as he made fewer assumptions about the background spacetime. Yet he reached the same conclusion as the others, namely, that the geometry of the internal space had to be Calabi-Yau. None other would do. "Approaching this question from two directions strengthened the finding," Horowitz explains. "Moreover, it suggested this was the most natural way of doing the compactification, because you were led to the same conditions from two different starting points."[13]

The foursome finished their work in 1984 and promptly distributed their findings to colleagues via preprints, though the paper did not formally come

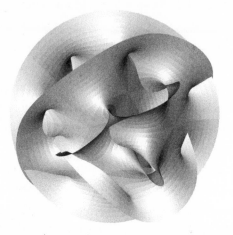

6.2—A rendering of the two-dimensional "cross-section" of a six-dimensional Calabi-Yau manifold (Andrew J. Hanson, Indiana University)

6.3—The Real Calabi-Yau: Eugenio Calabi (left) and Shing-Tung Yau (Photo of Calabi by permission from E. Calabi; photo of Yau by Susan Towne Gilbert)

out until the next year. The paper coined the term *Calabi-Yau space*, introducing the physics world to this strange, six-dimensional realm.

Prior to the publication of the 1985 paper, Calabi "hadn't expected there was any physical meaning to this work. It was purely geometric," he says. The paper, however, changed that, thrusting this mathematical construct into the heart of theoretical physics. It also brought unanticipated attention to the two mathematicians behind these spaces, Calabi recalls, "putting us on the news map. That sort of thing is always flattering, all the notoriety that came with this talk of Calabi-Yau spaces, but it was really not our doing."[14]

Our work, for a while at least, became all the rage in physics, but the "news map" spanned even broader terrain. There was an off-Broadway show (*Calabi-Yau*), the title of an electro/synthpop album by the band DopplerEffekt (*Calabi Yau Space*), the name of a painting (*Calabi-Yau Monna Lisa*) by the Italian artist Francesco Martin, and the butt of a joke in a *New Yorker* story by Woody Allen ("'My pleasure,' she said, smiling coquettishly and curling up into a Calabi-Yau shape").[15] It was a surprising run for such an abstruse idea, given that manifolds of this sort are hard to describe in words and harder still to visualize. A space with six dimensions, as one physicist commented, has "three more than I can comfortably imagine." The picture is further complicated by the presence of twisting, multidimensional holes running through the space, of which there could be a small number, or—in the deluxe Swiss cheese version—upward of five hundred.

Perhaps the simplest feature of Calabi-Yau spaces is their compactness. Rather than resembling an endless sheet of paper stretching to infinity in all directions, Calabi-Yau manifolds are more like a crumpled-up piece of paper that folds back on itself, only the crumpling-up has to be done in a meticulous way. A compact space contains no infinitely long or wide regions, and if you have a large enough suitcase, the space will fit inside. Another way of putting it, suggests Cornell's Liam McAllister, is that a compact space "can be covered with a quilt made out of a finite number of patches"—each patch, of course, being of finite size.[16] If you're standing on the surface of such a space and keep walking in the same direction, it may be possible to end up where you started. And even if you don't make it back exactly, you'll never get infinitely far away from the starting point, no matter how far you walk. Calling a Calabi-Yau space compact is by no means

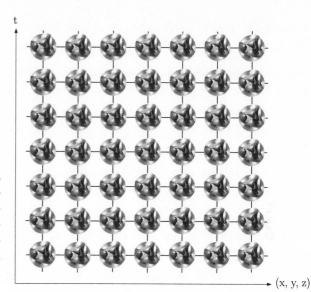

6.4—If string theory is correct, at any point in four-dimensional spacetime there's a hidden (six-dimensional) Calabi-Yau manifold. (Calabi-Yau images courtesy of Andrew J. Hanson, Indiana University)

an exaggeration. Although the exact size of such a manifold is still in question, it is thought to be exceedingly small, with diameters on the order of 10^{-30} centimeters (making it more than a quadrillion times smaller than an electron). Denizens of the four-dimensional realm like us can't ever see this six-dimensional realm, but it's always there, attached to every point in our space. We're just too big to go inside and look around.

That isn't to say that we don't interact with those unseen dimensions. As we walk around or sweep our arm through the air, we pass through the hidden dimensions without even realizing it—the movement, in a sense, cancels itself out. Imagine a herd of caribou, 100,000 strong, all headed in the same direction—going from, say, Alaska's coastal plane to the Brooks Range, where they'll find a nice valley to spend the winter in. Each animal, explains Allan Adams of MIT, takes a slightly different path over the 800-mile journey, but when looked at in aggregate, all the individual meanderings offset each other and the herd follows just one general path.[17] Our brief forays into Calabi-Yau space similarly cancel themselves out, rendering them inconsequential compared with the longer trajectories we follow in the four-dimensional domain.

Another way to think of it is that we live in a space without end; our horizons are vast even if we only manage to visit a tiny fraction of it. Yet everywhere we go in this big wide world, there's also this tiny, invisible realm, always within striking distance, that we can never fully access. Let's imagine an unusual x-y axis in which the x direction represents our infinite, four-dimensional space and the y direction represents the internal Calabi-Yau space. At every point along the x-axis, there is a hidden six-dimensional realm. Conversely, at every point along the y-axis, there is an additional four-dimensional space or direction that we can explore as well.

One of the most amazing things is that the concealed, internalized portion of the universe—a place we can never see, touch, smell, or feel—could have a more profound effect on the physics we experience than the concrete world of bricks and stones, cars and rocket ships, and billions and billions of galaxies. Or so string theory proclaims. "All of the numbers we measure in nature—all of the things we consider fundamental, such as the mass of quarks and electrons—all of these are derived from the geometry of the Calabi-Yau," explains the physicist Joe Polchinski of the University of California, Santa Barbara. "Once we know the shape, in principle, we'd have everything."[18] Or as Brian Greene has put it, "The code of the cosmos may well be written in the geometry of the Calabi-Yau shape."[19] If Einstein's relativity is proof that geometry is gravity, string theorists hope to carry that notion a good deal further by proving that geometry, perhaps in the guise of Calabi-Yau manifolds, is not only gravity but physics itself.

certainly not going to be the one to cast doubt on these sweeping assertions. a reasonable person might wonder, just the same, that if the Calabi conjecture was considered too good to be true, what do we call this? And how on earth can we explain it? I'm afraid the real explanation may strike some as unsatisfying and perhaps even circular—that being that Calabi-Yau manifolds can pull off the miraculous feat simply because that ability is, from the very start, built into string theory's basic machinery. Even so, it still may be possible to provide some inkling as to how that "machine"—which takes in ten-dimensional manifolds and spits out four-dimensional physics—actually works. Here's an admittedly simplistic picture of how the particles and their masses can be derived from a given Calabi-Yau manifold if we assume that the manifold in question is what is known as *non-simply connected*. A non-simply connected manifold is like a torus with one or more holes, where some loops sitting on the surface cannot be shrunk down to points. (This is in contrast to a sphere, a simply connected surface, where every loop can be shrunk to a point like an extremely taut rubber band sliding off the equator and wrapping around the north pole.) Starting with a complicated, six-dimensional manifold with some number of holes, you can figure out all the different ways in which strings can wrap around the manifold, going through various holes one or more times. It's a complicated problem, as there are many ways you can do the wrappings, and each loop or cycle you make can have different sizes that depend, in turn, on the size of the holes. From all these possibilities, you draw up a list of potential particles. The masses can then be determined by multiplying the length of the string by its tension, which is equivalent to its linear energy density, throwing in the kinetic energy of motion as well. The objects you fashion in this way can have any number of dimensions between zero and six. Some objects are allowed; some are not. If you take all the allowed objects and all the allowed motions, you end up with a list of particles and masses.

Another way to view this is that in the reigning picture of quantum physics, owing to the central tenet known as wave-particle duality, every particle can be thought of as a wave and every wave can be thought of as a particle. Particles in string theory, as has been stated before, correspond to a specific vibrational pattern of a string, and a string vibrating in a characteristic, well-prescribed way is also very much like a wave. The question then is to discern how the geometry of this space influences the waves that can form.

Let's suppose the space in question is the Pacific Ocean, and we're floating in the middle of it, thousands of miles from the nearest continent and far above the ocean floor. One can imagine that in general, waves forming on the surface here are relatively unaffected by the shape or topography of the seafloor many miles below. But the situation is entirely different in a confined space, such as a shallow and narrow cove—where a mild bump far out at sea becomes a relative tsunami in the shallow waters near shore or, to take a less extreme example, where the reefs and rocks beneath the surface have a major bearing on how and where waves form and break. In this example, the open ocean is like a noncompact (or extended) space, whereas the shallow coastal waters are more like the small, compact dimensions of string theory, where the geometry dictates the kind of waves that can form and so, by extension, the kind of particles that can form.

A musical instrument, such as a violin, is another example of a compact space that produces characteristic vibrational patterns or waves, corresponding to notes rather than particles. The sounds produced by plucking a string depend not only on the length and thickness of the string but also on the shape of the instrument's interior—the acoustic chamber—where waves of certain frequencies resonate at maximum amplitude. The strings are named for their fundamental frequencies, which on most violins happen to be G, D, A, and E. Physicists—like instrument makers seeking the right shape for the sounds they hope to produce—are hunting for the Calabi-Yau manifold with the proper geometry to give rise to the waves and particles that we see in nature.

The way physicists normally attack a problem of this sort is to find solutions to the wave equation, more formally known as the Dirac equation. The solutions to the wave equation are, not surprisingly, waves and their corresponding particles. But this is a very difficult equation to solve, and we normally cannot solve it for all the possible particles to be found in nature—only for the so-called massless ones that correspond to the lowest or fundamental frequencies of a particular string. The massless particles include all the particles we see, or intuit, in the world around us, including those spotted within our high-energy physics accelerators. Some of these particles, like electrons, muons, and neutrinos, do actually have mass in spite of the terminology "massless." But they acquire their mass through a mechanism totally different from how the so-called massive particles, which are expected to form at the much higher-energy "string scale," acquire their mass. The mass of ordinary particles like electrons, moreover, is so much lighter than that of these heavier particles—by a factor of roughly a quadrillion or more—that it is considered a fair approximation to call these ordinary particles massless.

While limiting ourselves to massless particles makes solving the Dirac equation somewhat more manageable, it is still not easy. Fortunately, Calabi-Yau manifolds have certain attributes that help. The first is supersymmetry, which reduces the number of variables, converting a second-order differential equation (in which some derivatives are taken twice) into a first-order differential equation (in which derivatives are taken just once). Another way supersymmetry helps is that it pairs each fermion with its own special boson. If you know all the fermions, then you automatically know all the bosons and vice versa. So you just have to work things out for one class of particles or the other, and you get to pick the one whose equations are easier to solve.

Another special feature of Calabi-Yau manifolds and their geometry in particular is that the solutions to the Dirac equation—in this case, the massless particles—are the same as the solutions to another mathematical formulation known as the Laplace equation, which is considerably easier to work with. The biggest advantage stems from the fact that we can get the answers to the Laplace equation—and identify the massless particles—without having to solve any differential equations at all. Nor do we need to know the exact geometry of the Calabi-Yau or its metric. Instead, we can get all that we need from topological "data" about the Calabi-Yau, all of which is contained in a 4-by-4 matrix called a Hodge diamond. (We'll be discussing Hodge diamonds in the next chapter, so I won't be saying more about this now, other than that this topological sleight of hand enables us to round up the massless particles rather effortlessly.)

Getting the particles, however, is just the beginning. Physics, after all, is more than just a bunch of particles. It also includes the interactions or forces between them. In string theory, loops of string moving through space may either join together or split apart, and their inclination to do one or the other depends on the *string coupling*, which is a measure of the force between strings.

Calculating the strength of particle interactions is a painstaking task—requiring almost the full arsenal of string theory tools—that can, in current practice, take a year to work out for a single model. Supersymmetry, again, makes the computations somewhat less taxing. Mathematics can help, too, since this kind of problem has long been familiar to geometers and, as a result, we have many tools to bring to bear on it. If we take a loop and let it move and vibrate in Calabi-Yau space, it can become a figure eight and then split into two separate circles or loops. Conversely, two circles can come together to form a figure eight. The passage of these loops through spacetime sweeps out a Riemann surface, and that is precisely the picture we have of string interactions, although mathematicians had not connected it to physics until string theory arrived on the scene.

Given those tools, how close can physicists come to matching their predictions to the world we see? This topic will be the focus of Chapter 9, but let's consider the 1985 paper of Candelas, Horowitz, Strominger, and Witten, which represented the first serious attempt to match string theory—by way of Calabi-Yau compactifications—to the real world.[20] Even then, the physicists were able to get a lot of things right. Their model produced, for example, the desired amount of supersymmetry in four dimensions (denoted as $N = 1$, meaning the space remains invariant under four symmetry operations, which can be thought of as four different kinds of rotations). That, in itself, is a great success. For if they'd instead come up with maximal supersymmetry ($N = 8$, implying the more demanding situation of invariance under thirty-two different symmetry operations), that would have constrained physics so much that our universe would have to be just flat space without any of the curvature that we believe to exist or any of the complexities, like black holes, that make life so interesting—at least for the theorists. If Candelas and his colleagues had failed on this front, and one could prove that this six-dimensional space could never furnish the appropriate supersymmetry, then compactification in string theory, at least in this instance, would have failed.

The paper was a great leap forward and is now considered part of the first string revolution, but it did miss the mark in other areas, such as the number of families of particles. In the Standard Model of particle physics—a model that has ruled particle physics over recent decades and incorporates the electromagnetic, weak, and strong forces—all the elementary particles from which matter is composed are divided into three families or generations. Each family consists of two quarks, an electron or one of its relatives (a muon or tau), and a neutrino, of which there are three varieties: the electron, muon, or tau neutrino. The particles in family 1 are the most familiar ones in our world, being the most stable and the least massive. Those in family 3 are the least stable and the most massive, whereas the members of family 2 lie somewhere in between. Unfortunately for Candelas and company, the Calabi-Yau manifold they worked with yielded four families of particles. They were off by just one, but in this case the difference between three and four was huge.

In 1984, Strominger and Witten started inquiring about the number of families, and I was eventually asked if I could come up with a Calabi-Yau that yielded three families. Horowitz had stressed the importance of this to me as well. What was needed was a manifold with an Euler characteristic of 6 or –6, as Witten had shown a few years before that for a certain class of Calabi-Yau manifolds (those with a nontrivial fundamental group—or noncontractible loop—among other features) the number of families equals the absolute value of the Euler characteristic divided by two. A version of this formula even appeared in the quartet's widely cited 1985 paper.

I found some time to work on the problem later that year while flying from San Diego to Chicago en route to Argonne National Laboratory, which was hosting one of the first major string theory conferences ever. I was giving a talk there and had set aside the time on the plane to prepare my remarks. It occurred to me that since my physics friends considered the three-family business so important, maybe I should give that matter some thought. And luckily, I did come up with a solution during the flight—a Calabi-Yau manifold whose Euler number was –6, making it the first manifold ever constructed that would yield the three families of particles the standard model demanded. While this wasn't a great leap forward, it was a "small breakthrough," as Witten put it.[21]

The method I used for constructing that manifold was rather technical, but later proved quite useful. For starters, I took the Cartesian product of two cubic *hypersurfaces*. A hypersurface is a submanifold—that is, a surface of one dimension lower than the ambient space in which it sits, like a disk tucked inside a sphere or a line segment tucked inside a disk. The hypersurface of a cubic surface of three complex dimensions has two complex dimensions overall. The product of two of these hypersurfaces has $2 \times 2 = 4$ complex dimensions. That's one dimension too many, and I cut that down to three complex dimensions (or six real dimensions), which is what you want in string theory, by taking a cross-section, or slice, of the four-dimensional product manifold.

Unfortunately, that procedure doesn't quite give us the manifold we wanted, because that manifold would yield nine families of particles rather than the desired three. But the manifold had threefold symmetry, which enabled me to create a so-called quotient manifold in which every point corresponds to three points in the original. Taking the quotient, in this case, was like dividing the original manifold into three equivalent pieces. In this way, the number of points was decreased by a factor of three, as was the number of families.

To my knowledge, this quotient manifold was the first Calabi-Yau ever constructed—and the only one available for a long time—that had an Euler

6.5—In geometry we can reduce the dimensions of an object by taking a slice of it and looking only at the revealed cross-section. For example, if you slice through a three-dimensional apple, you'll expose a two-dimensional surface—one of many such surfaces you could expose depending on where and how you choose to slice. An additional cut into that surface would pick out a one-dimensional line sitting on the surface. Cutting that line, in turn, would select a single (zero-dimensional) point. Therefore, each successive slice (up to that final point) reduces by one the dimensionality of the object at hand.

characteristic of 6 or −6, which could be exploited to generate three families of particles. In fact, I hadn't heard of anything like it until late 2009, when Candelas and two colleagues—Volker Braun of the Dublin Institute for Advanced Studies and Rhys Davies of Oxford—did something similar, constructing a Calabi-Yau manifold with an Euler characteristic of −72 and a quotient manifold with an Euler characteristic of −6. Ironically, in the late 1980s, Candelas had constructed the original (or "parent") Calabi-Yau—one of eight thousand manifolds so created at the time—with two colleagues, but did not recognize its potential applicability until more than twenty years later.[22]

I bring this up because back in 1986, when Brian Greene began trying to get realistic physics out of a Calabi-Yau manifold, there weren't a lot of manifolds around to choose from. To get the number of families right, he went with the Calabi-Yau that I had fabricated in 1984 en route to Argonne. Working on this problem first as an Oxford graduate student and then as my postdoc at Harvard, Greene—along with Kelley Kirklin, Paul Miron, and Greene's former Oxford adviser Graham Ross—came even closer to the Standard Model than Candelas, Horowitz, Strominger, and Witten had a year or so earlier. Greene's model provided more details—a step-by-step prescription—for generating physics from a Calabi-Yau manifold. He and his colleagues got supersymmetry right, as well as the number of families right, massive neutrinos (with extremely small mass), and pretty much everything else you wanted, except for a few extra particles that weren't supposed to be there. So this Calabi-Yau manifold came close— indeed closer than we'd ever been before—but didn't quite do the job, either. This shouldn't be taken as criticism of their work, since almost a quarter of a century later, no one has finished "the job" yet.

In those early days, physicists had hoped there was just one Calabi-Yau manifold to worry about—a unique solution from which we could calculate everything—or just a handful, in which case they could quickly weed out the less worthy examples and select the right one. When Strominger and Witten first started asking me about the number of known (and already constructed) Calabi-Yau manifolds, I only knew of two examples for certain. One, the *quintic threefold*, is arguably the simplest Calabi-Yau you can have. (It's called a quintic because it's described by a fifth-degree polynomial equation that takes the general form $z_1^5 + z_2^5 + z_3^5 + z_4^5 + z_5^5 = z_1 \times z_2 \times z_3 \times z_4 \times z_5$. It's called a threefold because it's a manifold with three complex dimensions.) The second Calabi-Yau manifold was made by combining (or taking the "product" of) three complex one-dimensional tori and then modifying the resultant manifold.

Around this time, Strominger asked me about the total number of Calabi-Yau manifolds possible. I told him there were probably tens of thousands—each representing a different topology and a different solution to the equations of string theory. Within each of those topologically distinct families, moreover, lay an infinite variety of possible shapes. I delivered the same news to a larger crowd of physicists during my 1984 Argonne lecture, and many of them were dismayed when I tossed out the round figure of ten thousand or so—an estimate that has held up pretty well to this day.

At first, the physicists weren't in a position to construct Calabi-Yau manifolds on their own, as the mathematics was still too unfamiliar to them, which meant they were dependent on people like me to tell them about such objects. But once they became conversant in the literature, the physicists moved fast and constructed many examples themselves, working independently of mathematicians.

Shortly after my Argonne lecture, Candelas and his students took the same general approach I'd used in constructing the first Calabi-Yau that gave rise to three families of particles, computerized the technique, and in this way generated many thousands of Calabi-Yau manifolds (as mentioned a moment ago). I'd only worked out a few of them myself and was never very good at computer calculations. But in light of Candelas's achievement and the output of his computer, the notion of a large number of manifolds was no longer abstract or merely an estimate made by a partisan mathematician. It was a fact, and if you had any doubts on the subject, you needed to look no further than Candelas's published database.

The upshot of this was that string theory was starting to look far more complicated than initially envisioned. It was no longer a matter of taking the Calabi-Yau manifold and wringing every last drop of physics out of it you could. Before doing that, you first had to answer the question "Which Calabi-Yau?" And, as will be seen in Chapter 10, the problem of too many Calabi-Yau manifolds has gotten worse over the years rather than better. Yet this problem had even become apparent in 1984, when, says Strominger, "the uniqueness of string theory was already in question."[23]

As if that weren't bad enough, there was another numbers problem afflicting string theory in the early days, and this one had to do with the number of string theories itself. There wasn't just one string theory. There were five separate theories—with the names Type I, Type IIA, Type IIB, Heterotic SO(32), and Heterotic $E_8 \times E_8$—that differed, for example, as to whether the strings must be closed loops or could be open strands as well. Each theory belonged to a different symmetry group, and each contained a unique set of assumptions about things like the chirality (or handedness) of fermions and so forth. A competition of sorts arose as to which of these five candidate theories might ultimately prevail and become the true Theory of Everything. In the meantime it was paradoxical—not to mention just plain embarrassing—to have five "unique" theories of nature.

In 1995, in an intellectual tour de force, Witten showed that all five string theories represented different corners of the same overarching theory, which he called *M-theory*. Witten never said what the *M* stood for, but speculation has included terms like *master, magical, majestic, mysterious, mother, matrix,* and *membrane.*

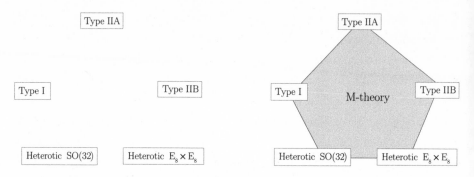

6.6—At first the five different string theories were regarded as competing theories that were studied separately and considered distinct from each other. Edward Witten and other architects of the "second string revolution" showed that the five theories are all related—tied together through a common framework called M-theory (though, apparently, no one knows what the *M* stands for).

The last word in that series is of special significance because the fundamental ingredients of M-theory are no longer just strings. They're more general objects, called membranes or branes, that could have anywhere from zero to nine dimensions. The one-dimensional version, a "one-brane," is the same as a familiar string, whereas the two-brane is more like what we think of as a membrane, and a three-brane is like three-dimensional space. These multidimensional branes are called p-branes, and a category of these objects called D-branes are subsurfaces within a higher-dimensional space to which open strings (as opposed to closed loops) attach. The addition of branes has made string theory richer and better equipped to take on a broader range of phenomena (as will be explored in later chapters). Furthermore, demonstrating that the five string theories were all connected to each other in fundamental ways meant that one could choose to work with whatever version made solving a particular problem easiest.

M-theory has another important feature that distinguishes it from string theory. The theory exists in eleven dimensions rather than ten. "Physicists claim to have a beautiful and consistent theory of quantum gravity, yet they can't agree on the number of dimensions," Maldacena notes. "Some say ten, and some say eleven. Actually, our universe may have both—ten and eleven dimensions."[24]

Strominger agrees that the "notion of dimension is not absolute." He compares string theory and M-theory to different phases of water. "If you put it

6.7—Edward Witten at the Institute for Advanced Study (Photo by Cliff Moore)

below freezing it's ice; above freezing it's liquid; above boiling it's steam," he says. "So it looks very different depending on what phase it's in. But we don't know which of these phases we're actually living in."[25]

Even the principal author of M-theory, Witten himself, concedes that the ten- and eleven-dimensional descriptions of the universe "can both be correct. I don't consider one more fundamental than the other," he says, "but at least for some purposes, one might be more useful than the other."[26]

As a practical matter, physicists have had better success so far in explaining the physics of our four-dimensional world when starting from a ten- rather than eleven-dimensional perspective. Researchers have tried to go from eleven dimensions directly to four by compactifying on a seven-dimensional so-called G_2 manifold—the first compact version of which was constructed in 1994 by Dominic Joyce, a mathematician currently at Oxford. So one might think that much of what we've talked about so far—such as deriving four-dimensional physics from a ten-dimensional universe by way of six-dimensional Calabi-Yau manifolds $(4 + 6 = 10)$—could have been suddenly rendered obsolete by Witten's eureka moment. Fortunately, at least for the purposes of the current discussion, that is not the case.

One drawback of the G_2 approach, as explained by Berkeley physicist Petr Horava, a Witten collaborator and a key contributor to M-theory, is that we can't recover four-dimensional physics by compactifying on a "smooth" seven-dimensional manifold. Another problem is that a seven-dimensional manifold, unlike Calabi-Yau manifolds, cannot be complex, because complex manifolds must have an even number of dimensions. That's probably the most important difference, Horava adds, "because complex manifolds are much better behaved, much easier to understand, and much easier to work with."[27]

Furthermore, there's still much to be learned about the existence, uniqueness, and other mathematical properties of seven-dimensional G_2 manifolds. Nor do we have a systematic way of searching for these manifolds or a general set of rules to draw on, as we do for Calabi-Yau manifolds. Both Witten and I have looked into developing something similar to the Calabi conjecture for G_2 manifolds, but so far neither he nor I nor anybody else has gotten very far. That's one reason M-theory is not as developed as string theory, because the mathematics is much more complicated and not nearly as well established.

Because of the difficulties with the G_2 manifolds, most efforts in M-theory have focused on indirect routes of compactifying from eleven dimensions to four. First, eleven-dimensional spacetime is treated as the product of ten-dimensional spacetime and a one-dimensional circle. We compactify the circle, shrinking it down to a minuscule radius, which leaves us with ten dimensions. We then take those ten dimensions and compactify on a Calabi-Yau manifold, as usual, to get down to the four dimensions of our world. "So even in M-theory, Calabi-Yau manifolds are still in the center of things," says Horava.[28] This approach—pioneered by Witten, Horava, Burt Ovrut, and others—is called heterotic M-theory. It has been influential in introducing the concept of brane universes, in which our observable universe is thought to live on a brane, and has also spawned alternative theories of the early universe.

So for now, at least, it appears that all roads lead through Calabi-Yau. To get realistic physics out of string theory and M-theory, and some cosmology as well, this is still the geometry that holds the "cosmic code"—the space wherein the master plan resides. Which is why Stanford physicist Leonard Susskind, one of the founders of string theory, claims that Calabi-Yau manifolds are more than just the theory's supporting structure or scaffolding. "They are the DNA of string theory," he says.[29]

Seven

THROUGH THE LOOKING GLASS

Though Calabi-Yau manifolds entered physics with a bang, these beguiling shapes were quickly at risk of becoming a whimper, and for reasons that had nothing to do with string theory's embarrassment of riches (in the form of the multiplicity of theories that Edward Witten ultimately helped sort out). The attraction of these geometric shapes was obvious: As Duke physicist Ronen Plesser describes it, "the hope was that we could classify these spaces, figure out the kind of physics they give rise to, exclude some of them, and then conclude that our universe is described by space number 476, and everything we might want to know could be derived from that."[1]

That simple vision has yet to be realized, even today, and as progress stalled and enthusiasm waned more than two decades ago, doubts inevitably crept in. By the late 1980s, many physicists felt that the role of Calabi-Yau manifolds in physics was doomed. Physicists like Paul Aspinwall (now at Duke), who was then finishing his Ph.D. work at Oxford, found it hard to get jobs and secure grants to pursue their research into Calabi-Yau manifolds and string theory. Disenchanted string theory students, including two of Brian Greene's former classmates and coauthors at Oxford, started leaving the field to become financiers. Those who remained, such as Greene, found themselves having to fend off charges of "pursuing calculations for their own sake—of learning math and regurgitating it as physics."[2]

Perhaps that was true. But given that Greene and Plesser would soon make critical contributions to an idea called mirror symmetry, which would revitalize Calabi-Yau manifolds and rejuvenate a somnolent branch of geometry, I must admit that I'm pleased they settled on that particular line of research rather than, say, stock futures. Before this renaissance came, however, these manifolds would hit a low point that, for a while at least, looked as if it might prove fatal.

Signs of trouble began to emerge through string theory investigations of a concept known as conformal invariance. A string moving through spacetime sweeps out a surface of two real dimensions (one of space and one of time) and one complex dimension called a world sheet. If the string is a loop, the world sheet takes the form of a continuously extending, multidimensional tube (or more precisely, a complex Riemann surface without a boundary), whereas a strand of open string traces an ever-elongating strip (that is, a complex Riemann surface with a boundary). In string theory, we study all possible vibrations of strings, and those vibrations are governed by a physical principle (or *action principle*) that depends on the conformal structure of the world sheet, which is itself an intrinsic feature of Riemann surfaces. In this way, string theory has conformal invariance automatically built into it. The theory is also scale invariant, meaning that multiplying distances by some constant shouldn't change the relationships between points. Thus, you should be able to change the surface—like putting air in a balloon or shrinking the surface by letting air out, or expanding it in other ways that change the distance between two points or change the shape—without changing anything important in string theory itself.

The problems arise when you insist on conformal invariance in a quantum setting. Much as a classical particle follows a geodesic, a path that minimizes the distance traveled (as described by the principle of least action we talked about in Chapter 3), a classical string follows a path of minimal distance. As a result, the world sheet associated with that moving string is a minimal-area surface of a special kind. The area of this two-dimensional world sheet can be described by a set of equations, a two-dimensional field theory, that tells us exactly how a string can move. In a field theory, all forces are described as fields that permeate spacetime. The movement of a string and its behavior in general are a result of the forces exerted on it, and the string moves in such a way as to minimize the area of the world sheet. Among the vast number of possible world sheets, encompassing the vast number of possible ways the string can move, this field theory picks out the one with minimal area.

The quantum version of this field theory must capture not only the gross features of a string moving in spacetime—and the surface that such an object traces out—but also some of the finer details that stem from the fact that the string oscillates as it moves. The world sheet, as a result, will have small-scale features that reflect those oscillations. In quantum mechanics, a particle or string moving in spacetime will take all possible paths. Rather than just selecting the one world sheet possessing minimal area, a quantum field theory takes a weighted average of all possible world sheet configurations and assigns more importance in its equations to surfaces with lower area.

But after all this averaging has been done—by integrating over all possible world sheet geometries—will the two-dimensional quantum field theory that emerges still retain scale invariance and other aspects of conformality? That depends on the metric of the space in which the world sheets exist; some metrics allow for a conformal field theory, and others do not.

To see whether scale invariance is maintained in a specific metric, we calculate something known as the beta function, which measures deviations from conformality. If the value of the beta function is zero, nothing changes when we deform the world sheet by blowing it up, stretching it, or shrinking it, which is good if you like your theories conformal. It had been assumed that the beta function would automatically vanish (or drop to zero) in a Ricci-flat metric such as those we can find in Calabi-Yau manifolds. Unfortunately, as with many of the complicated equations we've been discussing, the beta function cannot be calculated explicitly. Instead, we approximate it by an infinite sum known as a power series. The more terms we use from the series, the better our approximation.

You can get a sense of how we do this by imagining that you're trying to measure the surface area of a sphere by wrapping it in a wire mesh. If the wire consists of a single loop that can assume a single position around the sphere, you can't get a good area estimate. If you instead had four triangular loops, configured as a tetrahedron surrounding the sphere, you could get a much better area approximation. Increasing the number to twelve loops (shaped like pentagons in the case of a dodecahedron) and to twenty loops (shaped like triangles in the case of an icosahedron) would yield even more refined approximations. Much as in our example, the terms in the beta function's power series are known as loops. Working with just one term in the series gives us the one-loop beta function; working with the first two terms gives us the two-loop beta function, and so forth.

The problem with adding more loops to the wire mesh is that these beta function calculations, which are extraordinarily difficult to start with, get harder—and more computationally intensive—as the number of loops goes up. Calculations showed that the first three terms of the power series were zero, just as predicted, which was reassuring to physicists. But in a 1986 paper, Marcus Grisaru (a physicist now at McGill) and two colleagues, Anton van de Ven and Daniela Zanon, showed that the four-loop beta function does not vanish. A subsequent calculation by Grisaru and others showed that the five-loop beta function does not vanish, either. This seemed like a major blow to the position of Calabi-Yau manifolds in physics, because it suggested that their metric did not allow for conformal invariance to hold.

"As a believer in both string theory and supersymmetry, our findings were of some concern," Grisaru says. "We were happy that the result gave us much notoriety, but you don't always want the notoriety that comes from demolishing a beautiful building. Still, my attitude about science is that you have to accept whatever you get."[3]

But perhaps all was not lost. A separate 1986 paper by David Gross and Witten, both of whom were then at Princeton, argued that even though having an exactly Ricci-flat metric on a Calabi-Yau manifold does not work, the metric can be changed slightly so that the beta function will vanish, as required. Tweaking the metric in this way didn't mean making a single adjustment; instead it meant making an infinite number of adjustments, or quantum corrections. But in situations like this, when an infinite series of corrections are required, there's always the question as to whether that series will eventually converge on a solution. "Could it be that when you start adding in all the corrections, there'd be no solution at all?" Plesser asks.

In the best of circumstances, changing the metric by a little bit will yield a solution that itself changes by a little bit. We know, for example, how to solve the equation $2x = 0$, the answer being $x = 0$. "If I try to solve $2x = 0.100$, the answer changes just slightly ($x = 0.050$), which is the situation we like," Plesser explains. We can also solve the equation $x^2 = 0$ without much difficulty (again, $x = 0$). "But if you ask me to solve $x^2 = -0.100$, my solution (at least in real numbers) just vanished," he says. "So we see that a small modification can take you to a solution that is just a little bit different or to a solution that [for real numbers] doesn't even exist."[4]

In this case of the amended Calabi-Yau, Gross and Witten determined that the series would converge. They showed that you can correct the Calabi-Yau metric, term by term, and at the end of that procedure, you'll be left with a very complicated equation that can be solved nevertheless. In the process, all the loops in the beta function will go to zero.

As a result, explains Shamit Kachru of Stanford, "people didn't have to throw out Calabi-Yau's; they just had to modify them slightly. And since you couldn't write down the Calabi-Yau metric to begin with, the fact that the metric had to be modified slightly wasn't such a big deal."[5]

Further insights on how the Calabi-Yau metric had to be changed came in that same year from the work of Dennis Nemeschansky and Ashoke Sen, who were then based at Stanford. The resultant manifold is still a Calabi-Yau topologically, and its metric is almost Ricci flat but not quite. Nemeschansky and Sen offered an explicit construction, a precise formula, showing how the modified metric differs from Ricci flatness. Their work, along with that of Gross and Witten, "helped save Calabi-Yau's for physics, for without it, people would have had to abandon the whole approach," asserts Sen. What's more, Sen argues, we could have never gotten to the solution without the first assumption that the Calabi-Yau manifold used in string theory was Ricci flat. "For if we didn't start from a Ricci-flat metric, it's hard to imagine any procedure we could have used to take us to the corrected metric."[6]

I agree with Sen, but that doesn't mean the Ricci-flat assumption has been rendered useless. One way to think about it is that the Calabi-Yau with a Ricci-flat metric is like the solution to the equation $x^2 = 2$. What you really want, however, is the solution to the equation $x^2 = 2.0000000001$, because the manifold is almost Ricci-flat but not exactly. The only way to get to the modified, corrected metric is to start with the solution to $x^2 = 2$ and move from there to the one you want. For most purposes, however, the solution to $x^2 = 2$ is good enough. As a general matter, the Ricci-flat metric is also the easiest to use and captures most of the phenomena you're after anyway.

The next big steps in the resurrection of Calabi-Yau manifolds were made by Doron Gepner (then a Princeton physics postdoc) over a several-year period starting in 1986. Gepner devised conformal field theories, each of which bore striking similarities—in terms of the associated physics—to a single Calabi-Yau manifold of a distinct size and shape. Initially Gepner found that the physics related to his field theory—involving certain symmetries, fields, and particles—looked the same as that for a string propagating on a particular Calabi-Yau manifold. That got people's attention, as there seemed to be a mysterious link between two things, a conformal field theory and a Calabi-Yau, which ostensibly had nothing to do with each other.

One of the people whose interest was piqued was Brian Greene—my Harvard postdoc at the time, who had an expertise in the mathematical underpinnings of Calabi-Yau manifolds, having completed his doctoral dissertation on the subject, and who also had a strong grounding in conformal field theory. He started talking to people in the physics department who also worked on conformal theories, including two graduate students, Ronen Plesser and Jacques Distler. Distler and Greene began investigating *correlation functions* associated with the field theory and the corresponding Calabi-Yau. The correlation functions in this case involved "Yukawa couplings" that dictate how particles interact, including interactions that confer mass to a given particle. In a paper submitted in the spring of 1988, Distler and Greene found that the correlation functions (or Yukawa couplings) for the two cases numerically agreed—more evidence that the field theory and Calabi-Yau were somehow related if not the same.[7] Gepner arrived at a similar conclusion, regarding the match in Yukawa couplings, in a paper submitted shortly thereafter.[8]

More specifically, Distler and Greene, and Gepner separately, found that for precise settings of manifold size and shape, they could compute all the correlation functions, which are a set of mathematical functions that, taken in sum, completely characterize the conformal field theory. The result, in other words, spelled things out in starkly explicit terms, identifying an exact conformal field theory with all the correlation functions, as well as showing the exact size and shape of the associated Calabi-Yau. For a restrictive class of the Calabi-Yau manifolds known to date, we have since identified a corresponding Gepner model.

That link, which was firmly established by the late 1980s, helped turn around thinking regarding the usefulness of Calabi-Yau manifolds. As Kachru put it, "you couldn't doubt the existence of his [Gepner's] conformal field theories, because they were fully solvable, fully computable. And if you can't doubt those theories, and they show the same properties as his Calabi-Yau compactifications, then you can't doubt those compactifications, either."[9]

"Gepner's paper really saved Calabi-Yau's," claims Aspinwall, at least insofar as physics and string theory were concerned.[10] The connection, moreover, between a representative Gepner model and a specific Calabi-Yau compactification helped set the stage for the discovery of mirror symmetry—a finding compelling enough to remove any lingering doubts as to whether Calabi-Yau manifolds were spaces worth studying.

Some of the earliest hints of mirror symmetry came in 1987, when Stanford physicist Lance Dixon and Gepner observed that different K3 surfaces were linked to the same quantum field theory, thereby implying that these disparate surfaces were related through symmetry. Neither Dixon nor Gepner, however, produced a paper on the subject (though Dixon gave some talks), so the first written statement of mirror symmetry probably came in a 1989 paper by Wolfgang Lerche (of Caltech), Cumrun Vafa, and Nicholas Warner (of MIT); they argued that two topologically distinct Calabi-Yau threefolds—six-dimensional Calabi-Yau manifolds rather than four-dimensional K3's—could give rise to the same conformal field theory and hence the same physics.[11] This was a stronger statement than the Dixon-Gepner one because it linked Calabi-Yau manifolds with different topologies, whereas the earlier finding applied to surfaces with the same topology (though with different geometries), owing to the fact that all K3s are topologically equivalent. The problem was that no one knew how to construct the pairs of Calabi-Yau manifolds that might be tied together in this strange way. Gepner models turned out to be one of the keys to unraveling that puzzle, and those same models also helped bring Brian Greene and Ronen Plesser together for the first time.

In the fall of 1988, Brian Greene learned through a conversation with Vafa (whose offices were on the same "theory" floor of the Harvard physics building) about this suspected link between different Calabi-Yau manifolds. Greene realized, right off, that this idea would be hugely important if it could be proved.

He joined forces with Vafa and Warner to gain a better understanding of the link between Calabi-Yau manifolds and Gepner models. Greene, Vafa, and Warner spelled out the steps for going from a Gepner model to a particular Calabi-Yau, explains Greene.[12] The researchers provided "an algorithm that showed why they are connected and how they are connected. Give me a Gepner model, and I can show you in a flash which Calabi-Yau it's connected to."[13] The Greene, Vafa, and Warner paper explained why every Gepner model yields a Calabi-Yau compactification. Their analysis took the guesswork out of matching Gepner models with Calabi-Yau manifolds, as Gepner previously had to look through tables to find a Calabi-Yau that produced the desired physics.

Now that the relationship between Gepner models and Calabi-Yau manifolds had been solidified, Greene teamed up with Plesser in 1989 in the hopes of going further. One of the first things they realized, says Greene, "is that we now had a potent tool for analyzing very complicated [Calabi-Yau] geometry using a field theory that we had complete control over and, indeed, a complete understanding of."[14] What would happen, they wondered, if they changed the Gepner model a little bit? They thought the altered model would probably correspond to a somewhat different Calabi-Yau. So they tried it out, by performing a rotational symmetry operation (like rotating a square by 90 degrees) on the Gepner model. It left everything in the field theory unchanged. Performing the same symmetry operation on the Calabi-Yau, however, produced manifolds of different topology as well as geometry.

The symmetry operation, in other words, changed the Calabi-Yau topology while leaving the conformal field theory that goes along with it intact. The result, then, was two Calabi-Yau manifolds with two distinct topologies connected to the same physical theory. "And that, in a nutshell, is mirror symmetry," Gepner notes.[15] It is also, more generally, what we call a *duality*, in which two objects (in this case, the Calabi-Yau manifolds) that appear unrelated nevertheless give rise to the same physics.

Greene and Plesser's first paper on the subject of mirror symmetry contained ten so-called mirror partners or mirror manifolds in nontrivial (as in not totally flat) Calabi-Yau manifolds, starting with the simplest Calabi-Yau, the quintic threefold. Along with the other nine examples, their paper gave a formula that showed how to construct mirror pairs from any Gepner model—hundreds, if not thousands, of which are known to date.[16]

7.1—Brian Greene (Photo © Andrea Cross)

7.2—Ronen Plesser (Photo by Duke Photography)

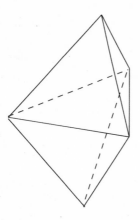
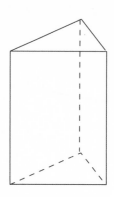

7.3—The double tetrahedron, which has five vertices and six faces, and the triangular prism, which has six vertices and five faces, are simple examples of mirror manifolds. These rather familiar-looking polyhedra can be used, in turn, to construct a Calabi-Yau manifold and its mirror pair, with the number of vertices and faces of the polyhedra relating to the internal structure of the associated Calabi-Yau. (The details of this "construction" procedure are quite technical, however, going well beyond the scope of this discussion.)

Mirror manifolds have some fascinating attributes that came to light by juxtaposing objects that previously seemed unrelated. Greene and Plesser found, for example, that one Calabi-Yau manifold might have 101 possible shape settings and one possible size setting; the mirror would be just the opposite, having one possible shape setting and 101 size settings. Calabi-Yau manifolds have holes of various dimensions (odd- and even-numbered), but Greene and Plesser found a curious pattern between the pairs: The number of odd-dimensional holes in one manifold equaled the number of even-dimensional holes in its mirror, and vice versa. "This means that the total number of holes . . . in each is the same even though the even-odd interchange means that their shapes and fundamental geometrical structures are quite different," Greene says.[17]

That still doesn't quite explain the "mirror" aspect of this symmetry, which might be easier to picture through topology. It was found, for example, that Calabi-Yau mirror pairs had Euler characteristics of opposite sign, thereby implying markedly different topologies, albeit crudely; the numbers themselves convey only a bit of information about a space, and as we've seen, many very different-looking spaces—such as a cube, tetrahedron, and sphere—can all have

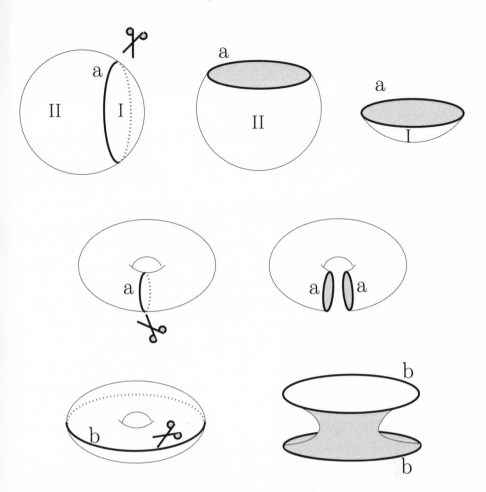

7.4—Surfaces (and again, we mean orientable or two-sided ones) can be distinguished from each other topologically by comparing their Betti numbers. In general, the Betti number tells you the number of ways you can cut all the way around a two-dimensional surface without dividing it into two pieces. There are no ways of making such cuts to a sphere, so its Betti number is 0. A donut, on the other hand, can be cut in two different ways without dividing it in half—as shown—so its Betti number is 2.

the same Euler characteristic. This crudeness can be refined, however, by breaking the Euler characteristic into sums and differences of integers called Betti numbers, which contain more detailed information about a space's inner structure.

Any given object has $n + 1$ Betti numbers, where n is the number of dimensions the object has. A zero-dimensional point, therefore, has a Betti number of 1; a one-dimensional circle has a Betti number of 2; a two-dimensional surface, such as a sphere, has 3; and so forth. The first Betti number is denoted b_1, the second b_2, and the last b_k, where the kth Betti number is the number of independent, k-dimensional cycles or loops that can be wrapped around (or threaded through) the space or manifold in question. (We'll say a bit more about cycles later in this chapter.)

In the case of two-dimensional surfaces, the first Betti number describes the number of cuts that can be made to a space without dividing it in two. If you take the surface of a sphere, a two-dimensional space, there's no way to cut all the way around the surface without dividing it into two pieces. That's equivalent to saying the first Betti number of a sphere equals 0.

Now we'll take a hollow donut. If you make an incision all the way around the "equator," you'll still end up with a single pastry, albeit an eviscerated one. Similarly, if you slice the donut the other way, the short way through the hole, you'll be left with a severed donut that's still just a single pastry (albeit an unsightly one). Since there are two different ways of cutting the donut, neither of which divides the space into two, we say its first Betti number is 2.

Next, let's consider a pretzel with two holes. We can slice through either hole or slice through from one hole to the other or make a slice all the way around the outside edge and still be left with a single pretzel. That makes four different ways of cutting the double-holed pretzel, none of which divide it into two pieces, so its first Betti number is 4. The same pattern holds for a pretzel with 18 holes; its first Betti number is 36.

We can get an even more refined description of the topology of different manifolds, however. Each of the Betti numbers is itself the sum of finer numbers called Hodge numbers, discovered by the Scottish mathematician W. V. D. Hodge. These numbers afford an even higher-resolution glimpse into a space's substructure. This information is encapsulated in a so-called Hodge diamond.

Hodge diamonds enable us to visualize the "mirror" in mirror symmetry. A given grid of sixteen numbers corresponds to a particular six-dimensional Calabi-Yau manifold, which we'll call M. To get the Hodge diamond for the mirror manifold, M', we draw a line that goes from the middle of the lower left edge of the diamond to the middle of the upper right edge. Then we flip the Hodge numbers over this diagonal line. The modified Hodge diamond, which characterizes the mirror partner, is literally a reflection or mirror image of the original.

$$
\begin{array}{ccccccc}
 & & & 1 & & & \\
 & & 0 & & 0 & & \\
 & 0 & & h^{1\cdot1}(V) & & 0 & \\
1 & & h^{2\cdot1}(V) & & h^{2\cdot1}(V) & & 1 \\
 & 0 & & h^{1\cdot1}(V) & & 0 & \\
 & & 0 & & 0 & & \\
 & & & 1 & & &
\end{array}
\qquad
\begin{array}{ccccccc}
 & & & 1 & & & \\
 & & 0 & & 0 & & \\
 & 0 & & h^{2\cdot1}(V) & & 0 & \\
1 & & h^{1\cdot1}(V) & & h^{1\cdot1}(V) & & 1 \\
 & 0 & & h^{2\cdot1}(V) & & 0 & \\
 & & 0 & & 0 & & \\
 & & & 1 & & &
\end{array}
$$

7.5—Detailed topological information about a Calabi-Yau manifold (of three complex dimensions) is contained in a four-by-four array of numbers known as a Hodge diamond. Although a Calabi-Yau manifold may not be uniquely characterized by its Hodge diamond, manifolds with different Hodge diamonds are topologically distinct. The two Hodge diamonds shown are mirror images of each other, corresponding to a Calabi-Yau manifold and its mirror partner.

The fact that the Hodge numbers for the manifolds are reversed is a consequence rather than an explanation of mirror symmetry, because this can also happen in two manifolds that are not actual mirror pairs. The reflected Hodge numbers seen by Greene and Plesser were merely the first hints, rather than proof, that they'd stumbled upon an interesting new symmetry. Far more compelling, says Plesser, was determining that the physics (or conformal field theories) associated with the mirror pairs was "literally identical."[18]

Outside corroboration came a few days after Greene and Plesser had submitted their paper in 1989. Candelas informed Greene that he and two students had found a striking pattern after surveying a large number of Calabi-Yau manifolds they'd created by computer. These manifolds, they noticed, seemed to come in pairs in which the number of even-dimensional holes of one equaled the number of odd-dimensional holes of the other. The observed interchange in number of holes, shape and size settings, and Hodge numbers between the two manifolds was certainly intriguing, but could have been mere mathematical coincidence, says Greene, "perhaps of no more consequence to physics than going to one store, where milk was a dollar and juice two dollars, and then finding another store, where milk was two dollars and juice just one dollar. What clinched the deal was the argument that Plesser and I gave establishing that pairs of different Calabi-Yau's yielded the same physics. That's the real definition of mirror symmetry—the one from which all the interesting consequences follow—and it is much more than the simple interchange of two numbers."[19]

The two efforts were not only parallel but "complementary," according to Greene. While he and Plesser had gone deeper by exploring the physical consequences of this correspondence, Candelas and his students—by virtue of their computer program—had found a much larger sample of Calabi-Yau shapes whose Hodge numbers fell into mirror pairs. With these two papers (both published in 1990), Greene proclaimed that the "mirror symmetry of string theory" had finally been established.[20]

Vafa, for one, was happy to see the conjecture he'd contributed to upheld, though he'd felt all along that mirror symmetry would prove correct. "I sometimes say it was more courageous of us to have formulated it without any known examples," he jokes.[21]

Initially, I had been skeptical of Vafa and Greene's program, as I told them, because almost all of the Calabi-Yau manifolds we'd found up to that point had negative Euler characteristics. If what they were saying was true, and these manifolds really came in pairs of opposite Euler characteristics, then we should have been finding about as many with positive Euler characteristics as those with negative values, given that the Euler characteristic of the manifold and its mirror must have opposite signs. Fortunately, those concerns were not sufficient to make Vafa, Greene, Plesser, and others abandon their investigations into the possibility of this new kind of symmetry. (The moral being it's often better to go ahead and look, rather than concluding in advance that something cannot be found.) And before long, we started finding more Calabi-Yau manifolds with positive Euler characteristics—enough, in fact, to put my earlier concerns to rest.

I soon arranged for Greene to talk about mirror symmetry to mathematicians, and several big shots—including I. M. Singer from MIT—were planning to be there. A physicist by training, Greene was nervous about speaking before such a crowd. I told him he should try to use the word *quantum* liberally in his lecture, as mathematicians tended to be impressed with that word. Perhaps, I suggested, he could describe mirror symmetry in terms of "quantum cohomology"—a term I coined at the time.

Cohomology has to do with the cycles, or loops, of the manifold and how they intersect. Cycles, in turn, are linked to subsurfaces within the manifold—

also called *submanifolds*—that have no boundaries. To get a better idea of what we mean by submanifold, imagine a hunk of Swiss cheese cut into the shape of a ball. You could think of that entire Swiss cheese ball as a three-dimensional space that could be wrapped along the outside with a plastic sheet. But inside you might also find hundreds of smaller holes—subsurfaces within the larger surface—that you might be able to wrap or string something like a rubber band through. The submanifold is a geometric object with a precisely defined size and shape. A cycle, to a physicist, is a less sharply defined loop based solely on topology, whereas most geometers don't see any difference between a cycle and a submanifold. Nevertheless, we tend to use cycles—such as a circle winding through a donut hole—to extract information about a manifold's topology.

Physicists have a way of associating a quantum field theory with a given manifold. But since that manifold normally has an infinite number of cycles, they use an approximation that drives that number down to a finite, and therefore manageable, value. This process is called *quantization*—taking something with an infinite number of possible settings (such as frequencies on the FM radio dial) and saying that only certain values are going to be allowed. Doing that involves making quantum corrections to the original equation—an equation that is all about cycles and therefore all about cohomology, too. Hence the name *quantum cohomology*.

It turns out that there is more than one way to do these quantum corrections. Thanks to mirror symmetry, we can take a Calabi-Yau manifold and produce its physically equivalent mirror partner. The partners are described by two apparently different but fundamentally equivalent versions of string theory, Type IIA and Type IIB, that describe the same quantum field theory. We can do these quantum correction calculations relatively easily for the B model, where the quantum corrections turn out to be zero. The A model calculation, where the quantum corrections are not zero, is practically impossible.

About a year after Greene and Plesser's paper, a further development in mirror symmetry commanded the attention of the mathematics community. That's when Candelas, Xenia de la Ossa, Paul Green, and Linda Parks showed how mirror symmetry could help solve mathematical puzzles, particularly in the fields of algebraic and enumerative geometry—some of which had resisted solution for decades or more. The problem Candelas and his colleagues had taken on, the so-called quintic threefold problem, had certainly been around for a while. Also called the Schubert problem, in honor of Hermann Schubert, the nineteenth-century German mathematician who solved the first part of this puzzle, the problem concerns the number of rational curves—that is, curves with genus 0, or no holes, such as a sphere—that can fit on a quintic (six-dimensional) Calabi-Yau.

Counting in this way might sound like an odd pastime, unless you're an enumerative geometer, in which case your days are filled with activities such as this. The task, however, is rarely as simple as emptying the contents of a jelly-bean bowl onto a table and counting the number of treats. Counting the number of objects on a manifold, and finding the right way to frame the problem so that the number you get is useful, has been a continuing challenge for mathematicians for a century or more. The number we seek at the end of this process must be finite, so we must restrict our search to compact space rather than to, say, an infinite plane. If we want, for example, to count the number of points of intersection between two curves, complications arise when these two curves are tangential or touch each other. Enumerative geometers have developed techniques to handle such situations and to arrive, hopefully, at a discrete number.

One of the earliest problems of this sort was posed around the year 200 B.C. by the Greek mathematician Apollonius, who asked how many circles can be drawn that are tangent to three given circles. The general answer to this question, which happens to be eight, can be obtained with a ruler and compass. But more sophisticated computational techniques are needed for the Schubert problem.

Mathematicians have attacked this problem in phases, taking it one degree at a time. A degree, in this context, refers to the highest exponent of any term in the polynomial equations that describe the term. For example, the degree of the polynomial $4x^2 - 5y^3$ is three, and $6x^3y^2 + 4x$ is of degree five (the x^3y^2 exponents get combined), whereas $2x + 3y - 4$ is of degree one and corresponds to a straight line. So the point of this exercise is to pick the manifold—the quintic threefold in this case—and pick the degree of the curves in question and then ask how many of these curves there are.

Schubert solved the problem for the first degree, showing that the number of lines on the quintic is exactly 2,875. In 1986, roughly a century later, Sheldon Katz (currently at the University of Illinois) demonstrated that the number of curves of degree two (such as circles) is 609,250. Candelas, de la Ossa, Green, and Parks tackled the degree-three case, which loosely translates to the number

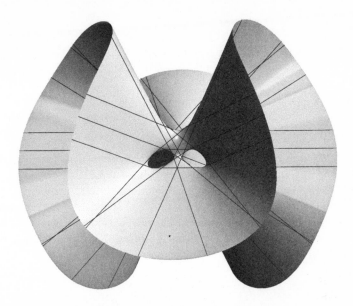

7.6—In a celebrated result of nineteenth-century geometry, the mathematicians Arthur Cayley and George Salmon proved that there are exactly 27 lines on a so-called cubic surface, as illustrated. Hermann Schubert later generalized this result, which is known as the Cayley-Salmon theorem. (Image courtesy of the 3D-XplorMath Consortium)

7.7—A common problem in algebraic and enumerative geometry is to count the number of lines or curves on a surface. To gain a sense of what it means to have lines on the surface, one can see that the doubly ruled hyperboloid shown is a surface made up entirely of lines. It's called doubly ruled because there are two distinct lines going through every point. Such a surface is not a good candidate for enumerative geometry, however, since an infinite number of lines can be drawn. (Photo by Karen Schaffner, Department of Mathematics, University of Arizona)

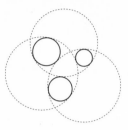

7.8—The Apollonius problem, one of the most famous problems in all of geometry, asks how many circles can be drawn that are tangent to three circles in a plane. The framing of this problem, as well as its original solution, has been attributed to the Greek geometer Apollonius of Perga (circa 200 B.C.). The eight solutions to this problem— eight different tangent circles—are shown. The mathematician Hermann Schubert extended that result two thousand years later, showing that there are sixteen spheres tangent to four given spheres.

of spheres that can fit inside a particular Calabi-Yau space. But they employed a trick based on mirror symmetry. Whereas solving the problem on the quintic itself was exceedingly difficult, the quintic's mirror manifold—which Greene and Plesser had constructed—offered a much easier setting in which to address the question.

In fact, Greene and Plesser's original paper on mirror symmetry had identified this basic approach, showing that a physical quantity, the Yukawa couplings, could be represented by two very different mathematical formulas—one corresponding to the original manifold and the other corresponding to its mirror pair. One formula, which involved the number of rational curves of various degrees that could be found on the manifold, was absolutely "horrendous" to deal with, according to Greene. The other formula, which depended on the shape of the manifold in a more general way, was much simpler to work with. But since both formulas described the same physical object, they had to be the same— just as the words *cat* and *gato* look different yet also describe the same furry creature. Greene and Plesser's paper had an equation that explicitly stated that these two different-looking formulas were equal. "You can have an equation that you know is abstractly correct, but it can nevertheless be a major challenge to evaluate that equation with adequate precision to extract numbers from it," says Greene. "We had the equation but we didn't have the tools to leverage it into the determination of numbers. Candelas and his collaborators developed the tools to do that, which was a huge accomplishment" that has had a major impact on geometry.[22]

This illustrates the potential power of mirror symmetry. You may not have to bother counting the number of curves in a Calabi-Yau space, when an entirely different calculation—which seems to have nothing to do with the original counting chore—yields the same answer. And when Candelas and company applied this approach to the number of degree-three curves on a quintic three-fold, they got 317,206,375.

Our interest, however, is not strictly in the number of rational curves but extends to the manifold itself. For in the course of this curve counting, we're essentially moving the curves around, using well-established techniques, until we've covered the whole space. In the course of doing that, we've defined that space—be it a quintic threefold or some other manifold—in terms of those curves.

The effect of all of this was to reinvigorate an area in geometry that had been largely moribund. The idea of using mirror symmetry to solve problems in enumerative geometry, which was pioneered by Candelas and his collaborators, led to the rebirth of that whole field, according to Mark Gross, a mathematician at the University of California, San Diego. "At the time, the field was more or less dead," Gross says. "As the old problems had been solved, people went back to check Schubert's numbers with modern techniques, but that was getting pretty stale." Then he adds, out of the blue, "Candelas brought in some new methods that went far beyond anything Schubert had envisioned."[23] The physicists had been eagerly borrowing from mathematics, but before mathematicians would borrow from physics, they were demanding more proof of the rigor of Candelas's method.

Fortuitously, at about this time—May 1991, to be exact—I had organized a conference at the Mathematical Sciences Research Institute (MSRI) at Berkeley for mathematicians and physicists to talk about mirror symmetry. I. M. Singer, one of the founders of MSRI, had originally selected a different topic for the conference, but I mentioned to him that some new developments in mirror symmetry were particularly exciting. Having attended Brian Greene's talk somewhat earlier, Singer agreed with me and asked me to chair the weeklong event.

My hope for the proceedings was to overcome barriers stemming from differences in language and assumed knowledge between the respective fields. Candelas discussed his result for the Schubert problem there, but his number disagreed with the number obtained through ostensibly more rigorous techniques by the Norwegian mathematicians Geir Ellingsrud and Stein Arild Strømme, who had arrived at 2,682,549,425. The algebraic geometers in the crowd tended to be arrogant, assuming the physicists must have made the mistake. For one thing, explains University of Kaiserslautern mathematician Andreas Gathmann, "mathematicians didn't understand what the physicists were doing, because they [the physicists] were using completely different methods—methods that didn't exist in math and didn't always appear to be justified."[24]

Candelas and Greene worried that they might have made an error, but they couldn't see where they had gone wrong. I spoke with both of them at the time, especially Greene, wondering whether something might have gone awry during the process of integrating over an infinite-dimensional space that then has to be reduced to finite dimensions. Choices have to be made in the course of doing that, and none of the options are perfect. While that made Candelas and Greene somewhat uneasy, we couldn't find any flaws in their reasoning, which was based on physics rather than an outright mathematical proof. Moreover, they remained confident about mirror symmetry in general, despite the mathematicians' criticism.

This all became clear a month or so later, when Ellingsrud and Strømme found an error in their computer program. Upon rectifying it, they got the same answer that Candelas et al. had. The Norwegian mathematicians showed great integrity in rerunning their program, checking their results, and publicizing their mistake. Many people in the same position try to conceal a mistake for as long as possible, but Ellingsrud and Strømme did the opposite, promptly informing the community of their error and subsequent correction.

This proved to be a big moment for mirror symmetry. The announcement of Ellingsrud and Strømme not only advanced the science of mirror symmetry, but also helped change attitudes toward the subject. Whereas many mathematicians had previously considered mirror symmetry rubbish, they now came to realize there was something to be learned from the physicists after all. As a case in point, the mathematician David Morrison (then at Duke University) had been one of the most vocal critics at the Berkeley meeting. But his thinking turned around, and before long, he would make many important contributions to mirror symmetry, string theory, and topology-changing transitions of Calabi-Yau manifolds.

Beyond solving the degree-three Schubert problem, Candelas and company had used their mirror symmetry method to calculate the solutions for degrees one through ten, while producing a general formula for solving every order of the quintic threefold problem and for predicting the number of rational curves for any and all degrees. In so doing, they'd taken major strides toward answering a centuries-old challenge cited in 1900 by the German mathematician David Hilbert as one of the twenty-three biggest problems in all of mathematics— namely, trying to establish a "rigorous foundation of Schubert's enumerative calculus" such that "the degree of the final equations and the multiplicity of their solutions may be foreseen."[25] Candelas's formula took many of us by surprise. The numerical solutions to the Schubert problem appeared to be just a string of numbers with no pattern and no apparent relationships between them. The work of Candelas and colleagues showed that rather than being random digits, these numbers were actually part of an exquisite structure.

The structure Candelas and his colleagues observed gave rise to a formula for doing the work, which was tested via mathematically intensive computations for polynomials of degrees one through four. The first three problems have already been discussed, whereas the fourth-degree problem was solved in 1995 by the mathematician Maxim Kontsevich (currently at the Institut des Hautes Études Scientifiques, IHES), who obtained the number 242,467,530,000. Although the Candelas group's formula agreed with all known data points, one had to wonder whether this proposition could actually be proved. Many mathematicians, including Kontsevich, took additional steps to help put the equations into the form of a full-fledged conjecture, mainly by supplying definitions for some terms in that equation. The resulting statement, which came to be known as the mirror conjecture, could then be put to the ultimate test: a mathematical proof. Proving the mirror conjecture would provide mathematical validation for the notion of mirror symmetry itself.

Here again we venture into one of those areas of controversy that pop up from time to time in mathematics. These things happen, I suppose, because we live in an imperfect world populated by imperfect beings, and because mathematics—despite the popular image—is not a strictly intellectual pursuit, done in total isolation, divorced from politics, ambition, competition, and emotion. It often seems that in matters like this, the smaller the stakes, the bigger the controversy.

My colleagues and I had been studying the mirror conjecture, and its generalizations, since 1991—pretty much as soon as the Candelas results were announced. In a paper posted on the math archives in March 1996, Alexander Givental of the University of California, Berkeley, claimed to have proved the mirror conjecture. We scrutinized his paper very carefully and were not alone in finding it hard to follow. That year, I personally invited an MIT colleague who is an expert on the subject (and who wishes to remain anonymous) to lecture on Givental's proof in my seminar. He politely declined, citing his serious misgivings about the arguments presented in that paper. My colleagues and I also failed in our attempt to reconstruct Givental's entire argument, despite our attempts to contact him and piece together the steps we found most puzzling. We consequently abandoned that effort and pursued our own proof of the mirror conjecture, which we published a year later.

Some observers, including Gathmann, called our paper the "first complete rigorous proof" of the conjecture, arguing that Givental's "proof was hard to understand and at some points incomplete."[26] Amherst mathematician David Cox, coauthor (with Katz) of *Mirror Symmetry and Algebraic Geometry*, similarly concluded that we had provided the conjecture's "first complete proof."[27] On the other hand, some people take a different view, maintaining that Givental's proof, published a year before ours, was complete, containing no significant gaps. While people are free to debate the matter further, I believe the best thing to say at this point is that collectively the two papers constitute a proof of the mirror conjecture and to leave it at that. Continuing to quarrel over credit makes little sense, especially when there are so many unsolved problems in math that we might instead devote our energy to.

Controversy aside, what did those two papers actually prove? For starters, the mirror conjecture proof showed that Candelas had the right formula for predicting the number of curves of a given degree. But our proof was a good deal broader. The Candelas formula applied only to counting curves on quintic threefolds, whereas our results applied to a much broader class of Calabi-Yau manifolds, including those of interest to physicists, and to other objects, such as vector bundles, about which we'll be saying more in Chapter 9. Moreover, in our generalization, the mirror conjecture involved not just counting curves but counting other geometric characteristics as well.

As I see it, proving this conjecture offered a consistency check on some of the ideas of string theory—a check that's grounded in rigorous mathematics, thereby putting the theory on firmer mathematical footing. But string theory has more than returned the favor, as mirror symmetry has helped create a new industry in algebraic geometry—enumerative geometry, a branch thereof, being one of the main beneficiaries—while contributing to the solution of long-standing problems in those fields. Indeed, many colleagues in algebraic geometry have told me they haven't done anything interesting in the past fifteen years except for the work inspired by mirror symmetry. The great windfall from string theory for mathematics suggests to me that the intuition of physicists must carry some weight. It means there must be some truth to string theory, even if nature doesn't work exactly the way the theory posits, because we have used it to solve classic problems that mathematicians had been unable to solve on their own. Even now, years later, it's hard to imagine an independent way of getting a formula like that of Candelas et al. were it not for string theory.

Ironically, one thing the proof of the mirror conjecture did not do was explain mirror symmetry itself. In many ways the phenomenon that physicists hit upon, and mathematicians have subsequently exploited, remains a mystery, although two main avenues of inquiry now under way—one called homological mirror symmetry, the other going by the acronym SYZ—are actively seeking such an explanation. The SYZ work attempts to provide a geometric interpretation of mirror symmetry, whereas homological mirror symmetry takes a more algebraic approach.

We'll start with the one I'm more invested in, the SYZ conjecture, perhaps because the acronym is short for the three authors of the original 1996 paper: Andrew Strominger is the S, Eric Zaslow of Northwestern University is the Z, and I'm the Y. Collaborations of this sort rarely have a formal starting point, and this one began in some sense with casual conversations I had with Strominger at a 1995 conference in Trieste, Italy. Strominger discussed a paper he'd recently written with Katrin Becker and Melanie Becker, physicist sisters now at Texas A&M. As D-branes were then entering string theory in a big way, almost to the point of taking over the field, their paper looked at how these branes fit in with Calabi-Yau geometry. The idea was that branes can wrap around submanifolds that sit inside a Calabi-Yau space. Becker, Becker, and Strominger were studying a class of submanifolds that preserve supersymmetry and that, as a result, have very desirable properties. Strominger and I were curious about the role these submanifolds might play in mirror symmetry.

I returned to Harvard tantalized by this possibility, which I immediately took up with Zaslow, a physicist-turned-mathematician who was my postdoc at the time. Before long, Strominger came from Santa Barbara to visit Harvard, which was actively recruiting him (though it took more than a year before he switched coasts). The three of us met then, which finally put S, Y, and Z in the same room at the same time and eventually on the same page, as we submitted our paper for publication in June 1996.

If SYZ is correct, it would offer a deeper insight into the geometry of Calabi-Yau spaces, while validating the existence of a Calabi-Yau substructure. It argues that a Calabi-Yau can essentially be divided into two three-dimensional spaces that are highly entangled. One of these spaces is a three-dimensional torus. If you separate the torus from the other part, "invert" it (by switching its radius from r to $1/r$), and reassemble the pieces, you'll have the mirror manifold of the original Calabi-Yau. SYZ, asserts Strominger, "provides a simple physical and geometrical picture of what mirror symmetry corresponds to."[28]

The key to understanding mirror symmetry, according to the SYZ conjecture, resides within the submanifolds of a Calabi-Yau and the way in which they're organized. You might recall our discussion earlier in this chapter of a Swiss cheese chunk as a surface containing many subsurfaces, or submanifolds, within. The submanifolds in this case are not any old random patch of surface but rather a discrete piece (of lower dimension than the manifold itself), defined by a single hole in the "cheese," that could be wrapped around or wound through individually. Similarly, the submanifolds in a Calabi-Yau of concern to the SYZ conjecture are wrapped by D-branes. (Not to confuse matters further, but one can also think of those D-branes as being the submanifolds rather than just wrapping them. Physicists tend to think in terms of branes, whereas mathematicians are more comfortable with their own terminology.) Subspaces of this sort that can satisfy supersymmetry are called special Lagrangian subman-ifolds, and as their name implies, they have special features: They have half the dimension of the space within which they sit, and they have the additional at-tribute of being length-, area-, or volume-minimizing, among other properties.

Let's consider as an example the simplest possible Calabi-Yau space, a two-dimensional torus or donut. The special Lagrangian submanifold in this case will be a one-dimensional space or object consisting of a loop through the hole of the donut. And since that loop must have the minimum length, it must be a circle, the smallest possible circle going through that hole, rather than some ar-bitrary, squiggly, or otherwise meandering loop. "The entire Calabi-Yau, in this case, is just a union of circles," explains Mark Gross, who has probably done more to follow up on the SYZ conjecture since it was first posed than anyone else. "There is an auxiliary space, call it B, that tells you about the set of all those circles, and it is a circle itself."[29] B is said to *parameterize* that set of circles, mean-ing that every point on B corresponds to a different circle, just as every circle looping through the donut hole corresponds to a different point on B. Another way to put it is that B, which is called the moduli space, contains an index of every subspace in the bigger manifold. But B is more than just a glorified list, as it shows how all these subspaces are arranged. B may, in fact, be the lynchpin to the whole SYZ conjecture, claims Gross. Accordingly, we're going to spend a bit of time trying to understand this auxiliary space.

If we go up by one complex dimension from two real dimensions to four, the Calabi-Yau becomes a K3 surface. Instead of being circles, the submanifolds in this case are two-dimensional tori—a whole bunch of them fitting together. "I can't draw this four-dimensional space," says Gross, "but I can describe B, which tells us the setting in which all these subobjects, these donuts, sit."[30] In this case, B is just a two-dimensional sphere. Every point on this sphere B corresponds to a different donut, except for twenty-four bad points corresponding to "pinched donuts" that have singularities—the significance of which shall be ex-plained shortly.

Now we'll go up one more complex dimension, so that the manifold in ques-tion becomes a Calabi-Yau threefold. B now becomes a *3-sphere*—a sphere, that is, with a three-dimensional surface (which is not one we can readily picture)—and the subspaces become three-dimensional donuts. In this case, the set of

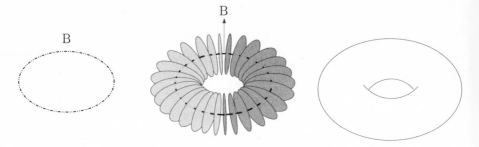

7.9 —The SYZ conjecture—named after its inventors, Andrew Strominger, the author (Shing-Tung Yau), and Eric Zaslow—offers a way of breaking up a complicated space such as a Calabi-Yau manifold into its constituent parts, or *submanifolds*. Although we cannot draw a six-dimensional Calabi-Yau manifold, we can draw the only two- (real) dimensional Calabi-Yau, a donut (with a flat metric). The submanifolds that make up the donut are circles, and all these circles are arranged by a so-called auxiliary space *B*, which is itself a circle. Each point on *B* corresponds to a different, smaller circle, and the entire manifold—or donut—consists of the union of those circles.

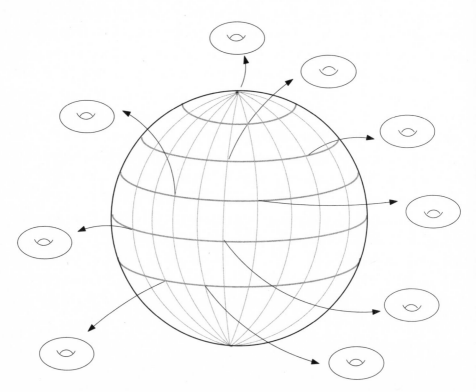

7.10—The SYZ conjecture offers a new way of thinking about K3 surfaces, which constitute a class of four-dimensional Calabi-Yau manifolds. We can construct a K3 surface, according to SYZ, by taking a two-dimensional sphere (the *auxiliary space*, B, in this example) and attaching a two-dimensional donut to every point on that sphere.

"bad" points, corresponding to singular donuts, fall on line segments that connect to each other in a netlike pattern. "Every point on this line segment is a 'bad' [or singular] point, but the vertices of the net, where the three line segments come together, represent the worst points of all," says Gross. They correspond, in turn, to the most scrunched-up donuts.[31]

This is where mirror symmetry comes in. Working with the initial idea of SYZ, the Oxford geometer Nigel Hitchin, Mark Gross, and some of my former students (Naichung Leung, Wei-Dong Ruan, and others) helped build the following picture: You have a manifold X made up of all kinds of submanifolds that are catalogued by the moduli space B. Next you take all those submanifolds, which have a radius of r, and invert them to a radius of $1/r$. (One of the strange though nice features of string theory—something that doesn't happen in classical mechanics—is that you can make a switch like this, flipping the radius of, say, a cylinder or a sphere, without affecting the physics at all.) If you have a point particle on a circle of radius r, its motion can be described strictly in terms of its momentum, which is quantized, meaning it can just assume certain (integer) values. A string, similarly, has momentum as it moves around a circle, but it can also wrap around the circle one or more times. The number of times it wraps around is called the *winding number*. So the motion of the string, unlike that of the particle, is described by two numbers: its momentum and its winding number, both of which are quantized. Let's suppose that on one hand we have a string of winding 2 and momentum 0 on a circle of radius r, and on the other hand we have a string of winding 0 and momentum 2 on a circle of radius $1/r$. Although the two situations sound different and conjure up quite different pictures, mathematically they are identical and so is the associated physics. This is known as *T duality*. "This equivalence extends beyond circles to products of circles or tori," says Zaslow.[32] In fact, the T in T duality stands for tori. Strominger, Zaslow, and I considered this duality so instrumental to mirror symmetry that we titled our original SYZ paper "Mirror Symmetry Is T Duality."[33]

Here's a simple way of seeing how T duality and mirror symmetry go hand in hand. Suppose manifold M is a torus—the product of two circles of radius r. Its mirror, M', is also a torus—the product of two circles of radius $1/r$. Let's further suppose that r is extremely small. Because M is so tiny, if you want to understand the physics associated with it, quantum effects must be taken into account. So your calculations necessarily become much more difficult. Getting the physics from the mirror manifold, M', is much easier because if r is very small, then $1/r$ is very big, and quantum effects can, therefore, be safely ignored. So mirror symmetry, under the guise of T duality, can simplify your calculations (as well as your life) tremendously.

Now let's see how all these notions fit together, starting with our two-dimensional example above. When you invert the radii of all the submanifolds (circles), the big manifold you're left with—which is made up of all these circles—has a different radius from what you started with, but it's still a torus. As a result, we call this example trivial, because the manifold and its mirror are topologically identical. Our four-dimensional example with K3 surfaces is also trivial in the same respect because all K3 surfaces are topologically equivalent. The six-dimensional example involving Calabi-Yau threefolds is more interesting. The components of this manifold include three-dimensional tori. Applying T duality will invert the radii of those tori. For a nonsingular torus, this radius change will not change the topology. However, Gross explains, "even if all the original submanifolds were of the 'good' [nonsingular] variety, changing the radius can still change the topology of the big manifolds because the pieces . . . can be put together in a non-trivial way."[34]

This might be best understood through analogy: You could take a stack of line segments—or, say, toothpicks—and use them to make a cylinder by sticking the toothpicks into a corkboard in a circular pattern. Instead of making a cylinder (which has two sides), you could also take those same toothpicks, introduce a twist, and assemble them into a single-sided Möbius strip. So you're using the same pieces, the same submanifolds, but the objects you're making have a very different topology.[35]

The point is that after the T duality shift and the different ways of assembling the submanifolds, we're left with two topologically distinct manifolds that are indistinguishable from a physics standpoint. That's part of what we mean by mirror symmetry but not quite the whole story, as another interesting feature of this duality is that mirror pairs should have opposite Euler numbers or characteristics. But all the submanifolds we're talking about here—the special Lagrangians—have an Euler characteristic of 0, which does not change when the radii are inverted.

Although what I just said is true with respect to the "good" (nonsingular) submanifolds, it does not hold for the "bad" (singular) ones. T duality interchanges the Euler characteristic of those submanifolds from +1 to –1, or vice versa. Suppose your original manifold includes thirty-five bad submanifolds, twenty-five of which have an Euler characteristic of +1, and ten with an Euler characteristic of –1. As Gross has shown, if you add all those numbers, you get the Euler characteristic of the manifold as a whole, which turns out to be +15 in this case. In the mirror manifold, everything is reversed: twenty-five submanifolds with a –1 Euler characteristic and ten with a +1 Euler characteristic, for a total of –15, which is the opposite of what we started with and which is just what we wanted.

These bad submanifolds, as discussed earlier, correspond to bad points on the moduli space *B*. "Everything interesting in mirror symmetry, all the topological changes, are happening at the vertices of *B*," Gross explains. So the emerging picture puts this space *B* at the center of mirror symmetry. Beforehand, the whole phenomenon had a mysterious air. "We had these two manifolds, *X* and *X'*, that were somehow related, but it was hard to see what they had in common," Gross adds. What they have in common is this other object, *B*, which nobody knew about at first.[36]

Gross thinks of *B* as a kind of blueprint. If you look at the blueprint from one perspective, you'll build one structure (or manifold), and if you look at it another way, you'll build a different structure. And those differences stem from the funny (singular) points on *B* where T duality doesn't work so well, and things change as a result.

That's pretty much where our current picture of mirror symmetry, as seen through the lens of SYZ, stands. One of the chief virtues of SYZ, claims Strominger, is that "mirror symmetry was demystified a little bit. Mathematicians liked it because it provided a geometric picture of where mirror symmetry comes from, and they can use that picture without reference to string theory."[37] Besides offering a geometric explanation of mirror symmetry, Zaslow says, "it also offers a process for constructing mirror pairs."[38]

It's important to bear in mind that SYZ is still a conjecture that has only been proved in a few select cases but not in a general way. Although the conjecture may not be provable as originally stated, it has been modified in light of new insights to its present incarnation, which, according to Gross, "is slowly absorbing all of mirror symmetry."[39]

Some would consider this last point debatable—and perhaps an overstatement. But SYZ has already been used—by Kontsevich and Yan Soibelman of

Kansas State University, for instance—to prove a specific example of homological mirror symmetry, the other leading attempt at a fundamental mathematical description of mirror symmetry.

Homological mirror symmetry was first unveiled by Kontsevich in 1993 and has been developing ever since, sparking considerable activity in both physics and math. Mirror symmetry, as originally formulated, didn't make much sense to mathematicians, as it concerned two different manifolds giving rise to the same physics. But as Soibelman explains, "in mathematics, there really is no concept of a physical theory associated with manifolds X and X'. So Kontsevich has tried to make that statement mathematically precise," expressing it in a manner that is not tied to physics.[40]

Perhaps the simplest way of describing homological mirror symmetry is in terms of D-branes, even though Kontsevich's idea predated the discovery of D-branes by a year or two. Physicists think of D-branes as subsurfaces upon which the endpoints of open strings must attach. Homological mirror symmetry anticipated D-branes, while providing a more refined description of these entities that became among the most basic ingredients of string theory, or M-theory, after the second string revolution. It's the familiar story in which physics (through the discovery of mirror symmetry) gave rise to mathematics, and mathematics repaid its debt generously.

One of the main ideas behind homological mirror symmetry is that there are two kinds of D-branes involved in this phenomenon, A-branes and B-branes (which are terms that Witten introduced). If you have a mirror pair of Calabi-Yau manifolds, X and X', A-branes on X are the same as B-branes on X'. This concise formulation, according to Aspinwall, "made it possible for mathematicians to clearly state what mirror symmetry is. And from that statement, you can produce everything else."[41]

It's as if we had two boxes of building blocks with different shapes, suggests Stony Brook University physicist Michael Douglas. "Yet when you stack them up, you can make the exact same set of structures."[42] This is similar to the correspondence between A-branes and B-branes posited by homological mirror symmetry.

A-branes are objects defined by what's called *symplectic geometry*, whereas B-branes are objects of *algebraic geometry*. We've already touched on algebraic geometry to some extent, which describes geometric curves in algebraic terms and solves geometric problems with algebraic equations. Symplectic geometry includes the notion of Kähler geometry that is central to Calabi-Yau manifolds, but it is more general than that. Whereas spaces in differential geometry are typically described by a metric that is symmetric across a diagonal line, the metric in symplectic geometry is antisymmetric along that same line, meaning that the signs change across the diagonal.

"These two branches of geometry used to be considered totally separate, so it was a big shock when someone came along and said the algebraic geometry of one space is equivalent to the symplectic geometry of another space," says Aspinwall. "Bringing two disparate fields together, finding that they are related in some sense through mirror symmetry, is one of the best things you can do in mathematics because you can then apply methods from one field to the other. Normally that really opens the floodgates, and many a Fields Medal has come from it."[43]

In the meantime, homological mirror symmetry is reaching out to other branches of mathematics, while also reaching out to SYZ. As of yet, however, there is "no strict mathematical equivalence between the two, [but] they each support the other," Gross claims. "And if both are correct, we should eventually find that on some level they are equivalent."[44]

It's an unfolding story. We're still figuring out what mirror symmetry means through our investigations of SYZ, homological mirror symmetry, and other avenues. The phenomenon has branched off, leading to new directions in mathematics that no longer involve mirror symmetry at all, and no one knows how far these explorations will take us, or where they will end up. But we do know where it started: with the discovery of an unusual property of compact Kähler manifolds bearing the name Calabi-Yau—spaces that were almost given up for dead more than two decades ago.

Eight

KINKS IN SPACETIME

The key to understanding the human mind, according to Sigmund Freud, is to study people whose conduct is out of the ordinary, anomalous—people with weird obsessions, for instance, including his famous patients like the "Rat man" (who had deranged fantasies involving loved ones and a pot of rats) and the "Wolf man" (who dreamed of being eaten alive by a pack of white wolves perched in a tree outside his bedroom window). Freud's premise was that our best hopes of learning about typical behavior would come from studying the most unusual, pathological cases. Through such investigations, he said, we might eventually come to understand both the norm and the deviations from the rule.

We often take a similar approach in mathematics and physics. "We seek out places where the classical descriptions fail, for these are the places where we are most likely to learn something new," explains Harvard astrophysicist Avi Loeb. Whether we are thinking about an abstract space in geometry or the somewhat more tangible space we call the universe, the places "where something bad happens to space, where things break down," as Loeb puts it, tend to be the places we call singularities.[1]

Contrary to what people might think, singularities abound in nature. They're all around us: A water droplet breaking off and falling from a leaky faucet is a common example (all too common in my household), the spot (well-known to surfers) at which an ocean wave breaks and crashes, the folds in a newspaper (which dictate whether your story is important or filler), or the pinch points on a slender balloon twisted into the shape of a French poodle. "Without sin-gularities, you cannot talk about shapes," notes the geometer Heisuke Hironaka, an emeritus professor at Harvard. He cites the example of a handwritten signa-ture: "If there is no crossing, no sharp point, it's just a squiggle. A singularity might be a crossing or something suddenly changing direction. There are many things like that in the world, and that's why the world is interesting."[2]

In physics and cosmology, two kinds of singularities stand out among count-less other possibilities. One is the singularity in time known as the Big Bang. I don't know how to make sense of the Big Bang as a geometer, because no one—physicists included—really knows what it is. Even Alan Guth, the inventor of the whole notion of cosmic inflation—the thing that "puts the bang in the Big Bang," as he explains it—admits that the term *Big Bang* has always suffered from "vagueness, probably because we still don't know (and may never know) what really happened."[3] So in this instance, I believe some humility is in order.

And though we're rather clueless when it comes to applying geometry to the precise moment of the universe's birth, we geometers have had somewhat better success in dealing with black holes. A black hole is basically a chunk of space crushed down to a single point under the weight of gravity. All that mass packed into a tiny area creates a superdense object whose escape velocity (a measure of its gravitational pull) exceeds the speed of light, trapping everything, light included.

Despite being a consequence of Einstein's general theory of relativity, black holes are still strange enough objects that Einstein himself denied their exis-tence as late as 1930—about fifteen years after the German physicist Karl Schwarzschild presented them as solutions to Einstein's famous equations. Schwarzschild didn't believe in the physical reality of black holes, either, but nowadays, the existence of such objects is widely accepted. "Black holes are now discovered with great regularity," as Andrew Strominger put it, "every time somebody at NASA needs a new grant increase."[4]

Even though astronomers have found large numbers of putative black holes while accumulating impressive data in support of that thesis, these objects are still enshrouded in mystery. General relativity provides a perfectly adequate de-

8.1—Twelve million light-years away, a supermassive black hole, approximately seventy million times more massive than the sun, is thought to reside in the center of the spiral galaxy M81. (Image courtesy of NASA)

scription of large black holes, but the picture falls apart when we move to the center of the maelstrom and consider a black hole's vanishingly small singular point of infinite curvature. Nor can general relativity contend with tiny black holes, smaller than a grain of dust—a regime in which quantum mechanics inevitably comes into play. General relativity's inadequacies become glaringly apparent in the case of such miniature black holes, where masses are large, distances small, and the curvature of spacetime off the charts. That's precisely where string theory and Calabi-Yau spaces have helped out, which is gratifying since the theory was invented, in part, to deal with that very clash between general relativity and quantum mechanics.

One of the highest-profile disputes between those two celebrated fields of physics revolves around whether information is destroyed by a black hole. In

1997, Stephen Hawking of Cambridge University and Kip Thorne of Caltech made a bet with John Preskill, also of Caltech. At issue was an implication of Hawking's theoretical finding in the early 1970s that black holes are not totally "black." These objects, he discovered, have a tiny but nonzero temperature, meaning that they must retain some thermal energy. Like any other "hot" body, the black hole will radiate its energy away until there's nothing left and the black hole completely evaporates. If the radiation given off by the black hole is strictly thermal and is thus lacking in information content, then the information originally stored within a black hole—say, if it swallows up a star with a particular composition, structure, and history—will disappear when the black hole evaporates. That would violate a fundamental tenet of quantum theory, which holds that the information of a system is always preserved. Hawking argued that, quantum mechanics notwithstanding, in the case of black holes, information can be destroyed, and Thorne agreed. Preskill maintained that the information would survive.

"We believe that if you throw two ice cubes into a pot of boiling water on Monday and inspect the water atoms on Tuesday, you can determine that two ice cubes were thrown in the day before," Strominger explains—not in practice, but in principle, yes.[5] Another way to think of it is to take a book, say, *Fahrenheit 451*, and toss it into a fire. "You may think the information is lost, but if you have infinite observational power and calculation capacity—if you measure everything about the fire and keep track of the ashes and enlist the services of 'Maxwell's demon' (or in this case 'Laplace's demon')—then you can reproduce the original state of the book," notes Caltech physicist Hirosi Ooguri.[6] If you were to toss that same book into a black hole, however, Hawking argued that the data would be lost. Preskill, on the other hand, like Gerard 't Hooft and Leonard Susskind before him, took the position that the two cases are not fundamentally different and that the black hole radiation must, in a subtle way, carry the information of the Ray Bradbury classic—information that, in theory, can be recovered.

The stakes were high, as one of the lynchpins of science—the principle of scientific determinism—was at risk. *Determinism* is the idea that if you have all possible data describing a system at a particular time and know the laws of physics, then you can, in principle, determine what will happen in the future, as well as deduce what happened in the past. But if the information can be lost or destroyed, determinism is no longer valid. You can't predict the future, nor can you deduce the past. If the information is lost, in other words, you're lost, too. The stage was thus set for a classic showdown. "This was the moment of truth for string theory, which had claimed it could consistently reconcile quantum mechanics and gravity," says Strominger. "So could it explain Hawking's paradox?"[7]

Strominger tackled this question with Cumrun Vafa in a groundbreaking 1996 paper.[8] They used the notion of black hole entropy as their way into the problem. Entropy is a measure of a system's randomness or disorder but also provides indications about a system's information content and storage. Think of a bedroom, for instance, that has a variety of shelves, drawers, and counters, as well as assorted artwork displayed on the walls and mobiles hanging from the ceiling. Entropy relates to the number of different ways you can organize, or disorganize, all the stuff—furniture, clothes, books, posters, and assorted knickknacks—in that room. To some extent, the number of possible ways of arranging these same items in this same space depends on the room's size or volume—its length, width, and height all factoring into the computation.

The entropy of most systems scales with the volume. In the early 1970s, however, the physicist Jacob Bekenstein, then a graduate student at Princeton, suggested that the entropy of a black hole is proportional to the area of the event horizon surrounding the black hole, rather than to the volume bounded up inside the horizon. The event horizon is often called the point of no return, and any object that crosses this invisible line in spacetime will succumb to the pull of gravity and fall, inexorably, into the black hole. But it is perhaps better thought of as a surface of no return, as it is indeed a two-dimensional surface rather than a "point." For a nonspinning (or "Schwarzschild") black hole, the area of that surface depends solely on the mass of the black hole: The bigger the mass, the bigger the area. The notion that the entropy of a black hole—reflecting all possible configurations of this object—depended only on the horizon area suggested that all the configurations resided on the surface and that all the information about the black hole was stored on the surface, too. (One could make a parallel here with the bedroom of our previous example, in which all the stuff is arranged along the surface—walls, ceiling, and floor—rather than floating in midair in the room's interior space.)

Bekenstein's work, coupled with Hawking's ideas on black hole radiation, yielded an equation for computing a black hole's entropy. The entropy, according to the so-called Bekenstein-Hawking formula, was indeed proportional to the horizon area. More specifically, the entropy was proportional to the area divided by four times Newton's gravitational constant (G). Using this formula, one could see that a black hole three times as massive as the sun would have an astoundingly high entropy, on the order of 10^{78} joules per Kelvin. The black hole, in other words, would be extremely disordered.

The fact that a black hole had such a staggeringly high entropy came as a total shock, given that in general relativity, a black hole is completely described by just three parameters: its mass, charge, and spin. A gigantic entropy, on the other hand, suggests a tremendous variability in a black hole's internal makeup that must go far beyond those three parameters. The question was: Just where does that variability come from? What other things inside a black hole can vary as well?

The trick apparently lies in breaking down a black hole into its microscopic constituents, just as the Austrian physicist Ludwig Boltzmann had done for gases in the 1870s. Boltzmann showed that you could derive the thermodynamic properties of a gas from the combined properties of its many individual molecular constituents. (These molecules can be abundant indeed, approximately 10^{20} of them per ounce of a typical gas under typical conditions.) Boltzmann's idea was remarkable for a number of reasons, including the fact that he hit upon it decades before there was firm evidence that molecules existed. Given the vast number of gas constituents, or molecules, Boltzmann argued that the average of motions and behaviors of the individual molecules would produce the bulk properties—the same volume, temperature, and pressure—for the gas as a whole. He thus provided a more precise view of the system in which the gas was no longer just a single entity but instead consisted of multitudinous parts. His perspective also led to a new definition of entropy: the number of possible arrangements of microscopic states (or microstates) that give rise to the same macroscopic features. Putting this relation into more quantitative terms, the entropy (S) equals the natural log of the number of microstates. Or, equivalently, the number of microstates equals e^S.

The approach that Boltzmann helped pioneer is called statistical mechanics, and roughly a century later, people were trying to devise a statistical mechanical interpretation of black holes. Two decades after Bekenstein and Hawking brought this problem to prominence, no one had yet succeeded. What was needed was a "microscopic theory of black holes," says Strominger, "a derivation of the laws of black holes from some fundamental principles—the analogue of Boltzmann's derivation of the thermodynamics of gases." It's been known since the nineteenth century that every system has an associated entropy, and we've known since Boltzmann that a system's entropy depends on the number of microstates contained therein. "It would be a deep and unnerving asymmetry if the relation between entropy and the number of microstates was valid for every system in nature except a black hole," Strominger adds.[9]

These microstates, moreover, are "quantized," according to Ooguri, which is the only way you can hope to get a countable number. You can put a pencil on a desk in an infinite number of ways, just as there's an infinite number of possible settings along the electromagnetic spectrum. But as mentioned in Chapter 7, radio frequencies are quantized in the sense that radio stations only broadcast at a select number of discrete frequencies. The energy levels of a hydrogen atom are similarly quantized, meaning that you can't pick an arbitrary value; only certain values are allowed. "Part of the reason Boltzmann had such difficulty convincing others of his theories was that he was way ahead of his time," Ooguri says, "half a century before quantum mechanics was invented."[10]

So this was the challenge Strominger and Vafa took up. It was a real test of string theory, since it involved the quantum states of black holes, which Strominger has called "quintessential gravitational objects." He felt it was "incumbent upon the theory to give us a solution to the problem of computing the entropy, or it wasn't the right theory."[11]

The plan he and Vafa concocted was to take the entropy derived from that quantum microstate calculation and compare it with the value obtained from the Bekenstein-Hawking area formula based on general relativity. Although the problem was not new, Strominger and Vafa brought new tools to bear on it, drawing not only on string theory but also on Joe Polchinski's discovery of D-branes and the emergence of M-theory—both of which occurred in 1995, the year before their paper came out.

"Polchinski pointed out that D-branes carry the same kind of charge as black holes and have the same mass and tension, so they look and smell the same,"

notes Harvard physicist Xi Yin. "But if you can use one to calculate the properties of the other, such as the entropy, then it's much stronger than a passing resemblance."[12] This is indeed the approach that Strominger and Vafa followed,
using these D-branes to construct new kinds of black holes, guided by string
theory and M-theory.

The possibility of building black holes out of D-branes and strings (the latter
being the one-dimensional version of D-branes) stems from the "dual" description of D-branes. In models in which the strength of all forces acting upon
branes and strings (including gravity) is low (what's called weak coupling),
branes can be thought of as thin, membrane-like objects that have little effect
on the spacetime around them and therefore bear little resemblance to black
holes. On the other hand, at strong coupling and high interaction strengths,
branes can become dense, massive objects with event horizons and a powerful
gravitational influence—objects, in other words, that are indistinguishable from
black holes.

Nevertheless, it takes more than just a heavy brane—or a stack of heavy
branes—to make a black hole. You also need some way of stabilizing it, which
is most easily accomplished—at least in theory—by wrapping the brane around
something stable, something that will not shrink. The problem is that an object
that has high *tension* (measured in terms of mass per unit length, area, or volume) could shrink so small that it almost disappears without some underlying
structure to stop it, much as an ultratight rubber band would shrink down to a
tight clump if left alone.

The key ingredient was supersymmetry, which, as discussed in Chapter 6,
keeps the ground or vacuum state of a system from sinking to lower and lower
energy levels. And supersymmetry in string theory often implies Calabi-Yau
manifolds, because such spaces have this feature automatically built in. So the
question then becomes finding stable subsurfaces within the Calabi-Yau to
wrap branes on. These subsurfaces or submanifolds, which are of lower dimensionality than the overall space itself, are sometimes referred to by physicists
as cycles—a notion introduced previously in this book—which can sometimes
be pictured as a noncontractible loop around or through part of the manifold.
(Technically speaking, a loop is just a one-dimensional object, but cycles come
in higher dimensions, too, and can be thought of as noncontractible, higher-

dimensional "loops.") As physicists tend to view it, a cycle only depends on the topology of some object or hole you can wrap around, regardless of the geometry of that object or hole. "If you change the shape, the cycle will stay the same, but you'll have a different submanifold," explains Yin. Being just a feature of the topology, he adds, the cycle on its own has nothing to do with a black hole. "It's only when you wrap a brane on a cycle—or several branes on a cycle—that you can begin to think in terms of a black hole."[13]

To ensure stability, the thing you're wrapping with—be it a brane, string, or rubber band—must be tight, without any wrinkles in it. The cycle you're wrapping around, moreover, must be of the minimum possible length or area. Placing a rubber band around a uniform, cylindrical pole, for instance, wouldn't be an especially stable situation, because the band could easily be nudged from side to side. If, on the other hand, the pole were of varying thickness, the stable cycles (which in this case are circles) could be found at the thinnest, or minimal, points, where a rubber band would be difficult to nudge. Rather than thinking of a smooth pole as the object to be wrapped around, a better analogy for a Calabi-Yau would be a grooved pole or donut of varying thickness, with minimal cycles (or circles) again to be found at the points of minimal thickness.

There are different kinds of cycles that a brane could wrap around inside a Calabi-Yau: They could be circles, spheres, and tori of various dimensions, or Riemann surfaces of high genus. Given that branes carry mass and charge, the point of this exercise is to figure out the number of ways of placing them in stable configurations inside a Calabi-Yau so that their combined mass and

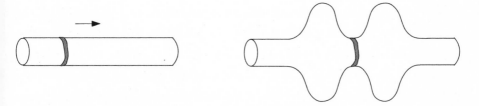

8.2—To make a black hole by wrapping a brane around an object, the object in question must be stable. In an analogous situation, one might consider wrapping a rubber band around a wooden pole. Of the two examples shown here, the right-hand figure represents by far the more stable arrangement because, in this case, the rubber band is wrapped around a minimal point, which holds it in place and keeps it from sliding sideways.

charge equals the mass and charge of the black hole itself. "Even though these branes are wrapped individually, they're all stuck together inside the internal [Calabi-Yau] space and can be thought of as parts of a bigger black hole," Yin explains.[14]

Here's an analogy, which I admit is rather unappetizing, so I'll have to blame it on an unnamed Harvard physicist who suggested it to me. (I'm sure that he, too, would pass the buck, claiming that he got the analogy from someone else.) The situation, in which individual wrapped branes glom together to form a bigger entity, can be compared to a wet shower curtain to which various strands of hair are stuck. Each strand of hair is like an individual brane that is bound to this bigger object, the shower curtain, that is like a brane itself. Even though each hair can be thought of as a separate black hole, they're all stuck together— all stuck to the same sheet—making them part of one big black hole.

Counting the number of cycles—counting the number of ways of arranging D-branes—is a problem in differential geometry, as the number you get from that counting corresponds to the number of solutions to a differential equation. Strominger and Vafa had thus converted the question of counting a black hole's microstates, and thereby computing its entropy, into a geometric question: How many ways can you put D-branes in Calabi-Yau manifolds to end up with the desired mass and charge? That, in turn, can be expressed in terms of cycles: How many spheres and other minimum-sized shapes—around which one can wrap a brane—can you fit inside a Calabi-Yau? The answer to both those questions obviously depends on the geometry of the given Calabi-Yau. If you change the geometry, you'll change the number of possible configurations or the number of spheres.

That's the general picture, but the calculation is still quite challenging, so Strominger and Vafa spent a good deal of time searching for a specific way of framing the problem—a way that they could actually solve. They settled on a very special case for their first go-round, selecting a five-dimensional internal space consisting of the product of a four-dimensional K3 surface and a circle. They also constructed a five-dimensional black hole sitting in flat five-dimensional space to which they could compare the structure they would build out of D-branes. This wasn't just any old black hole. It had special properties that were handpicked to make the problem manageable: It was both supersymmetric and *extremal*—the latter term meaning it had the minimum mass possible for a given charge. (We've already talked a fair amount about supersymmetry, but it only makes sense to talk about a supersymmetric black hole if the background vacuum in which it sits also preserves supersymmetry. That is not the case in the low-energy realm that we inhabit, where we don't see supersymmetry in the particles around us. Nor do we see it in the black holes that astronomers observe.)

Once Strominger and Vafa had their custom-built black hole, they could use the Bekenstein-Hawking formula to compute the entropy from the event horizon area. The next step was to count the number of ways of configuring D-branes in the internal space so as to match the designer black hole in total charge and mass. The entropy obtained in this fashion (equal to the log of the number of states) was then compared with the value obtained from the horizon area calculation, and the agreement was perfect. "They got it on the nose, including the factor of four, Newton's constant, and all those things," says Harvard physicist Frederik Denef. After twenty years of trying, Denef adds, "we finally had the first statistical mechanics derivation of black hole entropy."[15]

It was a major success for Strominger and Vafa and for string theory as well. The association between D-branes and black holes was significantly bolstered, while the two physicists also showed that the D-brane description itself was fundamental, Yin explains. "You might have wondered, can the brane be broken down further? Is it made up of smaller parts? We now know there are no additional structures of the brane because they got the right entropy, and entropy, by definition, counts *all* the states."[16] If, on the other hand, the brane was composed of different parts, this would add new degrees of freedom and, hence, more combinations that would have to be taken into account in the entropy calculation. But the 1996 result shows that this is not the case. The brane is all there is. Although branes of different dimensions look different from each other, none of them have subcomponents or can be broken down further. In the same way, string theory holds that the string (a one-dimensional brane in M-theory parlance) is all there is and cannot be subdivided into smaller pieces.

While the agreement between the two very different methods of obtaining the entropy was certainly gratifying, in some ways it was kind of surprising. "At first glance, the black hole information paradox appears to have nothing whatsoever to do with Calabi-Yau manifolds," claims Brown University physicist

8.3a—Harvard physicist Andrew Strominger (Kris Snibbe/Harvard University)

8.3b—Harvard physicist Cumrun Vafa (Stephanie Mitchell/Harvard University News Office)

Aaron Simons. "But the key to answering that question turned out to be counting mathematical objects inside a Calabi-Yau."[17]

The Strominger and Vafa result did not solve the information paradox outright, though the detailed description of a black hole they arrived at through string theory did show how information could be very precisely stored. They took the essential first step, Ooguri says, "by showing that the entropy of a black hole is the same as the entropy of other macroscopic systems," including the burning book from our previous example.[18] Both contain information that is at least potentially retrievable.

Of course, the 1996 results were really just the beginning, as the original entropy calculation had very little to do with real astrophysical black holes: The black holes in the Strominger-Vafa model, unlike those we see in nature, were supersymmetric—a stipulation made simply to make the calculation doable. Nevertheless, these results may still extend to non-supersymmetric black holes as well. As Simons explains, "regardless of supersymmetry, all black holes have a singularity. That is their central, defining feature, and that is the reason they are 'paradoxical.' In the supersymmetric black hole case, string theory can help us understand what happens around that singularity, and the hope is that the result is independent of the object being supersymmetric or not."[19]

Furthermore, the 1996 paper had described an artificial case of a compact five-dimensional internal space and a flat and noncompact five-dimensional space. That's not the way we normally view spacetime in string theory. The question was, did that same picture apply to the more conventional arrangement of a six-dimensional internal space and black holes sitting in flat, four-dimensional space?

The answer came in 1997, when Strominger, along with Juan Maldacena (then at Harvard) and Edward Witten, published a paper that carried the earlier work to the more familiar setting of a six-dimensional internal space (Calabi-Yau, of course) and an extended four-dimensional spacetime.[20] By duplicating the entropy calculation for Calabi-Yau threefolds, Maldacena says, "the space in which you put the branes has less supersymmetry"—and is thus closer to the world we see—"and the space in which you put the black holes has four dimensions, which is what we assume to be the case."[21] Moreover, the agreement with the Bekenstein-Hawking computation was even stronger because, as Maldacena explains, computing the entropy from the event horizon area is only accurate when the event horizon is very big and the curvature is very small. As the size of black holes shrinks and the surface area shrinks with it, the general-relativity approximation becomes worse and one therefore needs to make "quantum gravity corrections" to Einstein's theory. Whereas the original paper only considered "large" black holes (large compared with the diminutive Planck scale!), for which the general-relativity-derived number (the so-called first-order term) was sufficient, the 1997 calculation got the first quantum-corrected term in addition to the leading term. In other words, the agreement between the two very different ways of obtaining the entropy had just gotten better.

Ooguri, Strominger, and Vafa went even further in 2004, extending the 1996 result to any kind of black hole you could make by wrapping a brane around an object (such as a cycle) in a regular Calabi-Yau threefold, regardless of its size and therefore regardless of the degree to which quantum mechanics affects the system. Their paper showed how to compute the quantum gravity corrections to general relativity—not just the first few terms but the entire series, an infinite number of terms.[22] By including the other terms in the expansion, Vafa explains, "we get a more refined way of counting and a more refined answer and, fortunately, an even stronger agreement than before."[23] That is the approach we normally try to take in math and physics: If we find something that works under special circumstances, we try to generalize it to see if it's valid in broader circumstances and keep pushing it from there to see just how far we can go.

There is one last generalization of the original Strominger-Vafa work that I'd like to consider. For lack of a better phrase, it is even more general than anything we've discussed yet. The idea, which goes by the complicated-sounding name of the Anti-de Sitter Space/Conformal Field Theory (AdS/CFT) correspondence, was initially hit upon by Maldacena in 1997 and was elaborated on by Igor Klebanov of Princeton, Edward Witten, and others. Much like, as Maldacena puts it, a DVD and a 70-millimeter reel of film can both describe the same movie, the correspondence (technically still a conjecture) suggests that in some cases, a theory of gravity (such as string theory) can be completely equivalent to a standard quantum field theory (or conformal field theory, to be exact). The correspondence is surprising because it relates a theory of quantum gravity to a theory with no gravity at all.

AdS/CFT stems from the dual picture of D-branes that we talked about earlier. At very weak coupling, a network of D-branes wrapping cycles in a Calabi-Yau does not exert an appreciable gravitational pull and is best described by quantum field theory—a theory with no gravity in it. At strong coupling, however, this conglomeration of branes is better thought of as a black hole—a system that can only be described by a theory that includes gravity. Despite the integral role of Calabi-Yau manifolds in the work underlying the correspondence conjecture, Maldacena's idea did not originally involve these manifolds. However, subsequent attempts to frame the correspondence and understand it in greater detail (i.e., efforts by Klebanov and others, as well as some smaller contributions I've made in this area with James Sparks, my former Harvard postdoc who's now at Oxford) did involve Calabi-Yau manifolds quite directly, particularly Calabi-Yau singularities. "Calabi-Yau space is the setting in which the correspondence has been explored the most and is best understood," Sparks claims.[24]

Maldacena's original formulation, along with subsequent work on AdS/CFT, has provided another step toward resolving the black hole information paradox. Without getting into the details, the crux of the argument is that if black hole physics can be completely described by a quantum theory of particles—a setting without a black hole or its messy singularity in which we know information cannot be lost—then we can be sure that a black hole itself cannot lose information. So what happens to the information in an evaporating black hole? The idea is that Hawking radiation, which seeps out as a black hole evaporates, "is not random but contains subtle information on the matter that fell in," says Maldacena.[25]

Despite that insight, upon conceding his bet to Preskill in 2004, Hawking did not attribute his change of heart to string-theory-related ideas. Preskill, however, credits Strominger, Vafa, Maldacena, and others with building a "strong but rather circumstantial case that black holes really preserve information," noting that "Hawking has followed this work by the string theorists with great interest."[26] Strominger, for his part, believes this work "helped turn around Hawking's thinking on string theory and indeed turned around the whole world on string theory, because this was the first time that string theory had solved a problem that came from another area of physics and had been posed by someone outside of string theory."[27]

The work represented some validation that these crazy ideas involving strings, branes, and Calabi-Yau manifolds might be useful after all, but the scope of Maldacena's conjecture is not limited to the black hole paradox. Calling for a fundamental reconsideration of gravity, the correspondence has consequently engaged the efforts of a large fraction of the string theory community. One reason AdS/CFT has caught on to the extent it has is strictly pragmatic: "A computation that might be very difficult in one realm can turn out to be relatively straightforward in the other, thereby turning some intractable problems of physics into ones that are easily solved," explains Maldacena. "If true, the equivalence means that we can use a quantum particle theory (which is relatively well understood) to define a quantum gravity theory (which is not)."[28] The correspondence, in other words, can enable us to use our detailed knowledge of particle theories (without gravity) to improve our understanding of quantum gravity theories. The duality works in the other direction, too: When the particles in the quantum field theory are really strongly interacting, making calculations difficult, the curvature on the gravity side of the equation will be correspondingly low, making calculations there more manageable. "When one of the descriptions becomes hard, the other becomes easy, and vice versa," Maldacena says.[29]

Does the fact that string theory, according to AdS/CFT, can be equivalent to a quantum field theory—the kind of theory for which we have acquired substantial experimental proof—make string theory itself right? Maldacena doesn't think so, although some string theorists have tried to make that case. Strominger doesn't think so, either, but the work on black holes and AdS/CFT, which grew out of it, does make him think that string theory is on the right track. The insights that have sprung from both these fronts—the black hole entropy paradox and Maldacena's conjecture—"seem to argue for the inevitability of string theory," he says. "It seems to be a theoretical structure that you can't escape. It bangs you on the head everywhere you turn."[30]

Nine

BACK TO THE REAL WORLD

Upon meeting the Good Witch Glinda in *The Wizard of Oz*, Dorothy recounted the entire story of "how the cyclone had brought her to the Land of Oz, how she had found her companions, and of the wonderful adventures they had met with. 'My greatest wish now,' she added, 'is to get back to Kansas.'"[1]

Upon hearing this narrative so far, a tale frequented by visits from the Good Doctor Witten and others, during which we've heard of some wonderful adventures in the Land of Calabi-Yau—with its hidden dimensions, mirror partners, supersymmetry, and vanishing first Chern classes—some readers may, like Dorothy, be yearning for more familiar surroundings. The question, as always, is this: Can we get there from here? Can the combination of string theory and Calabi-Yau manifolds reveal secrets of a concealed, higher-dimensional domain that we can imagine but never set foot in—the theoretical equivalent of Oz— while also teaching us something new about the more concrete physical realm that some call Kansas?

"We can write physical theories that are interesting to look at from a mathematics standpoint, but at the end of the day, I want to understand the real world," says Volker Braun, a physicist at the Dublin Institute for Advanced Studies.[2] In our attempt to relate string theory and Calabi-Yau manifolds to the real world, the obvious point of comparison is particle physics.

The Standard Model, which describes matter particles and the force particles that move between them, is one of the most successful theories of all time, but as a theory of nature, it is lacking in several respects. For one thing, the model has about twenty free parameters—such as the masses of electrons and quarks—that the model does not predict. Those values must be put in by hand, which strikes many theorists as unforgivably ad hoc. We don't know where those numbers come from, and none of them seems to have a rationale in mathematics. String theorists hold out hope of providing this mathematical rationale, with the only free parameter—besides the string's tension (or linear energy density)—being the geometry of the compact internal space. Once you pick the geometry, the forces and particles should be completely fixed.

The aforementioned 1985 paper by Philip Candelas, Gary Horowitz, Andrew Strominger, and Edward Witten (see Chapter 6) "showed that you could bring all the key ingredients together to get a world that looked, at least in rough terms, like the Standard Model," says Candelas. "The fact that you could do this in a theory that included gravity led to a lot of interest in string theory."[3] One success of the Candelas et al. model is that it produced chiral fermions, a feature of the Standard Model whereby every matter particle has a kind of "handedness," with the left-handed version differing from its right-handed mirror image in important ways. As we saw earlier, their model also divided elementary particles into four families or generations, rather than the three that the Standard Model calls for. Although that number was off by one, Candelas says, "the main point was to show that you could get different generations—that you could get the repeated structure we see in the Standard Model."[4]

Strominger was equally sanguine, calling their pioneering Calabi-Yau compactifications "a spectacular jump from the basic principles of string theory to something close to the world we live in. It was like shooting a basketball from the far side of the court and having it sail through the hoop," he says. "In the space of all the things that could possibly happen in the universe, we came amazingly close. Now we want to do better, to find something that is not more or less right but exactly right."[5]

A year or so later, Brian Greene and colleagues produced a model that marked a step forward by getting the three generations that our theories require, chiral fermions, the right amount of supersymmetry (which we denote as $N = 1$), neutrinos with some mass (which is good) but not too much (which is even better), as well as reproducing the fields associated with the Standard Model forces (strong, weak, and electromagnetic). Perhaps the biggest shortcoming of their model was the presence of some extra (unwanted) particles that were not part of the Standard Model and had to be eliminated by various techniques. On the plus side, I was struck by the simplicity of their approach—that all they had to do, in essence, was pick the Calabi-Yau manifold (in this case opting for one that I had constructed), and with that single choice, they carried us closer toward generating the Standard Model. Although progress has been made in a number of areas in the decades since, string theory and string theorists have yet to realize the Standard Model. From our current vantage point, we don't even know whether string theory can reproduce the Standard Model.

Hard as this task now appears, proponents hope that string theory will not only take us there but eventually take us well beyond the Standard Model, which is where they believe we ultimately must go. For we already know the Standard Model is not the last word in physics. In the past decade or so, that model has been modified, or extended, a number of times to account for new experimental findings, such as the 1998 discovery that neutrinos, previously assumed to be massless, in fact have some mass. Furthermore, we now believe that 96 percent of the universe consists of dark matter and dark energy—mysterious forms of matter and energy about which the Standard Model says nothing at all. We expect there will be other discoveries to grapple with as well—whether they be the detection of supersymmetric particles (some of which are leading dark-matter candidates) or something completely unexpected—as the Large Hadron Collider goes about its business of smashing protons at high energies and seeing what comes out.

Even though Candelas and company and Greene and company weren't able to replicate the Standard Model, their compactifications surpassed it in at least one respect, as they made inroads toward achieving the Minimal Supersymmetric Standard Model. The MSSM, as it's called, is an extension of the conventional model, with supersymmetry added to it, meaning that it includes all the supersymmetric partners that are not included in the Standard Model itself. (Subsequent, string-theory-based efforts to realize the Standard Model, which will be discussed later in this chapter, encompass supersymmetry as well.)

For those who believe supersymmetry should be part of a theory of nature—a list that would probably include (though not be limited to) most string theorists—

the Standard Model is certainly lacking in that respect. But there's another major deficiency, which has come up repeatedly in these pages—namely, that the Standard Model, a theory of particle physics, overlooks gravity altogether, which is why it can never provide the ultimate description of our universe. Gravity is left out of the model for two reasons: For one thing, it is so much weaker than the other forces—strong, weak, and electromagnetic—that it is essentially irrelevant for the study of particle interactions on a small scale. (As a case in point, the gravitational force between two protons is roughly 10^{35} times weaker than the electromagnetic force they experience. Put in other terms, a magnet the size of a button is able—by virtue of the electromagnetic force—to pick a paper clip up off the ground, thereby overcoming the gravitational pull of the entire planet Earth.) The second reason, as has been amply discussed, is that no one yet knows how to tie gravity (described by general relativity) and the other forces (described by quantum mechanics) into a single, seamless theory. If string theory succeeds in reproducing the Standard Model while also bringing gravity into the fold, we'll be that much closer to having a complete theory of nature. And if string theory achieves this end, we'll have not only a Standard Model with gravity but also a supersymmetric Standard Model with gravity.

Various approaches to realizing the Standard Model are being tried—those involving orbifolds (which are like folds in flat space), intersecting branes, stacked branes, and the like—with advances being made on multiple fronts. This discussion, however, will focus on just one of those avenues, that of $E_8 \times E_8$ heterotic string theory, which is one of the theory's five varieties. We've made this choice not because it's necessarily the most promising (which I'm in no position to judge), but because efforts along these lines are closely tied to geometry, and it is the branch that arguably has the longest tradition of trying to go from Calabi-Yau geometry to the real world.

I'm not playing up the geometry angle simply because it happens to be the main thrust of this book. It's also vital to the endeavor about which we speak. For one thing, we cannot describe forces—an essential part of the Standard Model (and of any purported theory of nature)—without geometry. As Cumrun Vafa has said, "the four forces all have geometric underpinnings and three of them—the electromagnetic, weak, and strong—are related to each other by symmetry."[6] The Standard Model weaves together three forces and their associated symmetry (or *gauge*) groups: special unitary group 3, or SU(3), which corresponds to the strong force; special unitary group 2, or SU(2), which corresponds to the weak force; and the first unitary group, or U(1), which corresponds to the electromagnetic force. A symmetry group consists of the set of all operations, such as rotations, that you can do to an object and keep it unchanged. You take the object, apply the symmetry operation to it once or as many times as you want, and in the end, the object looks the same as when you started. In fact, you could not tell that anything was done to it.

Perhaps the simplest group to describe is U(1), which involves all the rotations you can do to a circle that is sitting in a two-dimensional plane. This is a one-dimensional symmetry group, because the rotations are around a single, one-dimensional axis perpendicular to the circle and lying at its center. SU(2) relates to rotations in three dimensions, and SU(3), which is more abstract, very roughly involves rotations in eight dimensions. (The rule of thumb here is that any group SU[n] has a symmetry of dimension $n^2 - 1$.) The dimensions of the three subgroups are additive, which means that the overall symmetry of the Standard Model is twelve-dimensional ($1 + 3 + 8 = 12$).

As solutions to the Einstein equation, Calabi-Yau manifolds of a particular geometry can help us account for the gravity part of our model. But can these manifolds also account for the other forces—those found in the Standard Model—and if so, how? In answering that question, I'm afraid we'll have to take a roundabout path. Particle physics, as presently cast, is a quantum field theory, meaning that all the forces, as well as all the particles, are represented by fields. Once we know these fields, which permeate four-dimensional space, we can derive the associated forces. These forces, in turn, can be described by vectors that have both an orientation and a size, meaning that at every point in space, an object will feel a push or pull in a certain direction and at a certain strength. At an arbitrary spot in our solar system, for instance, the gravitational force exerted on an object such as a planet is likely to point toward the sun, and the magnitude of that force will depend on the distance to the sun. The electromagnetic force exerted on a charged particle occupying a given spot, similarly, will depend on its location with respect to other charged particles.

The Standard Model is not only a field theory, but a special kind of field theory called a gauge theory—an approach that progressed greatly in the 1950s through the work of physicists Chen Ning Yang and Robert Mills (first mentioned in Chapter 3). The basic idea is that the Standard Model takes its various symmetries and combines them into a kind of composite symmetry group, which we denote as $SU(3) \times SU(2) \times U(1)$. What makes these symmetries special and unlike ordinary symmetries is that they are *gauged*. That means you can take one of the allowed symmetry operations, such as rotation on a plane, and apply it differently at different points in spacetime—rotating by, say, 45 degrees at one point, 60 degrees at another point, and 90 degrees at another. When you do that, despite the nonuniform application of the symmetry, the "equations of motion" (which control the dynamic evolution of fields) don't change, and neither does the overall physics. Nothing looks different.

Symmetries don't normally operate this way unless they are gauge symmetries. In fact, the Standard Model has four "global" symmetries relating to matter particles and charge conservation that are not gauged. (These global symmetries act on the matter fields of the Standard Model that we'll be talking about later in the chapter.) There's another global symmetry in the Standard Model, and in field theories in general, that is not gauged, either. Called the Poincaré symmetry, it involves simple translations (such as moving the whole universe one foot to the right, or doing the same experiment in two different laboratories) and rotations, where the final outcome looks the same.

However, if you want some of your symmetries to be gauged, Yang and Mills figured out that you need to put something else, something extra, into your theory. That "something else" happens to be gauge fields. In the Standard Model, the gauge fields correspond to the symmetries that are gauged—$SU(3) \times SU(2) \times U(1)$—which means, by association, that the gauge fields also correspond to the three forces that are incorporated in the model: the strong, weak, and electromagnetic. Incidentally, Yang and Mills were not the first to construct a $U(1)$ gauge theory describing electromagnetism; that had been achieved decades before. But they were the first to work out a gauge theory for $SU(2)$—a feat that showed the way to devise $SU(n)$ theories for any n of two or greater, including $SU(3)$.

The introduction of gauge fields enables you to have a theory with symmetries that are gauged, which in turn allows you to keep physics invariant, even when symmetry operations are differentially applied. Physicists didn't make the Stan-

9.1—Chen Ning Yang and Robert Mills, authors of the Yang-Mills theory (Courtesy of C. N. Yang)

dard Model that way just because it struck them as more elegant or more appealing. They made it that way because experiments told them that this is how nature works. The Standard Model is, therefore, a gauge theory for empirical reasons rather than for, say, aesthetic ones.

While physicists tend to speak in terms of gauge fields, mathematicians often express the same ideas in terms of bundles, which are a mathematical way of representing the fields that relate to the three forces. String theorists straddle the line between physics and math, and bundles play a key role in the heterotic constructions we'll be talking about momentarily.

Before getting to that, we need to explain how Calabi-Yau manifolds relate to these gauge fields (which mathematicians refer to as bundles). The fields that we see—four-dimensional gravity and the $SU(3) \times SU(2) \times U(1)$ gauge fields associated with the other three forces—unquestionably exist in the four-dimensional realm we inhabit, in keeping with our observations. Yet the gauge fields actually exist in all ten dimensions that string theory describes. The component that lies in the six compactified dimensions of the Calabi-Yau gives rise to the four-dimensional gauge fields of our world and hence to the strong, weak, and electromagnetic forces. In fact, it's fair to say that those forces are generated by the internal structure of the Calabi-Yau, or so string theory maintains.

We've said a bit about symmetry so far without mentioning the challenge that model builders face in what's called symmetry breaking. In the heterotic version of string theory we've been discussing, the ten-dimensional spacetime from which we start is endowed with what's called $E_8 \times E_8$ symmetry. E_8 is a 248-dimensional symmetry group that can be thought of, in turn, as a gauge field with 248 components (much as a vector pointing in some arbitrary direction in three dimensions has three components—described as the x, y, and z components). $E_8 \times E_8$ is an even bigger symmetry group of 496 (248 + 248) dimensions, but for practical purposes, we can ignore the second E_8. Of course, even the 248 symmetry dimensions pose a problem for re-creating the Standard Model, which has only twelve symmetry dimensions. Somehow, we've got to "break" the 248-dimensional symmetry of E_8 down to the twelve we want.

Let's go back to our example of a two-dimensional sphere or globe, which has rotational symmetry in three dimensions and belongs to the symmetry group $SO(3)$. (The term SO here is short for special orthogonal group, because it describes rotations around an orthogonal axis.) You can take that sphere and spin it around any of three axes—x, y, and z—and it will still look the same. But we can break that symmetry in three dimensions by insisting that one point must always stay fixed. On our planet, we could single out the north pole as that point. Now only one set of rotations, those that happen around the equator (on an axis that runs between the two poles), will keep that point (the north pole) fixed and unwavering. In this way, the threefold symmetry of the sphere has been broken and reduced to a one-dimensional symmetry, $U(1)$.[7]

In order to get down to four dimensions and the Standard Model with its twelve-dimensional symmetry group, we have to find some way of breaking the symmetry of the E_8 gauge group. In the E_8 case, we can break symmetry by choosing a particular configuration in which some of the 248 components of the big gauge field are turned on or turned off. In particular, we'll find a way to leave twelve of those little fields turned off, which is kind of like insisting that one spot on the sphere, the north pole, is not going to move. But they can't be just any twelve fields; they have to be the right ones to fit into the $SU(3) \times SU(2) \times U(1)$ symmetry groups. In other words, when you're done breaking down the massive

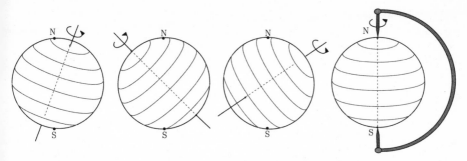

9.2—Owing to its great symmetry, a sphere remains unchanged under rotations under any axis running through the center. We can "break symmetry," however, by insisting that the north pole must remain fixed during rotation. Now rotation is only permitted on one axis (running between the north and south poles) rather than on any axis. Adherence to this condition breaks, or constrains, the unlimited rotational symmetry of the sphere.

E_8 group, what you'll have left in four dimensions are just the gauge fields of the Standard Model. The other fields, which correspond to the broken symmetries, don't disappear entirely. By virtue of being turned on, they'll reside at a high-energy regime that puts them far beyond our reach, totally inaccessible to us. You might say the extra symmetries of E_8 are hidden away in the Calabi-Yau.

Nevertheless, a Calabi-Yau manifold can't reproduce the Standard Model on its own. This is where bundles, which are literally extensions of the manifold, come in. Bundles are defined as groups of vectors attached to every point on the manifold. The simplest type of bundle is known as the tangent bundle. Every Calabi-Yau has one, but—as the tangent bundle of a Calabi-Yau is even harder to picture than the manifold itself—let's instead consider the tangent bundle of an ordinary two-dimensional sphere. If you pick a point on the surface of that sphere and draw two vectors tangent to that point, those vectors will define a plane (or a disk within a plane if you limit the vectors to an arbitrary length). If we do the same thing at every point on the surface and put all those planes (or disks) together, that collective entity will be the bundle.

Note that the bundle necessarily includes the manifold itself because the bundle contains, by definition, every single point on the manifold's surface. For that reason, the tangent bundle of a two-dimensional sphere is actually a four-dimensional space, because the tangent space has two degrees of freedom—or

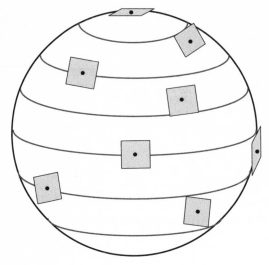

9.3—At each point on the surface of a sphere, there is a tangent plane intersecting the sphere at just that point and nowhere else. The *tangent bundle* for the sphere consists of all the planes tangent to every point on the sphere. Since the tangent bundle, by definition, includes every point on the sphere, it must also include the sphere itself. Although it would be impossible to draw a tangent bundle (with its infinite number of intersecting tangent planes), instead we show a sphere with patches of tangent planes at a few representative points.

two independent directions in which to move—and the sphere itself, being part of the bundle, has another two degrees of freedom that are themselves independent of the tangent space. The tangent bundle of a six-dimensional Calabi-Yau, similarly, is a twelve-dimensional space, with six degrees of freedom in the tangent space and another six degrees of freedom in the manifold itself.

Bundles are critical to string theorists' attempts to re-create particle physics in terms of Yang-Mills theory, under which gauge fields are described by a set of differential equations, unsurprisingly called the Yang-Mills equations. What we'd like to do here, specifically, is to find solutions to the equations for the gauge fields that live on the Calabi-Yau threefold. Since the principal reason Calabi-Yau manifolds entered string theory in the first place was to satisfy the requirement for supersymmetry, the gauge fields must also obey supersymmetry. That means we have to solve a special version of the Yang-Mills equations, the supersymmetric version, which are called the Hermitian Yang-Mills equations. In fact, these equations yield the least amount of supersymmetry you can have (what's known as $N = 1$ supersymmetry), which is the only amount of supersymmetry consistent with present-day particle physics.

"Before string theory forced us to get fancy, most physicists didn't think much about geometry and topology," says University of Pennsylvania physicist

Burt Ovrut. "We just wrote down equations like the Yang-Mills equations and tried to solve them." The only catch is that the Hermitian Yang-Mills equations are highly nonlinear differential equations that nobody knows how to solve. "To this day," Ovrut says, "there is not a single known [explicit] solution to the Hermitian Yang-Mills equations on a six-dimensional Calabi-Yau manifold. So you'd have to stop there, were it not for the work of some geometers who showed us another way to proceed."[8]

Bundles give us a way around this nonlinear differential logjam, because we can think of the bundle attached to the Calabi-Yau as an alternate description of the gauge fields that the Yang-Mills equations define. Exactly how that can be done is described by the DUY theorem, whose name is an acronym for Simon Donaldson (now at Imperial College), Karen Uhlenbeck at the University of Texas, and my name.

The idea behind the theorem is that the Hermitian Yang-Mills equations define a field that can be represented by a vector bundle. We proved that if you can construct a bundle on the Calabi-Yau that satisfies a specific topological condition—namely, that it is stable (or, to put it more technically, slope-stable)—then that bundle will admit a unique gauge field that automatically satisfies those equations. "This wouldn't have helped if you had traded in one infinitely difficult problem for another infinitely difficult problem," notes Ovrut. "But the second problem of constructing a stable bundle is much easier, and as a result, you don't have to solve those horrible differential equations at all."[9]

In other words, we found a geometric solution to a problem that we'd been unable to solve by other means. We showed that you don't have to worry about fields or differential equations. All you have to worry about is constructing a stable bundle. So what does it mean for a bundle to be slope-stable? The slope of a curve, as we've discussed, is a number that relates to curvature, and slope stability in this case relates to the curvature of the bundle. Putting it in loose terms, "the slope expresses a sense of balance," explains University of Pennsylvania mathematician Ron Donagi. "It says that the curvature in one direction can't be much bigger than the curvature in another direction. No matter which way you face, no direction can be too extreme in relation to the others."[10] Any bundle can be divided up into smaller pieces, or sub-bundles, and the stability condition means that the slope of any of these sub-bundles cannot be bigger than the slope of the bundle as a whole. If that condition is met, the bundle is slope-stable and the gauge fields will satisfy the Hermitian Yang-Mills equations. Supersymmetry will be satisfied as a result.

In some ways, the idea of slope stability, which is central to the DUY theorem, is a consequence of the Calabi-Yau theorem, because the theorem imposed specific curvature requirements on a Calabi-Yau manifold guaranteeing that its tangent bundle must be slope-stable. And the fact that the Calabi-Yau equations and Hermitian Yang-Mills equations are the same for the tangent bundle when the background metric is Calabi-Yau—another consequence of the Calabi conjecture proof—inspired me to think about the relation between slope stability and the Hermitian Yang-Mills equations. The idea that emerged was that a bundle will satisfy those equations if and only if it is stable.

That, in fact, is what Donaldson proved in his part of DUY—which he published in 1985, specifically regarding the special case of two complex dimensions. Uhlenbeck and I worked independently of Donaldson, and in a paper that came out a year later, we proved that the same result applied to any complex dimension and, consequently, to any space with an even number of real dimensions. I still consider this one of the most difficult theorems I've ever proved— or, in this case, co-proved. Our work, in combination with Donaldson's, is now collectively referred to as DUY.

The theorem is very much analogous to the proof of the Calabi conjecture, as both reduce a problem involving some nasty nonlinear system of equations that we don't know how to deal with to a problem of geometry, where we might have an inkling of how to proceed. In the Calabi case, I never solved the relevant differential equations explicitly. I just showed that if a manifold satisfies certain conditions (compact, Kähler, vanishing first Chern class) that can be checked through standard procedures of algebraic geometry, then a solution to those equations (in the form of a Ricci-flat metric) must exist. DUY works the same way, specifying the conditions on the bundle (namely, slope stability) such that a solution to the Hermitian Yang-Mills equations always exists. Algebraic geometers have also developed methods to assess a bundle's stability, although this turns out to be more complicated to check than whether a manifold's first Chern class is vanishing or not.

Some people, including physicists outside this particular area of mathematics, find DUY amazing because, on the face of it, the bundle conditions appear

9.4—Karen Uhlenbeck (Image courtesy of the University of Texas at Austin)

to have nothing to do with the differential equations you're hoping to solve. But it wasn't amazing to me, and if anything, it seemed like a natural extension of the Calabi conjecture. The Calabi proof is all about the manifold, the Calabi-Yau, whereas the DUY theorem is all about the bundle. You're looking for the metric of the bundle, but the metric for the manifold itself is given to you as part of the starting information. You can pick any "background" metric you want, including the Calabi-Yau metric.

The point of intersection between the Calabi conjecture and the DUY theorem is the tangent bundle. And here's why: Once you've proved the existence of Calabi-Yau manifolds, you have not only those manifolds but their tangent bundles as well, because every manifold has one. Since the tangent bundle is defined by the Calabi-Yau manifold, it inherits its metric from the parent manifold (in this case, the Calabi-Yau). The metric for the tangent bundle, in other words, must satisfy the Calabi-Yau equations. It turns out, however, that for the tangent bundle, the Hermitian Yang-Mills equations are the same as the Calabi-Yau equations, provided the background metric you've selected is the Calabi-Yau. Consequently, the tangent bundle, by virtue of satisfying the Calabi-Yau equations, automatically satisfies the Hermitian Yang-Mills equations, too.

The upshot of this is that the tangent bundle is really the first special case of the DUY theorem—the theorem's first solution—although that fact came out of the proof of the Calabi conjecture ten years before DUY was conceived. That's not the most interesting thing about DUY, however. The true power of DUY lies in prescribing the conditions (again regarding stability) that *other* bundles (not just the tangent bundle) must satisfy in order to ensure that a solution to the Hermitian Yang-Mills equations exists.

Before our paper on the subject came out in 1986, I told Edward Witten that Yang-Mills theory seemed to fit quite naturally with Calabi-Yau manifolds and therefore ought to be important for physics. Witten didn't see the relevance at first, but within about a year, he took my suggestion even farther, showing how this approach could be used in Calabi-Yau compactifications. Once Witten's paper came out, given his stature in the field, other people became interested in applying DUY to string theory. So this is another example of geometry's taking the lead even though it doesn't always go that way.

Now let's see how we can put some of this geometry (and topology) to use in order to generate particle physics from string theory. The first step is to pick a Calabi-Yau manifold, but not just any manifold will do. If we want to utilize certain approaches that have proved effective in the past, we need to pick a non-simply connected manifold—that is, a manifold with a nontrivial fundamental group. This, as I hope you recall, means you can find in that space a loop that cannot be shrunk to a point. The manifold, in other words, has to be like a torus rather than a sphere and have at least one hole. The presence of such a hole, cycle, or loop inevitably affects the geometry and topology of the bundle itself, which in turn affects the physics.

The second step is to construct a bundle that not only gives you the gauge fields of the Standard Model but also cancels the anomalies—the negative probabilities, unwanted infinities, and other irksome features—that had beset some of the earliest versions of string theory. When Michael Green and John Schwarz showed how to cancel anomalies in their momentous 1984 paper, their argument was framed in terms of gauge fields. Expressing this same idea in geometric and topological terms, one could say that a bundle will satisfy the anomaly cancellation requirement if its second Chern class equals the second Chern class of the tangent bundle.

We have already discussed the notion of Chern class, a technique for classifying topological spaces and measuring crude differences between them (see Chapter 4). The first Chern class, as noted, is vanishing (or zero) if you can orient all the tangent vectors on a manifold in the same direction. In some sense, this is like being able to comb a head of hair without getting a cowlick somewhere. This is impossible on a two-dimensional sphere, but you can avoid cowlicks on the surface of a two-dimensional torus. Thus we say the torus has a vanishing first Chern class, whereas the sphere's first Chern class is nonvanishing.

The second Chern class can be described in roughly the same way except that we need to consider two vector fields on some manifold rather than just one. (The vector fields we're talking about here are complex, meaning that the individual vector coordinates are described by complex numbers.) Assuming these two vectors fields are independent, at most points on the manifold they are likely to be pointing in different directions. But at certain points, a vector from each field may point in the same complex direction, or both vectors may go to zero. In fact, there may be a whole set of points where this is true. This set of points forms a closed, two-dimensional surface within our six-dimensional manifold, and collectively, these points represent the second Chern class.

So how does this tie in with anomaly cancellation? Green and Schwarz showed that no matter how bad the anomalies may be, if you can get them to cancel each other out and thereby disappear, you might have a viable theory after all. One way of getting rid of these bothersome anomalies is to make sure that the bundle you pick has the same second Chern class as that of the tangent bundle.

As for why that might work, we must remember that the bundles we're talking about here are, in a sense, stand-ins for the background fields, the gravitational and gauge fields, from which the forces of nature can be derived. The tangent bundle of the Calabi-Yau, for example, is a good facsimile of the gravitational field because the Calabi-Yau, as defined by a special metric, solves Einstein's gravity equations. Gravity, in other words, is literally encoded in that metric. But the metric also ties in with the tangent bundle, and here's why: The metric, as stated before, provides a function for computing the distance between any two points, A and B, on the manifold. We take all possible paths between A and B and break up each path into a set of tiny vectors, which are, in fact, tangent vectors; taken together, these vectors form a tangent bundle. That's why in our attempts at anomaly cancellation, we can use the tangent bundle of a Calabi-Yau to cover the gravity end of things.

We'll then choose an additional vector bundle for the purpose of reproducing the gauge fields of the Standard Model. So now we have our two bundles, one giving us the gravitational field and the other the gauge fields. Unfortunately, each field will inevitably have anomalies in it—there's no way to keep them out—but Green and Schwarz showed us there's no need for despair. They demonstrated, Donagi explains, "that the anomaly coming from gravity has the opposite sign as the anomaly coming from the gauge field. So if you can engineer things so that they have the same magnitude, they will cancel each other out."[11]

To find out if this works, we take the second Chern class of both the tangent bundle of the Calabi-Yau and the gauge field bundle. The answer we get for each bundle depends on those out-of-the-ordinary points where the vector fields align or vanish. However, you can't just count the number of such points, because there's actually an infinite number of them. What we can do, instead, is to compare the curves (of one complex dimension) that those points trace out. The curves corresponding to each of those bundles do not have to be identical for the second Chern classes to match, but they do have to be homologous.

Homology is a subtle concept, perhaps best defined through example, and I'm going to try to pick the simplest example possible—a donut with two holes. Each hole is cut out by a circle—a one-dimensional object—but each circle bounds a hole that is two-dimensional. And that's what we mean by homologous, the two circles of our double-donut being an example. Stating it in broader terms, we call two curves or cycles homologous if they are of the same dimension and bound a surface or manifold of one dimension higher. We use the term *Chern class* to indicate that there is a whole class of curves that are linked in this way through homology. The reason we brought this concept up in the first place is that if the curves for our two bundles are homologous—the tangent one representing gravity and the other one representing the gauge fields—then the second Chern class of these bundles will match. And as a result, the anomalies of string theory will magically cancel out, which is what we were after all along.

When people first began testing out these ideas—as Candelas, Horowitz, Strominger, and Witten did in their 1985 paper—they almost always used the tangent bundle, which was the only bundle that was well-known at the time. If you use the tangent bundle, then the second Chern class of your bundle cannot help but match the second Chern class of the tangent bundle. So you're covered on that score, but the tangent bundle will also satisfy the stability condition (which, as mentioned earlier, is a direct consequence of the Calabi conjecture proof). But investigators felt, nevertheless, that if other bundles met the above requirements—including stability—that could allow for more flexible options in terms of physics. Candelas says that even back in 1985, "we realized there were more general ways of doing things, bundles other than the tangent bundle that we might use. Although we knew it could be done, we didn't know then how to do it in a hands-on way."[12]

In the meantime, since the advent of the "second string revolution" in the mid-1990s, researchers saw that it was possible to loosen the restrictions on bundles further, thereby opening up many new possibilities. In M-theory, there is an extra dimension, and that gives you the freedom—and more elbow room—to accommodate extra fields that correspond to branes, the essential new ingredients ushered in by M-theory. With this extra ingredient, the brane now in the picture, the second Chern class of the gauge bundle no longer had to be equal to the second Chern class of the tangent bundle; it could instead be *less than or equal.* That's because the brane itself—or the curve upon which it's wrapped—has its own second Chern class, which can be added to the second Chern class of the gauge bundle to match that of the tangent bundle and thereby ensure anomaly cancellation. As a result, physicists now had a broader variety of gauge bundles to work with.

"Every time you weaken a condition—in this case, changing an equality into an inequality—you have more examples to draw on," Donagi explains.[13] Going back to our earlier example of a sphere or a beach ball, instead of attaching a tangent plane (or piece thereof) to every point on the ball's surface, we could attach a "normal" bundle with vectors pointing out from the surface. There are many other bundles one might construct by attaching a particular vector space to different points on the manifold and then gluing all these vector spaces together to make a bundle.

While this new freedom from M-theory has enabled researchers to explore a wider range of bundles, so far they haven't come up with many more examples that actually work. But at least the possibility now exists. The first step, again, is selecting a bundle that is stable and gets rid of the anomalies. From the DUY theorem, we know that such a bundle can give you the gauge fields, or forces, of the Standard Model.

Of course, the Standard Model is not just about forces. It's a theory of particle physics, so it's got to say something about particles, too. The question then is whether or how the particles of nature might be tied up with Calabi-Yau manifolds. There are two kinds of particles to talk about—matter particles, which are the things we can touch, and force-mediating particles, which include the photons that deliver light along with other particles we can't see, like weak bosons and gluons.

The force particles are in some sense easier to derive because if you got all the gauge fields right in the previous step, with all the right symmetry groups, then you already have these particles. They are literally part of the force fields, and the number of symmetry dimensions in each gauge field corresponds to the number of particles that communicate the force. Thus the strong force, which is endowed with eight-dimensional SU(3) symmetry, is mediated by eight gluons; the field of the weak force, which is endowed with three-dimensional SU(2) symmetry, is mediated by three particles, the W^+, W^-, and Z bosons; and the electromagnetic field, which is endowed with one-dimensional U(1) symmetry, is mediated by a single particle, the photon.

We can picture these particles in action fairly easily. Suppose two guys are roller-skating in parallel, and one throws a volleyball to the other. The guy who throws the ball will veer off in the direction opposite to that in which the ball travels, while the guy who catches the ball will veer off in the same direction as the ball is thrown. If you viewed this interaction from an airplane flying high enough above that you couldn't see the ball, it would appear as if there were a repulsive force pushing them apart. But if you looked extremely closely at that repulsive force and essentially "quantized" it, you'd see that the movements of the skaters were being caused by a discrete object, a volleyball, rather than by some invisible field. Quantizing the fields, either the matter fields or the gauge (force) fields, means that among all possible fluctuations or vibrations, you will only allow certain ones. Each specially selected fluctuation corresponds to a wave at a specific energy level and hence to a specific particle.

"That's what happens in the Standard Model," Ovrut says. "The matter particles are like the guys on roller skates, and the force particles are the volleyballs—

the photons, gluons, and the W$^+$, W$^-$, and Z bosons—that are exchanged between them."[14]

The ordinary matter particles will, however, take a bit more explaining. All normal matter particles, such as electrons and quarks, have spin-½, spin being an intrinsic, quantum mechanical property of all elementary particles that relates to a particle's internal angular momentum. These spin-½ particles are solutions to the Dirac equation, which was discussed in Chapter 6. In string theory, you have to solve this equation in ten dimensions. But when you fix the background geometry by selecting a Calabi-Yau manifold, the Dirac equation can be divided into a six-dimensional and a four-dimensional component. Solutions to the six-dimensional Dirac equation fall into two categories: heavy particles—many trillions of times heavier than anything observed in high-energy-accelerator experiments—and the particles of everyday life, whose mass is so small we can call it zero.

Regardless of the particle mass, finding the solutions to those component equations is quite difficult. Fortunately, geometry and topology again can save us from having to solve nearly impossible differential equations. In this case, we have to figure out the *cohomology* of the gauge bundle, as researchers at the University of Pennsylvania—including Braun (formerly at Penn), Donagi, Ovrut, and Tony Pantev—have shown. Cohomology is closely related to homology and, like homology, is concerned with whether two objects can be deformed into each other. The two concepts, as Donagi puts it, represent distinct ways of keeping track of the same properties.[15] Once you determine the cohomology class of a bundle, you can use it to find solutions to the Dirac equation, and generate the matter particles. "It's a beautiful mathematical approach," Ovrut claims.[16]

Employing these techniques and others, Vincent Bouchard of the University of Alberta and Donagi, as well as Ovrut and his colleagues, have produced models that appear to get a lot of things right. Both groups claim to get the right gauge symmetry group, the right supersymmetry, chiral fermions, and the right particle spectrum—three generations of quarks and leptons, plus a single Higgs particle, and no exotic particles, such as extra quarks or leptons that are not in the Standard Model.

But there is considerable debate regarding how close these groups have actually come to the Standard Model. Some questions have been raised, for instance, about methodologies and phenomenological details, such as the presence of "moduli particles," which will be discussed in the next chapter. Physicists I've heard from are of mixed opinion on this subject, and I'm not yet sold on this work or, frankly, on any of the attempts to realize the Standard Model to date. Shamit Kachru of Stanford considers the recent efforts another step forward, following on the advances of Candelas and Greene and their colleagues. "But no one," Kachru says, "has yet produced a model that hits it on the nose."[17] Michael Douglas of the Simons Center for Geometry and Physics at Stony Brook University agrees: "All of these models are kind of first cuts; no one has yet satisfied all the consistency checks of the real world. But even though both models are incomplete, we are still learning from all of this work."[18] Candelas credits the Bouchard-Donagi and Ovrut et al. models for showing us how to use bundles other than the tangent bundle. He believes this work will eventually point the way toward other models, noting that "it's likely there are other possibilities out there. But until you do it, you don't know what works."[19] Or if it works at all.

The next steps involve not only getting the right particles but also trying to compute their masses, without which we cannot make meaningful comparisons between a given model and the Standard Model. Before we can compute the mass, we need to determine the value of something called the *Yukawa coupling constant*, which describes the strength of interactions between particles—the interactions between the matter particles of the Standard Model and the Higgs field, and its associated particle (the Higgs boson), being of greatest relevance here. The stronger the coupling, the greater the mass of the particle.

Let's take one particle to start with, say, the down quark. As with other matter particles, the field description of the down quark has two components—one corresponding to the right-handed form of this particle and one corresponding to the left-handed form. Because mass, in quantum field theory, comes from interactions with the Higgs field, we multiply the two fields for the down quark— the right- and left-handed versions—by the Higgs field itself. The multiplication in this case actually corresponds to that interaction, with the size of that product, or *triple product*, telling you how strongly or weakly the down quark and Higgs field interact.

But that's just the first part of this complex procedure. One complication arises because the size of the triple product can vary as you move around on the "surface" of the Calabi-Yau. The Yukawa coupling constant, on the other hand, is not a variable quantity that depends on your location on the manifold. It is a global measure, a single number, and the way to compute that number is by integrating the size of the product of the down quark and Higgs fields over the entire manifold.

Remember that integration is really a process of averaging. You have some function (in this case, the product of three fields) that assumes different values at different spots on the manifold, and you want to get its average value. You have to do that because the Yukawa coupling is, in reality, a number rather than a function, just as the mass of a particle is a number as well. So what do you do? You chop the manifold up into tiny patches and determine the value of the function at each patch. Then you add up all those values and divide by the number of patches, and you'll get the average. But while that approach can carry you pretty far, it won't give you exactly the right answer. The problem is that the space we're working on here, the Calabi-Yau manifold, is really curvy, and if you were to take a tiny "rectangular" patch (assuming, for the moment, that we're in two dimensions) of size dx by dy, the size of that patch will vary depending on how flat or curved it is. So instead, you take the value of the function at a point in a particular patch and then weight that number by the size of the patch. In other words, you need a way of measuring how big the patch is. And for that we need the metric, which tells us the manifold's geometry in exacting detail. The only catch here, as we've said many times before, is that no one has yet figured out a way to calculate the Calabi-Yau metric explicitly, which is to say, exactly.

That could be the end of the line: Without the metric, we can never get the mass and thus will never know whether the model we have is anywhere near the Standard Model. But there are some mathematical methods—numerical (as in computer-based) techniques—we can use to approximate the metric. Then it becomes a question of whether your approximation is good enough to yield a reasonable answer.

Two general approaches are presently being tried out, and both rely on the Calabi conjecture in some way. That conjecture, as noted many times before, says that if a manifold satisfies certain topological conditions, a Ricci-flat metric exists. Without producing the metric itself, I was able to prove that such a metric exists. The proof employed a so-called deformation argument, which basically involves showing that if you start with something—let's say, some sort of metric—and keep deforming it in a certain way, that process will eventually converge on the metric you want. Once you can prove that this deformation process converges on the desired solution, there is a good chance that you can find a numerical scheme that converges as well.

Recently, two physicists—Matt Headrick of Brandeis University and Toby Wiseman of Imperial College—have performed numerical computations along these lines, working out an approximate metric for a K3 surface, the four-dimensional Calabi-Yau we've touched on often. The general strategy they used, called discretization, takes an object with an infinite number of points, such as the points tracing out a continuous curve, and represents it by a finite (discrete) number of points, with the hope being that this process will eventually converge on the curve itself. Headrick and Wiseman believe their process does converge, and while their results look good, they have not yet proved this convergence. One drawback of this approach has nothing to do with their analysis per se, but instead relates to the limitations of present-day technology: Available computers simply do not have the capacity to produce a detailed metric for six-dimensional Calabi-Yau manifolds. It all boils down to the fact that the calculation in six dimensions is much bigger, with many more numbers to be crunched, than the four-dimensional problem.

Computers will, no doubt, continue to improve and may eventually become powerful enough to bring the six-dimensional calculation within reach. In the meantime, there's another way to proceed that faces fewer computational constraints. The approach dates back at least to the 1980s, when I proposed that a Ricci-flat metric can always be approximated by placing (or, more technically, "embedding") a Calabi-Yau manifold in a very high-dimensional background space. This background space is called *projective space*, which is like a complex version of flat Euclidean space except that it's compact. When you put something like a manifold in a bigger space, the subspace automatically inherits a metric (what we call an *induced metric*) from the background space. A similar thing happens when you put a sphere in ordinary Euclidean space; the sphere adopts the metric of the background space. Drawing on a familiar analogy, we

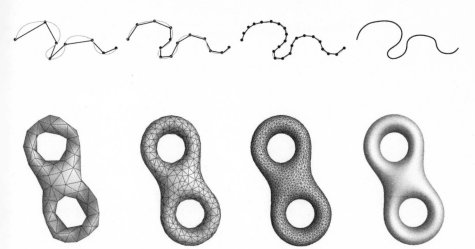

9.5—Through the process of *discretization*, you can approximate a one-dimensional curve and a two-dimensional surface with a finite number of points. The approximation, naturally enough, gets better and better as you increase the number of points.

can also think of a hole in Swiss cheese as being embedded in the larger space. And assuming we know how to measure distances in that larger space—the "Big Cheese," as it were—then we know how to measure the size of that hole as well. In that sense, the embedded space, or hole, inherits a metric from the cheesy background space in which it sits.

In the 1950s, John Nash had proved that if you put a Riemannian manifold in a space of high enough dimensions, you can get any induced metric that you want. The Nash embedding theorem, which is one of this illustrious mathematician's greatest works (among a long and diverse list, I might add), only applies to real manifolds sitting in real space. In general, the complex version of Nash's theorem is not true. But I suggested that a complex version of the theorem might be true under certain circumstances. I argued, for example, that a large class of Kähler manifolds can be embedded into a higher-dimensional projective space in a manner such that the induced metric is arbitrarily close to the original metric, provided the induced metric is suitably scaled or "normalized"— meaning that all its vectors are multiplied by a constant. Being a special case of Kähler manifolds, Calabi-Yau manifolds with a Ricci-flat metric satisfy this topological condition. That means the Ricci-flat metric can always be induced,

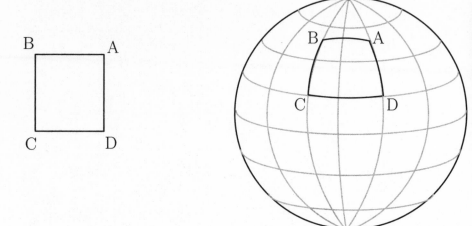

9.6—In geometry, we often talk about "embedding" an object or space in a higher-dimensional "background space." Here we embed a square—a one-dimensional object, as it consists of a line segment bent several times—in a two-dimensional background space, a sphere.

and can always be approximated, by embedding the manifold in a background or projective space of sufficiently high dimension. My graduate student at the time, Gang Tian, proved this in a 1990 paper, which was in fact his thesis work. Several important refinements of my original statement have been made since. This includes the thesis of another of my graduate students, Wei-Dong Ruan, who proved that an even better, more precise approximation of a Ricci-flat metric was possible.

The main refinement has to do with how you embed the Calabi-Yau in the background space. You can't just plop it in some haphazard way. The idea is to pick a proper embedding so that the induced metric will be arbitrarily close to the Ricci-flat metric. To do this, you put the Calabi-Yau in the best place possible, the so-called balanced position, which is the one position, among all possibilities, where the inherited metric comes closest to Ricci flat. The notion of a balanced position was introduced in 1982 by Peter Li and me for the case of submanifolds (or subsurfaces) in a sphere sitting in real space. We later extended that result to the more general case of submanifolds in a complex, high-dimensional background (or projective) space. Jean-Pierre Bourguignon, the current director of IHES, then joined our collaboration, coauthoring a 1994 paper with us on the subject.

I had previously conjectured—at a geometry conference at UCLA—that every Kähler manifold that admits a Ricci-flat metric (including a Calabi-Yau) is stable, but the word *stable* is hard to define. In subsequent geometry seminars, I continued to stress the relevance of the Bourguignon-Li-Yau work, as it's now called, to the notion of stability. Finally, some years later, my graduate student Wei Luo (who was based at MIT) made the connection between the stability of a Calabi-Yau and the balance condition. With the link provided by Luo, I was able to recast my conjecture to say that if you embed a Calabi-Yau in a very high-dimensional space, you can always find a position where it is balanced.

Simon Donaldson proved that conjecture to be true. His proof also validated the main thrust of this new approximation scheme—namely, that if you embed the Calabi-Yau in a background space of higher and higher dimension and satisfy the balance condition, the metric will get closer and closer to Ricci flat. Donaldson proved this by showing that the induced metrics form a sequence, in background spaces of increasing dimension, and that the sequence converges, approaching perfect Ricci flatness at infinity. The statement only holds, however, because the Calabi conjecture is true: When Donaldson demonstrated that the metric converges to the Ricci-flat metric, his conclusion hinged on the existence of a Ricci-flat metric.

Donaldson's proof had practical ramifications as well, because he showed that there was a best way of doing the embedding—the balanced way. Framing the problem in this way gives you a means of attacking it and a possible computational strategy. Donaldson utilized this approach in 2005, numerically computing the metric for a K3 surface, and he showed there were no fundamental barriers to extending the technique to higher dimensions.[20] In a 2008 paper, Michael Douglas and colleagues built on Donaldson's result, deriving a numerical metric for a family of six-dimensional Calabi-Yau manifolds, the aforementioned quintic. Douglas is now collaborating with Braun and Ovrut on a numerical metric for the Calabi-Yau manifold in their model.

So far, no one has been able to work out the coupling constants or mass. But Ovrut is excited by the mere prospect that particle masses might be computed. "There's no way to derive those numbers from the Standard Model itself," he says, "but string theory at least offers the possibility, which is something we've never had before." Not every physicist considers that goal achievable, and Ovrut admits that "the devil is in the details. We still have to compute the Yukawa couplings and the masses, and that could turn out completely wrong."[21]

It's doubtful, says Candelas, that the models we have in hand now will turn out to be the ultimate model of the universe. In trying to construct such a theory, he says, you have "an awful lot to get right. As you dig deeper into these models, sooner or later we're likely to come to things that don't work."[22] So rather than regarding the current models as the last word on the subject, we should view these efforts as part of a general learning process during which critical tools are being developed. Similar caveats apply to parallel efforts to realize the Standard Model that involve branes, orbifolds, or tori, none of which has yet reached that end, either.

There has been progress, says Strominger. "People have found more and more models, and some of the models are getting closer to what we observe around us. But there hasn't been a repetition of making that basketball shot from clear across the court. And that's what we're still waiting for."[23]

Invoking another sports analogy, Strominger compared the original Calabi-Yau compactification paper of 1985, which he coauthored with Candelas, Horowitz, and Witten, to hitting a golf ball from two hundred yards away and coming close to the hole. "There was a feeling that it was going to take only one more shot to get it in," he recalls. But a couple of decades have passed, he says, and "physicists have yet to pick up that gimme."[24]

"Twenty-five years is a long time in theoretical physics, and it's only now that groups are making substantial headway," says Candelas. "We're finally reaching a stage where people can do something practical with these new ideas."[25]

While acknowledging that researchers have made noteworthy strides, MIT's Allan Adams argues that "it's not correct to assume that the nearness to the Standard Model means we're almost done." On the contrary, he says, it's hard to know how far we still have to go. Although we may appear to be close to our goal, there's still a "great gulf" between the Standard Model and where we stand today.[26]

At the end of Dorothy's adventures in the Land of Oz, she learned that she had the powers to get back home all along. After some decades of exploring the

Land of Calabi-Yau, string theorists and their math colleagues (even those equipped with the penetrating powers of geometric analysis) are finding it hard to get back home—to the realm of everyday physics (aka the Standard Model)—and, from there, to the physics that we know must lie beyond. If only it were as easy as closing our eyes, tapping our heels together, and saying "There's no place like home." But then we'd miss out on all the fun.

Ten

BEYOND CALABI-YAU

Crafting a successful theory is like running an obstacle course that's never been run before. You get past one hurdle—going over it, around it, or maybe even under—knowing there are many more to come. And even though you've successfully cleared all the hurdles behind you, you don't know how many lie ahead, or whether one coming up might stop you for good. Such is the case with string theory and Calabi-Yau manifolds, where we know that at least one hurdle still looming is of sufficient magnitude to potentially topple this whole glorious enterprise. I'm talking about the moduli problem, which has been the subject of many talks and papers, as well as the source of much grief and consternation. As we'll see, what begins with the relatively simple goal of addressing this problem can take us far afield from where we began, at times leaving us without any goalposts in sight.

The size and the shape of any manifold with holes in it are determined by parameters called moduli. A two-dimensional torus, for example, is in many ways defined by two independent loops or cycles, one going around the hole and another going through. The moduli, by definition, measure the size of the cycles, which themselves govern both the size and shape of the manifold. If the cycle going through the donut hole is the smaller of the two, you'll have a skinny donut; if it's the larger, you'll have a fat donut with a relatively small hole in the middle. A third modulus describes the degree to which the torus is twisted.

So much for the torus. A Calabi-Yau, which, as we've noted, can have upwards of five hundred holes, comes with many cycles of various dimensions and hence many more moduli—anywhere from dozens to hundreds. One way to picture them is as a field in four-dimensional spacetime. The field for the size moduli, for instance, assigns a number to every point in ordinary space corresponding to the size (or radius) of the unseen Calabi-Yau. A field of this sort—which is completely characterized by a single number at each point in space, with no direction or vector involved—is called a scalar field. One can imagine all sorts of scalar fields around us, such as those measuring the temperature at every point in space or the humidity, barometric pressure, and so forth.

The catch here is that if nothing constrains the manifold's size and shape, you're going to run headlong into the aforementioned moduli problem, which will dash any hopes you might have of eliciting realistic physics from this geometry. We're faced with this problem when the scalar fields that relate to a manifold's size and shape are massless, meaning that no energy is required to alter them. They are, in other words, free to change value without impediment. Trying to compute the universe under these ever-shifting circumstances is "like running in a race, and the finish line is always moving an inch away from you," as University of Wisconsin physicist Gary Shiu puts it.[1]

There's an even bigger problem: We know that such fields cannot exist in nature. For if they did, there'd be all kinds of massless moduli particles—associated with the scalar (moduli) fields—flying around at the speed of light. These moduli particles would interact with other particles with roughly the same strength as gravitons (the particles thought to mediate the gravitational force), thereby wreaking havoc with Einstein's theory of gravity. Because that theory, as described in general relativity, works so well, we know these massless fields and particles simply cannot be there. Not only would their presence conflict with well-established gravitational laws, it would also give rise to a fifth force and perhaps additional forces that have never been seen.

So there's the rub. Given that much of string theory now hinges on compactifications of Calabi-Yau manifolds, which have these moduli with their associated massless scalar fields and particles that don't appear to exist, is string theory itself doomed? Not necessarily. There might be a way around this problem, because there are other elements of the theory—things that we already knew

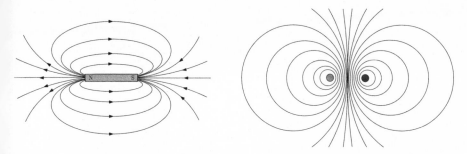

10.1—Fluxes can be thought of as lines of force that are not unlike the magnetic field lines shown here, although string theory includes fields that are more exotic and that point in the six compact directions we can't see.

about but left out to make our calculations simpler. When included, these elements make the situation look considerably different. These additional ingredients include items called *fluxes*, which are fields like electric fields and magnetic fields, although the new fields from string theory have nothing to do with electrons or photons.

Again, let's consider a two-dimensional torus, in this case, a particularly malleable donut whose shape is constantly shifting between skinny and fat. We can stabilize this torus into a fixed shape by wrapping wires through and around it. That is essentially the role of flux. Many of us have seen an effect like this when, say, a magnetic field is suddenly switched on, and iron filings, which were easily scattered about before, now assume a fixed pattern. The flux holds the filings in place, which is where they'll stay unless additional energy is applied to move them. In the same way, the presence of fluxes means that it now takes energy to change the manifold's shape, as the massless scalar fields have thus become scalar fields with mass.

Six-dimensional Calabi-Yau manifolds are more complex, of course, since they can have many more holes than a donut and the holes themselves can be of higher dimensions (up to six). That means there are more internal directions in which the flux can point, leading to many more possible ways of threading the field lines through those holes. Now that you have all these fluxes running through your manifold, you might want to know how much energy is stored in the accompanying fields. To calculate the energy, explains Stanford's Shamit

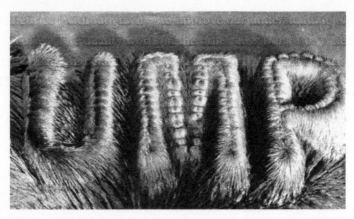

10.2—Just as we can fix and stabilize the arrangement of iron filings by applying a magnetic "flux," so too can we, in principle, stabilize the shape or size of a Calabi-Yau manifold by turning on the various fluxes of string theory. (Image courtesy of TechnoFrolics [www.technofrolics.com])

Kachru, you need to take an integral of the field strength squared "over the precise shape of the compact dimensions"—or, you might say, over the surface of the Calabi-Yau. So you divide the surface into infinitesimally small patches, determine the square of the field strength at each patch, add up all those contributions, divide by the number of patches, and you'll have your average value or integral. "Since varying that shape will vary the magnitude of the total field energy," Kachru says, "the shape the manifold chooses is the one that minimizes the flux energy of this field."[2] And that's how bringing fluxes into the picture can stabilize the shape moduli and in that way stabilize the shape itself.

That's part of the story, though we've neglected an important aspect of this stabilization process. Just as the magnetic or electric fields are quantized, the fluxes of string theory are also quantized, assuming integer values only. You can put in 1 unit of flux or 2 units of flux but not 1.46 units of flux. When we say that fluxes stabilize the moduli, we mean they restrict the moduli to particular values. You can't set the moduli to any value you choose—only to values that correspond to discrete fluxes. In that way, you've restricted the manifold—the Calabi-Yau—to a discrete set of shapes as well.

Although we spent the previous chapter exploring the heterotic version of string theory, it turns out that incorporating fluxes into heterotic models is rather difficult. Happily, the process is better understood in Type II string theory (a category that includes both Types IIA and IIB), which is dual to heterotic theory in some circumstances. I'll now mention an important 2003 analysis, performed in the Type IIB setting, that stands out in this regard.

So far we've only talked about stabilizing the shape moduli of a manifold with fluxes. The paper in question (dubbed KKLT after its authors, Shamit Kachru, Renata Kallosh, and Andrei Linde—all from Stanford—and Sandip Trivedi of the Tata Institute in India) is generally considered the first publication to show a consistent way of stabilizing all the moduli of the Calabi-Yau, both the shape moduli and size moduli. Stabilizing size is crucial for any string theory based on Calabi-Yau manifolds, because otherwise, there's nothing to keep the six hidden dimensions from unwinding and becoming infinitely large—bringing them to the same infinite size that we assume the other four dimensions have as well. If the small, invisible dimensions suddenly sprang free and expanded, we'd then be living in a spacetime of ten large dimensions, with ten independent directions to move in or to search for our missing keys, and we know our world doesn't look like that. (Which gives us some hope for finding lost keys.) Something has got to hold those dimensions back, and that something—according to the KKLT authors—turns out to be D-branes.[3]

Stabilizing the six-dimensional Calabi-Yau with branes is something like constraining an inner tube with a steel-belted radial tire. Just as the tire will hold back the tube as you pump air into it, the branes can curb the tiny manifold's inclination to expand.

"You say the shape and size are stabilized if you try to squash it and something pushes back, and if you try to expand it and something pushes back," explains Johns Hopkins physicist Raman Sundrum. "The goal was to make a compact, stable spacetime, and KKLT showed us how to do that—not just one way of doing so but many different ways."[4]

Having a stabilized volume and size is essential if we hope to explain phenomena like cosmic inflation—an idea that holds that almost all of the features we see in the universe today are the result of a brief though explosive period of exponential growth at the time of the Big Bang. This growth spurt, according to the theory, is fueled by the presence of a so-called inflaton field that endows the universe with a positive energy that drives expansion. "In string theory, we assume that positive energy must come from some kind of ten-dimensional sources, which have the property that as you make the compact [Calabi-Yau] space bigger, the associated energy gets smaller," says Cornell physicist Liam McAllister. When given a chance, all fields will try to spread out and get dilute. "What this means is that the system is 'happier' when the internal space gets bigger and the energy becomes lower," he says. "The system can reduce its energy by expanding, and it can reduce it to zero by expanding an infinite amount."[5] If there's nothing to keep the internal space from expanding, it will expand. When that happens, the energy that would otherwise drive inflation dissipates so quickly that the process would stop before it even got started.

In the KKLT scenario, branes provided a possible mechanism for realizing the universe we see—a universe that's influenced by inflation to a large degree. The goal of this exercise was not to reproduce the Standard Model or to get into the details of particle physics, but rather to go after some broader, qualitative features of our universe, including aspects of cosmology, which is arguably the broadest discipline of all.

For in the end, we want a theory that works on all scales—a theory that gives us both particle physics and cosmology. In addition to providing hints as to how inflation might work in string theory, the KKLT paper, as well as a 2002 paper by Kachru, Steve Giddings, and Joe Polchinski—the latter two of the University of California, Santa Barbara—showed how string theory might account for the apparent weakness of gravity, which is about a trillion trillion trillion times weaker than the electromagnetic force. Part of the explanation, according to string theory, is that gravity permeates all ten dimensions, which dilutes its strength. But in the Giddings-Kachru-Polchinski (GKP) scenario, the effect gets amplified exponentially by the geometrical concept of warping, which will be discussed later in this chapter. This explanation builds on a warped-geometry model first achieved in field theory—by Lisa Randall of Harvard and Sundrum—and later incorporated into string theory by GKP, as well as in the subsequent KKLT work.

Another milestone achieved by KKLT was providing a string theory description of how our universe might be endowed with a positive vacuum energy—sometimes called dark energy—the existence of which has become evident through measurements since the late 1990s. We won't provide an elaborate description of that mechanism, which gets rather technical and involves the placement of something called an antibrane (the antimatter counterpart to branes) in a warped region of the Calabi-Yau, such as the tip of a so-called conifold singularity—a noncompact, cone-shaped protrusion extending from the "body" of the manifold. At any rate, the exact details are not all that important here, since their study was never supposed to supply the definitive answer to any of these questions, Kachru explains. "KKLT is really intended to be a toy model— the kind of thing that theorists play with in order to study phenomena—although many other constructions are possible."[6]

The point, then, is that if the work on the moduli stabilization front keeps progressing and the work on the particle physics front keeps progressing, there is at least the potential, according to McAllister, "of having it all. If you take a Calabi-Yau manifold and throw in D-branes and fluxes, you may have all the ingredients you need in principle to get the Standard Model, inflation, dark energy, and other things we need to explain our world."[7]

The upshot of the KKLT paper was that in showing how the moduli can be stabilized, the authors showed how a Calabi-Yau manifold itself can be restricted to a distinct set of stable, or quasi-stable, shapes. That means you can pick a Calabi-Yau of a specific topological type, figure out the ways you can dress it up with fluxes and branes, and literally count the possible configurations. The trouble is that when you do the counting, some people may be unhappy with the result, because the number of possible configurations appears to be preposterously large, upwards of 10^{500}.

That figure, far from being exact, is meant to provide a rough indication of the number of arrangements (or shapes) you can get in a Calabi-Yau with many, many holes. Consider again a torus with flux winding through a particular hole to stabilize it. Because the flux is quantized, we'll further suppose it can take on one of ten integer values from 0 to 9. That's equivalent to saying there are ten stable shapes for the torus. If we had a torus with two holes instead of one and could run flux through each of them, there'd be 10^2, or 100, possible stable shapes. A six-dimensional Calabi-Yau can, of course, offer many more options. "The number ten to the five hundredth [10^{500}] was obtained by taking from mathematicians the maximum number of holes a manifold could have—on the order of five hundred—and assuming that through each hole, you could place fields or fluxes that have any of ten possible states," explains Polchinski, one of the people to whom this number is ascribed. "The counting here is really crude. The number could be much larger or much smaller, but it's probably not infinite."[8]

What does a number like this mean? For starters, it means that owing to the topological complexity of a Calabi-Yau manifold, the equations of string theory have a large number of solutions. Each of these solutions corresponds to a Calabi-Yau with a different geometry that, in turn, implies different particles, different physical constants, and so forth. Moreover, because Calabi-Yau manifolds are, by definition, solutions to the vacuum Einstein equations, each of those solutions, which involve different ways of incorporating fluxes and branes, corresponds to a universe with a different vacuum state and, hence, a different vacuum energy. Now here's the kicker: A fair number of theorists believe that *all* these possible universes might actually exist.

There's a picture that goes with this. Imagine a ball rolling on a vast, smooth, frictionless landscape. There being no preferred position, it can go anywhere without costing any energy. This is like the situation of unstabilized moduli and massless scalar fields. Now let's imagine that this surface is not entirely smooth but instead has little dips in it, in which the ball can get stuck without the input of some energy to dislodge it. This is the situation you get when the moduli are stabilized; each of the dips on the surface corresponds to a different solution to string theory—a different Calabi-Yau occupying a different vacuum state. Because we have such a large number of possible solutions, this "landscape" of different vacuum states is enormous.

Of course, this whole notion—the so-called landscape of string theory—has become extremely controversial. Some people embrace the picture it implies of multiple universes, some abhor it, and others (myself included) regard it as speculative. There is a question in some people's minds as to the practical value of a purported theory of nature that offers more solutions than we can ever sort through. One also has to wonder, in view of all the possible universes strewn across this landscape, if there's any conceivable way of finding ours.

More worrisome to others, the landscape idea has become closely tied to so-called anthropic arguments, some of which go like this: The cosmological

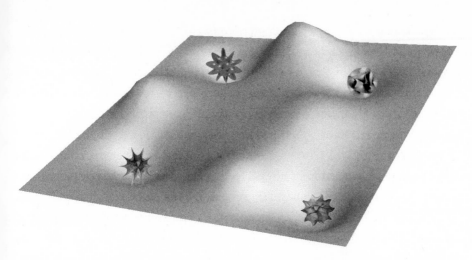

10.3—The energy of empty space, also called the vacuum energy, can assume a vast number of possible values that represent stable, or semistable, solutions to the equations of string theory. The concept of the "landscape" of string theory was invented, in part, to illustrate the idea that the theory has many possible solutions, corresponding to many possible vacuum states, or *vacua*—each of which could represent a different universe. The stable vacua in this figure are represented by dips or valleys on a sloping, hilly landscape—places in which balls, for instance, might get stuck as they rolled down a mountainside in different directions. The elevation of these troughs corresponds to the energy the vacuum assumes at that particular spot on the landscape. Some theories suggest there might be on the order of 10^{500} different solutions, each corresponding to a different Calabi-Yau manifold and hence a different geometry for the compact dimensions. Calabi-Yau spaces are an integral part of this picture because it is thought that the bulk of the vacuum energy is used to keep the six extra dimensions of string theory curled up in such spaces rather than allowing them to expand to infinity. (Calabi-Yau images courtesy of Andrew J. Hanson, Indiana University)

constant for our universe, as calculated from recent measurements in astronomy, appears to be very small, a factor of about 10^{120} lower than that predicted by our best physical theories. No one has been able to explain this discrepancy or this constant's exceedingly small size. But what if all the 10^{500} or so possible vacua in the landscape are actually realized somewhere—each representing a separate universe or subuniverse with a different internal geometry (or Calabi-Yau) and a different cosmological constant? Among all those choices, at least one of those subuniverses is bound to have an extremely low cosmological constant just like ours. And since we have to live somewhere, maybe, by chance, that's where we ended up. But it's not entirely dumb luck. For we couldn't live in a universe with a large cosmological constant, because the expansion there would have been so fast that stars, planets, and even molecules would never have a chance to form. A universe with a large negative cosmological constant would have quickly shrunk down to nothing—or to some violent singularity that would likely ruin your whole day. In other words, we live in the kind of universe in which we can live.

The physicist David Gross has compared anthropic-style reasoning of this sort to a virus that ought to be eradicated. "Once you get the bug, you can't get rid of it," he complained at a cosmology conference.[9] Stanford physicist Burton Richter claims that landscape enthusiasts such as his Stanford colleague Leonard Susskind have "given up. To them the reductionist voyage that has taken physics so far has come to an end," Richter wrote in the *New York Times*. "Since that is what they believe, I can't understand why they don't take up something else—macramé, for example."[10] Susskind has not taken statements like that lying down: There's no way around the multiple solutions of string theory, he contends, so, like it or not, the landscape is here to stay. Since that's the case, we'd better make peace with it and see if there's anything useful to be learned. "The field of physics is littered with the corpses of stubborn old men who didn't know when to give up," he wrote in his book, *The Cosmic Landscape*, while admitting that he too may be a "crusty old [curmudgeon], battling to the very end."[11]

It's fair to say that things have gotten a little heated. I haven't really participated in this debate, which may be one of the luxuries of being a mathematician. I don't have to get torn up about the stuff that threatens to tear up the physics community. Instead, I get to sit on the sidelines and ask my usual sorts of questions—how can mathematics shed light on this situation?

Some physicists had originally hoped there was only one Calabi-Yau that could uniquely characterize string theory's hidden dimensions, but it became clear early on that there were a large number of such manifolds, each having a distinct topology. Within each topological class, there is a continuous, infinitely large family of Calabi-Yau manifolds. This is perhaps easiest pictured with the torus. A torus is the topological equivalent of a rectangle. (If you roll a rectangular sheet up into a cylinder and smoothly connect the tips, you'll have a torus.) A rectangle is defined by its height and width, each of which can assume an infinite number of possible values. All of these rectangles and their associated

10.4—Two sides of the landscape debate: (a) Santa Barbara physicist David Gross and (b) Stanford physicist Leonard Susskind (Photo of Susskind by Anne Warren)

tori are topologically equivalent. They're all part of the same family, but there's an infinite number of them. The same holds for Calabi-Yau manifolds. We can take a manifold, modify its "height," "width," and various other parameters, and end up with a continuous family of manifolds, all of the same topological type. So KKLT and the related landscape concept didn't change that situation at all. At best, imposing the constraints that come from physics—by insisting that the flux be quantized—has led to a very large but finite number of Calabi-Yau shapes rather than an infinite number. I suppose that might be viewed as some progress.

Personally, I never shared the dream that some physicists once harbored of there being a single, god-given Calabi-Yau or even just a few. I always assumed that things were going to be more complicated than that. To me, that's just common sense. After all, who ever said that getting to the bottom of the universe and charting out its intrinsic geometry was supposed to be easy?

So what can we make of this landscape idea that has proved so unsettling to some? One course, I suppose, is to ignore it, as nothing has been settled, nothing proved. Some physicists consider the concept useful for addressing the cosmological constant problem, while others see no utility in it whatsoever. Since the whole notion of the string theory landscape emerged from looking at numerous vacuum states, many if not all of which relate to Calabi-Yau manifolds (depending on which version of this concept you're talking about), I suggest that one of the things the landscape might be telling us—if we're to put any stock in the idea at all—is that we need to understand Calabi-Yau manifolds better.

I realize that this statement may be somewhat naive. There are many possible solutions to string theory, and many possible geometries upon which one might compactify the theory's extra dimensions, with Calabi-Yau manifolds representing just the tip of the iceberg. I'm well aware of this situation and am even looking into some of these new areas myself. Nevertheless, most of the progress we've made so far in string theory, and most of the insights gleaned, have come from using Calabi-Yau manifolds as the test case—the model of choice. What's more, even some of the alternative geometries now being investigated—such as non-Kähler manifolds, which we'll get into in a moment—are produced by deforming or warping Calabi-Yau manifolds. There's no shortcut that can take us directly to non-Kähler geometries, which means that we need to understand Calabi-Yau manifolds before we can have a chance of understanding things like non-Kähler manifolds.

This is a common strategy in all areas of exploration: You establish a base camp, which serves as a familiar point of departure, before venturing into the unknown. Remarkably, despite the amount of study that's gone into the subject since I proved the existence of these manifolds in 1976, there are many simple questions—shockingly simple, in fact—that we cannot yet answer: How many topologically distinct Calabi-Yau manifolds are there? Is that number finite or infinite? And might all Calabi-Yau manifolds be related in some way?

We'll start with the first question: How many distinct topological varieties, or families, do Calabi-Yau manifolds come in? The short answer is that we don't know, though we can do a little better than that. More than 470 million Calabi-Yau threefolds have been created by computer so far. For those, we have constructed more than 30,000 Hodge diamonds, which means there are at least 30,000 distinct topologies. (Hodge diamonds, as you may recall from Chapter 7, are four-by-four arrays that sum up basic topological information about a threefold.) The number could be significantly greater than 30,000, however, as two manifolds can have the same Hodge diamond and still be topologically distinct. "No systematic effort has been made to estimate the number of topological types, mostly because no practically calculable numerical test is known to unambiguously distinguish between such threefolds," explains Howard University physicist Tristan Hubsch. "We still don't have an unequivocal ID number for a Calabi-Yau manifold. We know the Hodge diamond is part of it, but that doesn't uniquely define the manifold. It's more like a partial registration number of a car."[12]

It's not just that we don't know whether the number of Calabi-Yau manifolds is a little bit more than 30,000 or a lot more; we don't even know whether the number is finite or not. In the early 1980s, I conjectured that the number is finite, but University of Warwick mathematician Miles Reid holds the opposite opinion, arguing that the number is infinite. It would be nice to find out which view is right. "I suppose that for physicists hoping there were only a handful of Calabi-Yau manifolds, finding out there's an infinite number would only make things worse," says Mark Gross of the University of California, San Diego. "From a mathematical viewpoint, it doesn't really matter. We just want to know the answer. We want to understand the totality of Calabi-Yau manifolds. The conjecture that there is a finite number—be it right or wrong—serves as a kind of measuring stick for our understanding."[13] And from a purely practical standpoint, if the set of manifolds is finite, no matter how large it is, you can always take an average. But we really don't know how to take the average of an infinite number of objects, which therefore makes it harder for us to characterize those objects.

There have been no contradictions to my conjecture so far. It appears that all the methods we presently know of for constructing Calabi-Yau manifolds will lead to only a finite number of manifolds. It could be a matter of overlooking some kinds of construction, but after a couple of decades of searching, no one has come up with a new method that would lead to an infinite set.

The closest anyone has come to settling this question is a 1993 proof by Mark Gross. He proved that if you think of a Calabi-Yau, loosely speaking, as a four-dimensional surface with a two-dimensional donut attached to each point, there's only a finite number of these surfaces. "The vast majority of known Calabi-Yau manifolds fall into this category, which happens to be a finite set," Gross says. That's the main reason he supports the "finite" hypothesis. On the other hand, he notes, plenty of manifolds don't fall into this category and we haven't had any real success in proving anything about these cases.[14] That leaves the matter unresolved.

Which brings us to the second question, originally posed by Reid in 1987, and it is equally unsettled: Might there be a way in which all Calabi-Yau manifolds are related? Or as Reid put it: "There are all these varieties of Calabi-Yau's with all kinds of topological characteristics. But if you look at it from a wider perspective, these could all be the same thing. Basically it's a crazy idea— something that can't possibly be true. Nevertheless . . . " In fact, Reid considered the idea so outlandish, he never referred to it as a conjecture, preferring to call it a "fantasy" instead. But he still believes that someone might be able to prove it.[15]

Reid speculated that all Calabi-Yau manifolds might be related through something called a conifold transition. The idea—developed in the 1980s by the mathematicians Herb Clemens (then at the University of Utah) and Robert Friedman of Columbia—involves what happens to Calabi-Yau manifolds when you move them through a special kind of singularity. As always, the concept is much easier to picture on a two-dimensional torus. Remember that a torus can be described as a series of circles arranged around a bigger circle. Now we'll take one of those smaller circles and shrink it down to a point. That's a singularity because every other spot on the surface is smooth.

So at this pinch point—a so-called conifold singularity—it's as if you had two little cones or party hats coming together on an otherwise unremarkable donut. One thing you can now do is what geometers call surgery, which involves cutting out the offending point and replacing it with two points instead. We can then separate those two points, pulling apart the donut at that spot until it assumes a crescent shape. Next, we reconfigure that crescent into its topological equivalent, a sphere. We don't have to stop there. Suppose we next stretch out the sphere so that it, again, looks more like a crescent. Then we attach those ends to fashion a torus, only this time we've been a bit careless, with an extra fold somehow having been introduced into our shape. This gives us a torus with a different topology and two holes instead of one. If we were to continue this process indefinitely—introducing extra folds, or holes, along the way by virtue of our sloppiness—we'd eventually get to all possible two-dimensional tori. The conifold transition is thus a way of connecting topologically distinct tori by way of an intermediary (in this case, a sphere), and this general procedure works for other (nontrivial) kinds of Calabi-Yau manifolds as well.

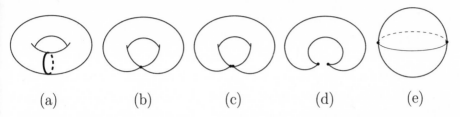

(a)   (b)   (c)   (d)   (e)

10.5—The conifold transition is an example of a topology-changing process. In this greatly simplified case, we start with a donut, which is made up of little circles, and shrink one of those circles down to a point. That point is a kind of singularity where two shapes that resemble cones come together. A cone-shaped singularity of this sort is called a *conifold*. Through a mathematical version of surgery, we replace that singular point with two points and then pull apart those points, so that the donut becomes more of a croissant. We then inflate the croissant to make it like a sphere. In this way, we've gone from a donut to a topologically distinct object, a sphere.

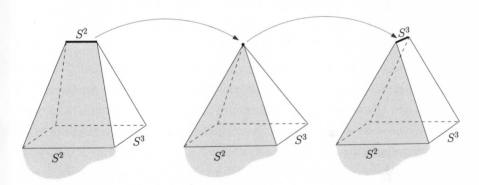

10.6—Here's another way of picturing the conifold transition. We'll start with the Calabi-Yau manifold on the left. It's a six-dimensional object because it has a five-dimensional base—being the "product" of a two-dimensional sphere (S^2) and a three-dimensional sphere (S^3)—plus an added dimension for the height. This Calabi-Yau surface is nice and smooth because it has a two-dimensional sphere (S^2) on top. Shrinking that sphere down to a point brings us to the middle picture, the pyramid. The point on the very tip of the pyramid is a singularity, the conifold. If we smooth out that pointy tip, by blowing up the point into a three-dimensional sphere (S^3), rather than the two-dimensional sphere (S^2) we started with, we'll arrive at the third panel, manifold M. So the idea here is that the conifold singularity serves as a kind of bridge from one Calabi-Yau to another. (Adapted, with permission, from a figure by Tristan Hubsch)

Six-dimensional Calabi-Yau manifolds aren't so simple. In our picture of the conifold transition, as suggested by Clemens, instead of shrinking a circle down to a point, we shrink down to a two-dimensional sphere. We're assuming here that every compact Kähler manifold, and hence every Calabi-Yau manifold, has at least one two-dimensional sphere of a special sort sitting inside. (The Japanese mathematician Shigefumi Mori proved that Kähler manifolds with positive Ricci curvature have at least one such subsurface, and we expect that this condition applies to the Ricci-flat Calabi-Yau case as well. Every Calabi-Yau manifold we know of has a two-dimensional sphere, so our intuition has held up so far. But we still don't have a proof for Ricci-flat Calabi-Yau manifolds.) After shrinking our two-dimensional sphere down to a point, we can replace that point with a shrunken three-dimensional sphere that can then be reinflated.

If our previous assumption is correct, after this surgery the manifold is no longer Kähler, since it no longer has a two-dimensional sphere, and therefore cannot be a Calabi-Yau. It's something else, a non-Kähler manifold. Continuing the conifold transition, we can take this non-Kähler manifold, insert a different two-dimensional sphere (where we had previously inserted the three-dimensional sphere), and end up with a different Calabi-Yau.

Although Reid did not invent the conifold transition, he was the first to see how it might be used to forge a link between all Calabi-Yau manifolds. A critical aspect of the conifold transition is that in getting from one Calabi-Yau surface to a different one, the geometry must pass through an intermediary stage—that being a non-Kähler manifold. But what if all these non-Kähler manifolds are connected in the sense that they can be molded into one another by means of squeezing, stretching, or shrinking? That, indeed, is the crux of Reid's fantasy.

Imagine a giant chunk of Swiss cheese, filled with numerous tiny holes or bubbles. If you live in one bubble, you can't go far before you hit a boundary, says Allan Adams. "But if you don't mind going through the cheese, you can get from one bubble to another. Reid conjectured that a conifold transition can take you through the cheese [the non-Kähler part] and into another bubble," the Kähler part, which is the Calabi-Yau.[16] The analogy works in another sense, too, for in this picture, the bulk of space is basically the cheese, save for the tiny bubbles scattered here and there. Those little bubbles are like little bits of Kähler spaces scattered amid a much larger non-Kähler background. And that's pretty much how we think of it: There's a vast number of non-Kähler spaces, with Kähler manifolds constituting just a tiny subset.

The general strategy underlying Reid's conjecture makes sense to Mark Gross. Since, he says, non-Kähler manifolds represent a much bigger set of objects, he says, "if you want to say things are related, allowing them to be part of this much bigger non-Kähler set certainly makes it easier."[17]

The situation is kind of like the Six Degrees of Kevin Bacon game, in which players try to show how everyone in Hollywood is connected to the prolific actor. It's the same with Calabi-Yau manifolds, Adams says. "Are they all neighbors? Can you smoothly deform one into the other? Definitely not. But Reid's conjecture says that every Calabi-Yau can be deformed into something else [a non-Kähler manifold] that knows all the other Calabi-Yau manifolds." Say you have a bunch of people and you're trying to see if they have something in common, he adds. "We just have to show that they know the same gregarious fellow, in which case they're all part of the same group—the group of this guy's acquaintances."[18]

How does Reid's proposition, or fantasy, about the connectedness of Calabi-Yau manifolds square with reality? In 1988, Tristan Hubsch and University of Maryland mathematician Paul Green proved that Reid's conjecture applied to about 8,000 Calabi-Yau manifolds, which included most of the manifolds known at the time. Subsequent generalizations of this work have shown that more than 470 million constructions of Calabi-Yau manifolds—almost all the known threefolds—are connected in the way that Reid suggested.[19]

Of course, we won't know it's true for all cases until it's proven. And more than two decades after Reid posed his conjecture, this is turning out to be a difficult fantasy to prove. A big part of the challenge, I believe, is that non-Kähler manifolds are not well understood by mathematical standards. We'll have a better chance of proving this proposition when we understand these manifolds better. As a general matter, we cannot say for sure that such manifolds—the non-Kähler ones—are even real (or mathematically viable). There is no broad existence proof, such as that pertaining to Calabi-Yau manifolds, and existence, so far, has only been established in a few isolated cases.

If we are intent on learning all we can about the manifolds that have given rise to the landscape conundrum and its attendant cosmological puzzles, it would be helpful to determine whether all Calabi-Yau manifolds are related. A key to doing this, as we've just established, may lie in the new frontier of non-Kähler manifolds. These manifolds are of keen interest to physicists not just for the insight they may give us on Calabi-Yau manifolds, but also because they may offer the compactification geometry needed to compute particle masses—one aspect of the quest to realize the Standard Model that has eluded us so far, while physicists have pursued strategies that rely exclusively on Calabi-Yau manifolds.

My colleague Melanie Becker, a physicist at Texas A&M University, believes that non-Kähler approaches may hold the answer. "The way to get particle content and masses," Becker says, "just might be through the compactification of non-Kähler manifolds." It could be the geometry we're looking for—the one that leads us to the promised land of the Standard Model. The reason she thinks so goes back to our discussion in the beginning of this chapter. String theorists introduced fluxes to get rid of massless scalar fields and thereby stabilize the size and shape of a Calabi-Yau manifold. But turning on these powerful fields or fluxes can have another consequence: It can distort the geometry of the manifold itself, changing the metric so that it's no longer Kähler. "When you turn on the flux, your manifold becomes non-Kähler, and it's a whole different ball game," Becker says. "The challenge is that this is really a whole new topic of mathematics. A lot of the math that applies to Calabi-Yau manifolds does not apply to non-Kähler manifolds."[20]

From the standpoint of string theory, one of the chief roles intended for these manifolds, be they Calabi-Yau or non-Kähler, is compactification—reducing the theory's ten dimensions to the four of our world. The easiest way to partition the space is to cut it cleanly, splitting it into four-dimensional and six-dimensional components. That's essentially the Calabi-Yau approach. We tend to think of these two components as wholly separate and noninteracting. Ten-dimensional spacetime is thus the Cartesian product of its four- and six-dimensional parts, and as we've seen, you can visualize it with the Kaluza-Klein-style model we discussed in Chapter 1: In this picture, our infinite, four-dimensional spacetime is like an infinitely long line, except that this line has some thickness—a tiny circle wherein the extra six dimensions reside. So what we really have is the Cartesian product of a circle and a line—a cylinder, in other words.

In the non-Kähler case, the four- and six-dimensional components are not independent. As a result, the ten-dimensional spacetime is not a Cartesian prod-

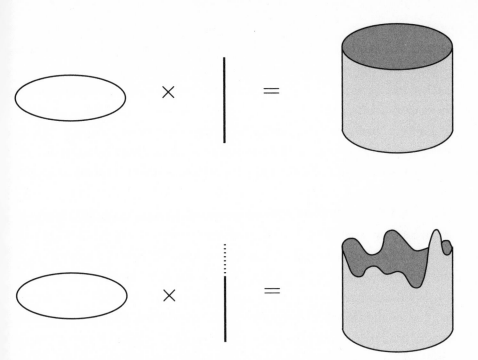

10.7—Taking the so-called Cartesian product of a circle and a line segment is like attaching that same line segment to every single point on the circle. The result is a cylinder. Taking the warped product is different. In this case, the length of the line segment does not have to be constant; it can vary depending on where you are on the circle. So in this case, the result is not an actual cylinder; it's more of a wavy, irregular cylinder.

uct but rather a *warped product*, signifying that these two subspaces do interact. Specifically, distances in the four-dimensional spacetime are influenced—and continually rescaled or warped—by the six-dimensional part. The extent to which the four-dimensional spacetime gets expanded or shrunk depends on a number called the warp factor, and in some models the effect, or warping, can be exponential.

This is perhaps easiest to picture in our cylinder example. Again, we'll represent the six-dimensional space with a circle. The four-dimensional part is a line perpendicular to that circle, and we'll represent it with a unit line segment (rather than an infinite line) to show how distances are affected. If there were no warping, as you moved the line segment to every point on the circle, you'd trace out a perfect (and solid) cylinder. Because of warping, however, the length of that line segment can vary over the surface of the circle. At one point, it may be 1, at another ½, at another 1½, and so forth. What you'll end up with is not a perfect cylinder but an uneven, wavy cylinder that's distorted by the warping.

This can all be expressed in more rigorous terms by a set of equations issued in 1986 by the physicist Andrew Strominger. In the earlier (1985) paper he wrote with Candelas, Horowitz, and Witten, which presented the first serious attempt at a Calabi-Yau compactification, they made the simplifying assumption that the four-dimensional and six-dimensional geometries were autonomous, notes Strominger. "And we found solutions in which they were autonomous, even though string theory didn't demand that. A year later, I worked out the equations you get when you don't make those assumptions." These are the so-called Strominger equations, which concern the situation when the fluxes are turned on and the four- and six-dimensional spaces interact. "The possibility of their not being autonomous is interesting because there are some really good consequences," Strominger adds. Prominent among these consequences is that warping might help explain important phenomena such as the "hierarchy problem," which relates to why the Higgs boson is so much lighter than the Planck mass and why gravity is so much weaker than the other forces.

The Strominger equations, which apply to non-Kähler manifolds, describe a bigger class of solutions than the equations in the 1985 paper, which only applied to Calabi-Yau manifolds. "In trying to understand the many ways string theory could be realized in nature, one needs to understand the more general solutions," Strominger says. "It's important to understand all the solutions to string theory, and Calabi-Yau space does not contain them all."[21] Harvard physicist Li-Sheng Tseng (my current postdoc) compares the Calabi-Yau manifold to a circle, "which, of all the smooth and closed one-dimensional curves you can draw, is the most beautiful and special." The Strominger equations (sometimes referred to as the Strominger system), he says, "involve a relaxation of the Calabi-Yau condition, which is like relaxing the circle condition to the ellipse condition." If you have a closed loop of string of fixed length, there is only one circle you can possibly make, whereas you can make an endless variety of ellipses by taking that circle and squashing or elongating it to varying degrees. Of all the curves you can make out of that loop, the circle is the only one that remains invariant to rotations around the center.

To see that a circle is just a special case of an ellipse, we need look no further than the equation that defines an ellipse on the Cartesian $(x\text{-}y)$ plane:

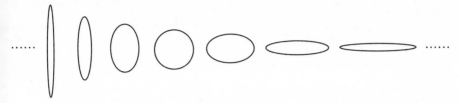

10.8—If you have a loop of fixed length, you can make an infinite number of ellipses— some pointier, some rounder—but you can only make one circle of that circumference. In other words, by relaxing some of the properties that make a circle so special, you can end up with any number of ellipses. Similarly, a Calabi-Yau manifold, which has Kähler symmetry by definition, is (like the circle) much more special than a non-Kähler manifold, which satisfies less rigorous conditions and encompasses a much broader class of objects.

$x^2/a^2 + y^2/b^2 = 1$, where a and b are positive, real numbers. A curve so described is never a circle except when $a = b$. Moreover, it takes two parameters, a and b, to define an ellipse, and just one (since $a = b$) to define a circle. That makes an ellipse a somewhat more complicated system than a circle, just as the Strominger (non-Kähler) system is more complicated to deal with than a Calabi-Yau, which can be described with fewer parameters.

Even though going from a circle to an ellipse, or from a Calabi-Yau to a non-Kähler manifold, may represent a step down in symmetry and beauty, Tseng notes, "it's clear that nature doesn't always choose the most symmetric configuration. Think, for example, of the elliptical orbits of the planets. So it's also possible that the internal six-dimensional geometry that describes our natural universe may not be quite so symmetric as the Calabi-Yau, but just slightly less so, as in the Strominger system."[22]

The system that Strominger proposed is no picnic to deal with, because it consists of four differential equations that have to be solved simultaneously— any one of which can be nightmarish to solve. There are two Hermitian Yang-Mills equations, which have to do with the gauge fields (see Chapter 9). Another equation ensures that the whole geometry is supersymmetric, while the last is designed to make the anomalies cancel, which is essential for the consistency of string theory.

As if this weren't challenging enough, each of these four equations is really a system of equations rather than a single equation. Each can be written as a single matrix (or tensor) equation, but since the matrix itself has many variables, you can split that single equation into separate component equations. For this same reason, the famous Einstein equation that encapsulates the theory of general relativity is really a set of ten field equations that describe gravity as the curvature of spacetime caused by the presence of matter and energy, even though it can be written as a single tensor equation. In the proof of the Calabi conjecture, solving the Einstein equation in the vacuum condition reduced to a single equation, albeit a rather imposing one. Non-Kähler manifolds are harder to work with than Calabi-Yau manifolds, because the situation has less symmetry and therefore more variables—all of which results in more equations to be solved. Furthermore, we don't really have the mathematical tools to understand this problem well at the moment. In the Calabi case, we drew on algebraic geometry, which has developed tools over the previous two centuries for dealing with Kähler manifolds but not their non-Kähler counterparts.

Still, I don't believe these two classes of manifolds are that different, from a mathematical standpoint. We've used geometric analysis to build Calabi-Yau manifolds, and I'm confident these techniques can help us build non-Kähler manifolds as well, assuming that first we can either solve the Strominger equations or at least prove that solutions exist. Physicists need to know whether non-Kähler manifolds can exist and whether it's possible to satisfy all four equations at once, for if it's not possible, people working on them may be wasting their time. I looked at the problem for nearly twenty years after Strominger proposed it and couldn't find a solution. Or, I should say, a smooth solution, a solution without singularities, as Strominger did find some solutions with singularities (but those are messy and extremely difficult to work with). After a while, people began to believe that a smooth solution did not exist.

Then a minor breakthrough occurred. Some colleagues and I found smooth solutions in a couple of special cases. In the first paper, which I completed in 2004 with Stanford mathematician Jun Li (a former graduate student of mine), we proved that a class of non-Kähler manifolds was mathematically possible. In fact, for each known Calabi-Yau manifold, we proved the existence of a whole family of non-Kähler manifolds that are similar enough in structure to be from the same neighborhood. This was the first time the existence of these manifolds was ever mathematically confirmed.

Although solving the Strominger equations is extraordinarily difficult, Li and I did about the easiest thing you could do in this area. We proved that those equations could be solved in the limited case, where the non-Kähler manifold is very close to the Calabi-Yau. In fact, we started with a Calabi-Yau manifold and showed how it could be deformed until its geometry, or metric, was no longer Kähler. Although the manifold could still support a Calabi-Yau metric, its metric was now non-Kähler, hence offering solutions to the Strominger system.

Probably more significantly, Li and I generalized the DUY theorem (as mentioned in Chapter 9, the acronym for the theorem's authors, Donaldson, Uhlenbeck, and me) to cover basically all non-Kähler manifolds. Having DUY in hand was of great practical value because it automatically took care of two of the four Strominger equations—those relating to Hermitian Yang-Mills theory—leaving the supersymmetry and anomaly cancellation equations to be solved. Given that DUY has been instrumental for Calabi-Yau compactifications (in terms of reproducing the gauge fields), we're hoping it will do the same for non-Kähler compactifications as well.

One promising avenue for generating non-Kähler manifolds, as suggested by Reid's conjecture, is to start with an already-known Calabi-Yau manifold and take it through a conifold transition. I recently looked into this with Jun Li and Ji-Xian Fu, a former Harvard postdoc of mine now based at Fudan University in Shanghai. The basic manifold we started with came from Herb Clemens, one of the architects of the conifold transition, but he provided just a general topology—a manifold without a metric and hence no geometry. Fu, Li, and I tried to give this manifold some shape by showing the existence of a metric that would satisfy the Strominger equations.

Those equations seemed appropriate here, because they apply not only to non-Kähler manifolds but also to Calabi-Yau manifolds, which are a special case. And Reid's conjecture, too, involves a procedure that takes you from Calabi-Yau manifolds to non-Kähler ones and back. So if you want a set of equations that can cover both these geometries, Strominger's formulations might be just the thing. So far, my colleagues and I have proved that Clemens's manifold does indeed satisfy three of the four Strominger equations, though we haven't found a solution to the most difficult one, the anomaly cancellation equation. I'm still pretty confident that this manifold exists. After all, in most human endeavors, three out of four is pretty good. But until we solve that last equation, we won't have proved a thing.

Fu and I went farther with another example, showing how to construct a topologically diverse class of non-Kähler manifolds that satisfies the Strominger equations. Built from scratch, rather than constructed by modifying known Calabi-Yau manifolds, these manifolds are intrinsically non-Kähler. They consist of K3 surfaces (four-dimensional Calabi-Yau manifolds) with a two-dimensional torus attached at every point. Solving the Strominger equation, in this case, involved solving a Monge-Ampère equation (a class of nonlinear differential equations discussed in Chapter 5) that was more complicated than the one I had to solve for the Calabi proof. Fortunately, Fu and I were able to build from the earlier argument. Our method, as with the Calabi proof, involved making a priori estimates, which means that we had to make guesses regarding the approximate values of various parameters.

Fu and I found a special method that enabled us to solve not just one equation but all four. Whereas in the case of the Calabi conjecture, I was able to obtain all possible solutions to the Monge-Ampère equation, this time Fu and I obtained just a subset of possible solutions. Unfortunately, we don't understand the system well enough to know how large or small that subset is.

At least we've taken some preliminary steps. Most physicists who have begun working on non-Kähler compactifications are assuming that the Strominger equations can be solved without bothering to prove it. Li, Fu, and I have shown that these equations can be solved in the isolated approaches we've identified so far, which is another way of saying that these particular manifolds—a fraction of all non-Kähler manifolds—really do exist. This is just the starting point for the bigger problem I want to tackle: finding a metric that satisfies the Strominger system, and all its equations, in the most general terms. While no one has come close to accomplishing this yet—and all signs suggest that a proof will not come easily—my colleagues and I, with our more modest steps, have at least raised the possibility.

Becker tells me that if I succeed in this venture, it will be even more important than the proof of the Calabi conjecture. She could be right, but it's hard to tell. Before I solved the Calabi conjecture, I didn't know its full significance. And even after I solved it, physicists did not recognize the importance of the proof and accompanying theorem until eight years after the fact. But I continued to explore Calabi-Yau spaces because, to me, they looked pretty. And these spaces characterized by the Strominger system have a certain allure as well. Now we'll have to see how things pan out.

In the meantime, Fu and I have offered the manifolds we've produced so far to fellow physicists through a collaboration with Melanie Becker, Katrin Becker, Tseng, and others—maybe even Strominger if we can bring him into the fold. Since then, this group has constructed more examples of the original Fu-Yau model, while exploring the physics in a preliminary way. Unlike some of the heterotic string theory compactifications described in the last chapter, this team has not been able to get the right "particle content" or the three generations of particles we see in the Standard Model. "What we do have," says Melanie Becker, "are stabilized moduli, which is a prerequisite to everything else, as well as an actual way to compute masses."[23]

At this stage, it's hard to know what exactly will come from the efforts of physicists currently toying with non-Kähler compactifications and the many other alternatives to Calabi-Yau manifolds (including an area called nongeometric compactifications) that are currently being investigated. It's fair to ask whether Calabi-Yau compactifications are the right description of our universe or merely the simplest example from which we've learned—a fantastic experiment for letting us discover how string theory works and how we can have supersymmetry, all the forces, and other things we want in an "ultimate" theory. In the end, though, this exercise may yet lead us to a different kind of geometry altogether.

For now, we're simply trying to explore some of the many possibilities lying before us on the string theory landscape. But even amid all those possibilities, we still live in just one universe, and that universe could still be defined by Calabi-Yau geometry. I personally think Calabi-Yau manifolds are the most elegant formulation, as well as the most beautiful manifolds constructed so far among all the string vacua. But if the science leads us to some other kind of geometry, I'll willingly follow.

"In the past twenty years, we've uncovered many more solutions to string theory, including non-Kähler ones," says Joe Polchinski. "But the first and simplest solutions—Calabi-Yau manifolds—still look the closest to nature."[24]

I'm inclined to agree, though there are plenty of top-notch researchers who have a different opinion. Melanie Becker, for one, is a champion of the non-Kähler approach. Strominger, who has made major contributions in both the Calabi-Yau and non-Kähler realms, doesn't think Calabi-Yau spaces will ever become obsolete. "But we want to use everything we encounter as stepping-stones to the next level of understanding," he says, "and Calabi-Yau manifolds have been stepping-stones in many directions."[25]

Before long, hopefully, we'll have a better sense of where they might lead us. Despite my affection for Calabi-Yau manifolds—a fondness that has not diminished over the past thirty-some years—I'm trying to maintain an open mind on the subject, keeping to the spirit of Mark Gross's earlier remark: "We just want to know the answer." If it turns out that non-Kähler manifolds are ultimately of greater value to string theory than Calabi-Yau manifolds, I'm OK with that. For these less-studied manifolds hold peculiar charms of their own. And I expect that upon further digging, I'll come to appreciate them even more.

University of Pennsylvania physicist Burt Ovrut, who's trying to realize the Standard Model through Calabi-Yau compactifcations, has said he's not ready to take the "radical step" of working on non-Kähler manifolds, about which our mathematical knowledge is presently quite thin: "That will entail a gigantic leap into the unknown, because we don't understand what these alternative configurations really are."[26]

While I agree with Ovrut's statement, I'm always up for a new challenge and I don't mind the occasional plunge into uncharted waters. But since we're often told not to swim alone, I'm not averse to dragging a few colleagues along with me.

Eleven

THE UNIVERSE UNRAVELS

*(or Everything You Always Wanted to Know About
the End of the World but Were Afraid to Ask)*

A man walks into a laboratory where he is greeted by two physicists, a senior scientist and her younger male protégé, who show him a roomful of experimental apparatus—a stainless steel vacuum chamber, insulated tanks filled with cryogenic nitrogen and helium, a computer, various digital meters, oscilloscopes, and the like. The man is handed the controls to the machinery and told that the fate of the experiment—and, perhaps, the fate of the universe—lies in his hands. If the younger scientist is correct, the device will successfully extract energy from the quantum vacuum, providing humankind with unlimited bounty—"the energy of creation at our fingertips," as it's described. But if he's wrong, the elder researcher cautions, the device could trigger a phase transition whereby the vacuum of empty space decays to a lower energy state, releasing all of its energy at once. "It would be the end, not only of the earth, but of the universe as we know it," she says. The man anxiously grips the switch, as sweat from his palms spreads across the device. Seconds remain until the moment of truth. "You'd better decide fast," he's told.

Although this is science fiction—a scene from the short story "Vacuum States," by Geoffrey Landis—the possibility of vacuum decay is not total fantasy.[1] The issue has, in fact, been explored for decades in journals more scholarly than *Asimov's Science Fiction—Nature, Physical Review Letters, Nuclear*

Physics B, etc.—by noted researchers like Sidney Coleman, Martin Rees, Michael Turner, and Frank Wilczek. Many physicists today, and perhaps the majority of those who think about such things, believe the vacuum state of our universe—empty space devoid of all matter save for the particles that spring in and out of existence by virtue of quantum fluctuations—is metastable rather than permanently stable. If these theorists are right, the vacuum will eventually decay, and the effect on the universe will be devastating (at least from our point of view), although these worrisome consequences may not happen until long after the sun has disappeared, black holes have evaporated, and protons have disintegrated.

While no one knows exactly what will happen in the long run, there seems to be agreement, at least in some quarters, that the current arrangement is not permanent—that eventually, some sort of vacuum decay is in order. The usual disclaimers apply, of course: While many researchers believe that a perfectly stable vacuum energy, or cosmological constant, is not consistent with string theory, we should never forget that string theory itself—unlike the mathematics underlying it—is in no way proven. Furthermore, I should remind readers that I'm a mathematician, not a physicist, so we're venturing into areas that extend far beyond my expertise. The question of what may ultimately happen to the six compact dimensions of string theory is one for physicists, not mathematicians, to settle. As the demise of those six dimensions may correlate with the demise of our chunk of the universe, investigations of this sort necessarily involve treading on uncertain ground because, thankfully, we haven't yet done the definitive experiment on our universe's end. Nor do we have the means—outside of a fertile imagination like Landis's—of doing so.

With that in mind, please take this discussion with a big grain of salt and, if you can, try to approach it in the spirit I'm approaching it—as a wild, whimsical ride into the realm of maybe. It's a chance to find out what physicists think may become of the six hidden dimensions we've talked so much about. None of this has been proven, and we're not even sure how it might be tested, yet it's still an opportunity to see how these ideas might play out and to see how far informed speculation can take us.

Imagine that the man in Landis's story pulls the lever, suddenly initiating a chain of events that result in vacuum decay. What would happen? The short answer is, nobody knows. But no matter what the outcome—whether we go the way of fire or ice, to paraphrase Robert Frost—our world would almost certainly be changed beyond recognition in the process. As Andrew Frey of McGill University and his colleagues wrote in *Physical Review D* in 2003, "the kind of [vacuum] decays considered in this paper in a very real sense would represent the end of the universe for anyone unfortunate enough to experience one."[2] There are two main scenarios under consideration. Both involve radical alterations of the status quo, though the first is more severe as it would spell the end of space-time as we know it.

For starters, think back to that picture of a little ball rolling on a gently sloping surface, with each elevation point corresponding to a different vacuum energy level—a picture that we discussed in Chapter 10. For the moment, our ball is sitting in a semistable situation called a *potential well*, which is analogous to a small dip or hole on an otherwise hilly landscape. We'll assume that even at the bottom of that hole, the elevation is still, so to speak, above sea level: The vacuum energy, in other words, remains positive. If this landscape were a classical one, the ball would sit there indefinitely. Its resting place, in other words, would become its final resting place. But the landscape isn't classical. It's quantum mechanical, and with quantum mechanics in play, a funny thing can happen: If the ball is exceedingly tiny, which is the setting in which quantum phenomena become apparent, it can literally bore through the side of the hole to reach the outside world. That's the result of an absolutely real phenomenon known as quantum tunneling. This is possible because of the fundamental uncertainty built into quantum mechanics. According to the uncertainty principle formulated by Werner Heisenberg, location—contrary to the mantra of realtors—is not the only thing, and it's not even an absolute thing. Although a particle is perhaps most likely to be found in one spot, there's a chance it could be found in more improbable locales. And if there's a chance it could happen, the theory states, it eventually will, provided we wait long enough. This principle holds, in fact, regardless of the size of the ball, although the probability of a large ball's doing so would be even smaller.

Strange as it sounds, the real-world effects of quantum tunneling have been seen. This well-tested phenomenon provides the basis, for example, for scanning tunneling microscopes, whose operation depends on electrons making their way through seemingly impenetrable barriers. Microchip manufacturers, similarly, cannot make transistors too thin, or the performance of these devices will suffer from electron leakage due to tunneling effects.

The idea of particles like electrons tunneling through a wall—metaphorical or real—is one thing, but what about spacetime as a whole? The notion of an entire vacuum's tunneling from one energy state to another is, admittedly, harder to swallow, yet the theory has been pretty well worked out by Coleman and others, starting in the 1970s.[3] The barrier in this case is not a wall so much as a kind of energy field that is preventing the vacuum from reconstituting itself into a lower-energy, more stable, and therefore more favored state. Change in this case comes by way of a phase transition, similar to liquid water turning into ice or vapor, except that a large swath of the universe is transformed—perhaps a swath that includes our home.

This brings us to the punch line of the first scenario, in which our current vacuum state tunnels from its slightly positive energy value—a fact of life we now call dark energy or the cosmological constant—to a negative value instead. As a result, the energy now driving our universe apart at an accelerated clip would instead compress it to a point, thereby carrying us toward a cataclysmic event known as the Big Crunch. At this cosmic singularity, both the energy density and the curvature of the universe would become infinite—the same thing that, in principle, we'd encounter at the center of a black hole or if we ran the universe backward to the Big Bang.

As for what might follow the Big Crunch, all bets are off. "We don't know what happens to spacetime, let alone what happens to the extra dimensions," notes physicist Steve Giddings of the University of California, Santa Barbara.[4] It's beyond our realm of experience and grasp in almost every respect.

Quantum tunneling isn't the only way to trigger a change in vacuum state; another is through so-called thermal fluctuations. Let's go back to our minuscule ball in the bottom of the potential well. The higher the temperature, the faster that all atoms, molecules, and other particles in this system are moving around. And if particles are moving around, some will randomly crash into the ball, jostling it one way or another. On average, these jostles will cancel each other out and the ball will remain more or less stationary. But suppose, through some statistical fluke, multiple atoms slam into it, successively, from the same general direction. Several such jostles in a row could knock the ball clear out of the hole. It will then wind up on the hilly surface and perhaps roll all the way down to zero energy unless it gets stuck in other wells or holes on the way.

Evaporation might be an even better analogy, suggests New York University physicist Matthew Kleban. "You don't ever see water crawling out of a cup," he explains. "But water molecules keep getting bumped, especially when water is heated, and occasionally they're bumped hard enough to make it out of the cup. That's similar to what happens in this thermal process."[5]

There are two important differences, however. One difference is that the processes we're discussing here take place in a vacuum, which traditionally means no matter and hence no particles. So what's doing the knocking? Well, for one thing, the temperature never quite gets to zero (this fact turns out to be a feature of an expanding universe), and for another, space is never quite empty, because pairs of virtual particles—a particle and its antiparticle—are continually popping into existence and then disappearing through annihilation in an interval so brief, we've never been able to catch them in the act. The other difference is that this process of virtual particle creation and annihilation is a quantum one, so what we've been calling thermal fluctuations necessarily include some quantum contributions as well.

We're now ready to take up the second scenario, which may be more benign than the first, but only somewhat. Through quantum tunneling or perhaps a thermal or quantum fluctuation, our universe may wind up at another metastable spot (most likely at a slightly lower vacuum energy level) on the string theory landscape. But this, as with our current situation, would be merely a temporary way station, or a metastable rest stop en route to our final destination. This issue is related to how Shamit Kachru, Renata Kallosh, Andrei Linde, and Sandip Trivedi (the authors of the KKLT paper) explained string theory's great vanishing act—providing us with a universe of only four large dimensions, rather than ten, and doing so while simultaneously incorporating the notion of inflation into string cosmology. Even though we now see only four dimensions, "in the long run, the universe doesn't want to be four-dimensional," claims Stanford cosmologist Andrei Linde. "It wants to be ten dimensional."[6] And if we're really patient, it will be. Compactified dimensions are fine in the short run, but it's not the ideal state of affairs for the universe over the long haul, according to

Linde. "Where we are now is like standing on top of a building, but we haven't jumped yet. If we don't do it by our own will, quantum mechanics will take care of it for us, throwing us down to the lowest energy state."[7]

The reason a universe with ten large dimensions is ergonomically preferred comes down to this: In the most well-developed models we have today, the energy of the vacuum is a consequence of the compactification of the extra dimensions. Put in other terms, the dark energy we've heard so much about isn't just driving the cosmos apart in some kind of madcap accelerative binge: Some, if not all, of that energy goes into keeping the extra dimensions wound up tighter than the springs of a Swiss watch, although in our universe, unlike in a Rolex, this is done with fluxes and branes.

The system, in other words, has stored potential energy that is positive in value. The smaller the radius of the extra dimensions, the tighter the spring is wound and the greater the energy stored. Conversely, as the radius of those dimensions increases, the potential energy declines, reaching zero when the radius becomes infinite. That's the lowest energy state and hence the only truly stable vacuum, the point at which the dark energy drops to zero and all ten dimensions become infinitely large. The once-small internal dimensions, in other words, become decompactified.

Decompactification is the flip side of compactification, which as we've discussed, is one of the biggest challenges in string theory: If the theory depends on our universe's having ten dimensions, how come we only see four? String theorists have been hard-pressed to explain how the theory's extra dimensions are so well concealed, because, as Linde has noted, all other things being equal, the dimensions would rather be big. It's like trying to hold an increasing volume of water in an artificial reservoir with fixed walls. In every direction, in every corner of the structure, the water is trying to get out. And it won't quit trying until it does. When that happens, and the sides suddenly give way, water confined to a compact area (within the perimeter of the reservoir) will burst out and spread over an extended surface. Based on our current understanding of string theory, the same sort of thing will happen to the compact dimensions, be they curled up in Calabi-Yau spaces or in some other more complicated geometries. No matter what configuration is selected for the internal dimensions, they will eventually unwind and open up.

Of course, one might ask why, if it's so advantageous from an energy standpoint for the dimensions to expand, this hasn't happened already. One solution that physicists have proposed, as discussed in the last chapter, involves branes and fluxes. Suppose, for instance, you have a badly overinflated bicycle inner tube. Any weak spot in the tire would give rise to a bubble that would eventually burst. We could shore up a weak spot by applying a patch, which is somewhat like a brane, or bind up the whole tire with rubber bands to help it maintain its shape, as we think fluxes do with the Calabi-Yau spaces. So the idea is that we've got two opposing forces here—a natural tendency for an overinflated shape to expand that's kept in check by branes, fluxes, and other structures that wrap around the object and hold it in. The net result is that these countervailing forces are now perfectly balanced, having achieved some kind of equilibrium.

It's an uneasy peace, however. If we push the radii of the extra dimensions to larger values through modest quantum fluctuations, the branes and fluxes provide a restorative force, quickly bringing the radii back to where they started. But if you stretch the dimensions too far, the branes or fluxes can snap. As Giddings explains, "eventually a rare fluctuation will take you out to the threshold radius for decompactification and"—keeping in mind the slope on the right side of Figure 11.1—"it's all downhill from there."[8] We're off on the merry road to infinity.

Figure 11.2 tells a similar story with a nuance thrown in. Instead of tunneling out of our present situation straight to a universe with ten large dimensions, there will be an intermediary stop—and perhaps a series of them—in the landscape along the way. But in either case, whether you fly nonstop or make connections in Dallas or Chicago, the endpoint will be the same. And inevitable.

The landing, however, is not likely to be gentle. Remember that change, when it comes, is actually a phase transition of the vacuum, rather than a ball climbing out of a hole or burrowing through a wall. The change will start small, as a tiny bubble, and grow at an exponential clip. Inside that bubble, the compactification—which had been keeping six of the dimensions almost Planck-scale small—will start to undo itself. As the bubble spreads, a spacetime of four large dimensions and six tiny curled ones will become desegregated, in a sense. Where the dimensions had once been partitioned into

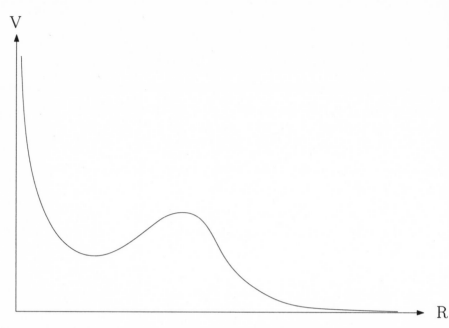

11.1—One theory holds that our universe sits in the little dip in the left-hand side of the graph, which locks the vacuum's potential energy (V) to a specific level as well as fixing the radius (R) of the compact extra dimensions. However, that arrangement may not be permanent. A little nudge could push us over the hill on the right—or we might just quantum-tunnel clear through the barrier—which would send us down the sloping curve toward infinitely large extra dimensions. The process, whereby previously tiny dimensions unwrap to become big, is called *decompactification*. (Adapted, with permission, from a figure by Steve Giddings)

compact and extended form, there will now be ten large dimensions all thrown together, with no barriers keeping them apart.

"We're talking about a bubble that expands at the speed of light," Shamit Kachru notes. "It starts in a certain location of spacetime, kind of like the way bubbles nucleate in water. What's different is that this bubble doesn't just rise and leave. Instead, it expands and removes all the water."[9]

But how can a bubble move so fast? One reason is that the decompactified state inside the bubble lies at a lower potential energy than exists outside the bubble. Because systems naturally move in the direction of lower energy— which in this case also happens to be the direction of increased dimension size—the resultant gradient in potential energy creates a force on the edge of

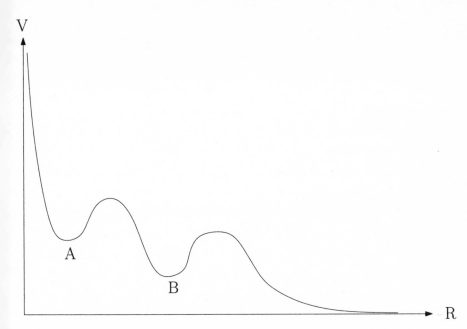

11.2—The story is pretty much the same as in Figure 11.1. Our universe is still headed toward decompactification and the realm of infinitely large extra dimensions, only this time, we're going to make an additional stop in the "landscape" along the way. In this scenario, our universe can be thought of as a marble that temporarily gets stuck in a trough (A) as it rolls down the hill. In principle, the marble could make many intermediary stops as it continues its descent, even though only one additional trough (B) is shown on this graph. (Adapted, with permission, from a figure by Steve Giddings)

the bubble, causing it to accelerate outward. The acceleration, moreover, is both sustained and high, which drives the bubble to the speed of light within a tiny fraction of a second.

Linde describes the phenomenon in more colorful terms. "The bubble wants to go as fast as possible because if you have the possibility of leading a great life in a lower vacuum energy, why would you want to wait?" he asks. "So the bubble moves faster and faster, but it cannot move faster than the speed of light."[10] Though, considering his description of the rewards ahead, it probably would if it could.

Because the bubble spreads outward at light speed, we'd never know what hit us. The only advance warning we'd get would be the shock wave that would arrive a fraction of a second earlier. The bubble would then smack into us head-on, carrying a lot of kinetic energy in its wall. That's just the first round of a double whammy. Because the bubble wall has some thickness, it would take a bit of time—albeit a mere fraction of a second—for the worst to come. The place we call home has four-dimensional laws of physics, whereas the bubble's interior obeys ten-dimensional laws. And those ten-dimensional laws will take over just as soon as the inside of the bubble infiltrates our world. As the playwright/screenwriter David Mamet once put it: "Things change."

In fact, everything you can imagine, from the tiniest particle to elaborate structures like galactic superclusters, would instantly explode into the six expanding dimensions. Planets and people would revert to their constituent parts, and those parts would be obliterated as well. Particles like quarks, electrons, and photons would cease to exist altogether, or they would reemerge with completely different masses and properties. While spacetime would still be there, albeit in an altered state, the laws of physics would change radically.

How long might we have before such an "explosion" in dimensions occurs? We're pretty sure the vacuum of our present universe has been stable ever since inflation ended approximately 13.7 billion years ago, notes Henry Tye of Cornell. "But if the expected shelf life is just fifteen billion years, that only gives us a billion or so years left."[11] Or just enough time to start packing.

But all signs suggest there's no need to hit the panic button just yet. It could take an extraordinarily long time—on the order of $e^{(10^{120})}$ years—for our spacetime to decay. That number is so big it's almost hard to fathom, even for a mathematician. We're talking about e—one of the fundamental constants in nature, a number that's approximately 2.718—multiplied by itself 10^{120} (a one with 120 zeros after it) times. And if that rough guess is correct, our waiting period is, for all practical purposes, infinitely long.

So how does one come up with a number like $e^{(10^{120})}$, anyway? The initial premise is that our universe is evolving into something called de Sitter space—a space dominated by a positive cosmological constant in which all matter and radiation eventually become so dilute as to be insignificant, if not absent altogether. (Such a space was first proposed in 1917 by the Dutch astrophysicist Willem de Sitter as a vacuum solution to the Einstein field equations.) If our universe, with its small cosmological constant, is in fact "de Sitter," then the entropy of such a space is prodigious, on the order of 10^{120} (more on where that number comes from in a minute). This kind of universe has a large entropy because its volume is so great. Just as there are more places you can put an electron in a big box than in a small box, a big universe has more possible states—and hence a higher entropy—than a small one.

De Sitter space has a horizon in the same way that a black hole has an event horizon. If you get too close to a black hole and cross the fateful line, you'll be sucked in and won't be coming home for supper. The same is true for light, which can't escape, either. And the same thing holds for a de Sitter horizon, as well. If you go too far in a space that's undergoing accelerated expansion, you'll never make it back to the neighborhood you started in. And light, as in the case of the black hole, won't make it back, either.

When the cosmological constant is small, and accelerated expansion relatively slow (which happens to be our present circumstance), the horizon is far away. That's why the volume of such a space is big. Conversely, if the cosmological constant is large, and the universe is racing apart at breakneck speed, the horizon—or point of no return—may be close at hand (quite literally) and the volume correspondingly small. "If you extend your arm too far in such a space," Linde notes, "the rapid expansion might tear your hand away from you."[12]

Although the entropy of de Sitter space is related to the volume, it's even more closely correlated with the surface area of the horizon, which scales with the distance to the horizon squared. (We can actually apply the same reasoning—and the same Bekenstein-Hawking formula—that we applied to black holes in Chapter 8, with the entropy of de Sitter space being proportional to the horizon area divided by four times Newton's gravitational constant G.) The distance to the horizon or, technically, the distance squared, in turn, depends on the cosmological constant: The greater the constant's value, the smaller the distance. Since the entropy scales with the distance squared, and the distance squared is inversely proportional to the cosmological constant, the entropy is also inversely proportional to the cosmological constant. The upper limit for the cosmological constant in our universe, according to Hawking, is 10^{-120} in the "dimensionless units" that physicists use.[13] (This number, 10^{-120}, is a rough approximation, however, and should not be taken as an exact figure.) The entropy, being the inverse or reciprocal of that, is therefore extremely large—roughly on the order of 10^{120}, as noted a moment ago.

The entropy, by definition, is not the number of states per se, but rather the log—or, to be precise, the natural log—of the number of states. So the number of states is, in fact, e^{entropy}. Going back to our graph in Figure 11.1, the number of possible states in our universe with a small cosmological constant, which is represented by the dip or minimum in the curve, is $e^{(10^{120})}$. Let's suppose, on the other hand, that the summit of the mountain, from which one would roll down toward dimensions of infinite radius, is such an exclusive place that there is just one state that puts you exactly on top. Therefore, the odds of landing at that particular spot, among all the other possibilities, are vanishingly small—roughly just $1/e^{(10^{120})}$. And that's why, conversely, the amount of time it will take to tunnel through the barrier is so mind-numbingly large we can't even call it astronomical.

One other point: In Figure 11.2, we presented a decompactification scenario in which our universe tunnels to a state of lower vacuum energy (and smaller cosmological constant), making a stopover in the landscape on its journey to the ultimate makeover, infinite dimensionality. But might we take a detour, tunneling up to a spot with higher vacuum energy instead? Yes, but it's much easier, and more probable, to go downhill. One can, however, make a somewhat more involved argument. Suppose there's a potential minimum at point A and a separate minimum at point B, with A being higher than B in elevation and hence in vacuum energy. Since A sits at higher energy, its gravity will be stronger, which means the space around it will be more strongly curved. And if we think of that space as a sphere, its radius will be smaller because smaller spheres bend more sharply than larger ones and thus have greater curvature. Since B sits at lower energy, its gravity will be weaker. Consequently, the space around it will have less curvature. If we think of that space as a sphere, it will have a larger radius and hence be less curved.

We've illustrated some aspects of this idea in Figure 11.3 (using boxes for A and B, rather than spheres) to show that it's more likely to travel "downhill" on the landscape toward lower energy—from A to B, in other words—than it is to go uphill. To see why, we can connect the two boxes with a thin tube. The two boxes will come into equilibrium, with the same concentration, or density, of

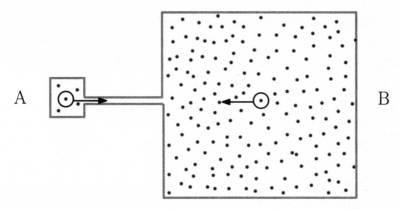

11.3—As discussed in the text, this figure tries to show why it's easier to "tunnel down" from A to B (in Figure 11.2) rather than to "tunnel up" from B to A. The analogy presented here is that it is more likely, on average, for a given molecule to make the trip from A to B than the other way around, simply because there are far fewer molecules in A than in B.

gases or molecules filling both boxes and the same number of molecules migrating from A to B as there are going from B to A. However, since B is much bigger than A, it will have many more molecules to begin with. So the odds of any individual molecule in A migrating to B is much greater than the odds of any individual molecule in B making the reverse trip.

Similarly, the probability of the appearance of a bubble that will transport you to a lower-energy spot in the landscape is substantially greater than that of the appearance of a bubble taking you the other way (uphill), just as any given molecule is more likely to make the trip from box A to B ("downhill," in other words).

In 1890, Henri Poincaré published his so-called recurrence theorem, which states that any system with a fixed volume and energy that can be described by statistical mechanics has a characteristic recurrence time equal to e^{entropy} of that system. The idea here is that a system like this has a finite number of states—a finite number of particle positions and velocities. If you start in a particular state and wait long enough, you'll eventually access all of them, just as a particle or molecule in our proverbial box will wander around, bounce off walls, and move in a random way, over time landing at every possible location in the box. (If we were to put it in more technical terms, instead of talking about possible locations in a box, we'd talk about possible states in "phase space.") The time it will take for spacetime to decompactify, then, is the Poincaré recurrence time—e^{entropy}, or $e^{(10^{120})}$ years. But there's one possible weak point in this argument, Kleban points out: "We don't yet have a statistical mechanical description of de Sitter space." The underlying assumption, which may or may not be borne out, is that such a description exists.[14]

There's not much more to be said about the subject today, nor that much more to be done, other than perhaps to refine our calculations, redo the numbers, and recheck our logic. It's not surprising that few investigators are strongly inclined to carry this much further, since we're talking about highly speculative events in model-dependent scenarios that are not readily testable and are expected to happen on a timescale just shy of forever. That's hardly the ideal prescription for getting grant money or, for younger researchers, gaining the admiration of their elders and, more importantly, securing tenure.

Giddings, who has given the subject more attention than most people have, is not letting the doomsday aspects of this picture drag him down. "On the positive side," he writes in his paper "The Fate of Four Dimensions," "the decay can result in a state that does not suffer the ultimate fate of infinite dilution," such as would occur in an eternally expanding universe endowed with a positive cosmological constant that is truly constant. "We can seek solace both in the relatively long life of our present four-dimensional universe, and in the prospect that its decay produces a state capable of sustaining interesting structures, perhaps even life, albeit of a character very different from our own."[15]

Like Giddings, I too am not losing any sleep over the fate of our four dimensions, six dimensions, or even ten dimensions. Inquiries into this area, as I mentioned before, are thought-provoking and entertaining, but they're also wildly conjectural. Until we obtain some observational data that tests the theory, or at least come up with practical strategies for verifying these scenarios, I'll have to regard them as closer to science fiction than to science. Before we spend too much time worrying about decompactification, however, we first ought to think about ways of confirming the existence of the extra dimensions themselves. Such a success, to my mind, would be more than enough to outweigh the potential downside of the various decay scenarios, which might eventually bring our universe to a bad end—not that any of the other ends, when you get right down to it, look particularly good, either.

As I see it, the unfurling of the hidden dimensions could be the greatest visual display ever witnessed—if it could be witnessed, though that seems extremely doubtful. Allow me one more flight of fancy here, and suppose that this scenario is eventually realized and the great unraveling of spacetime does occur at some point in the distant future. If that ever came to pass, it would be a spectacular confirmation (albeit belated) of the idea to which I've devoted the better part of my career. It's a pity that when the universe's best hiding places are finally exposed, and the cosmos opens up to its full multidimensional glory, no one will be around to appreciate it. And even if someone did survive the great transformation, there'd be no photons around to enable them to take in the view. Nor would there be anyone left with whom they could celebrate the success of a theory dreamed up by creatures who called themselves humans in an era known as the twentieth century, though it might be more properly referred to as the 137 millionth century (Big Bang Standard Time).

The prospect is especially dismaying to someone like me, who has spent decades trying to get a fix on the geometry of the six internal dimensions and, harder still, trying to explain it to people who find the whole notion abstruse, if not absurd. For at that moment in the universe's history—the moment of the great cosmic unwinding—the extra dimensions that are now concealed so well would no longer be a mathematical abstraction, nor would they be "extra" anymore. Instead, they'd be a manifest part of a new order in which all ten dimensions were on equal footing, and you'd never know which ones were once small and which ones were large. Nor would you care. With ten spacetime dimensions to play with, and six new directions in which to roam, life would have possibilities we can't even fathom.

Twelve

THE SEARCH FOR
EXTRA DIMENSIONS

Given that a decade has passed without a major revelation on the theoretical front, string theory partisans now find themselves under increasing pressure to link their ethereal ideas to something concrete. Hovering above all their fantastical notions is one question that won't go away: Do these ideas actually describe our universe?

That's a legitimate issue to raise in view of the provocative ideas broached here, any one of which might give the average citizen pause. One such claim is that everywhere in our world, no matter where we go, there is a higher-dimensional space always within reach, yet so small we'll never see it or feel it. Or that our world could implode in a violent Big Crunch or explode in a fleeting spurt of cosmic decompactification, during which the realm we inhabit instantly changes from four large dimensions to ten. Or simply that everything in the universe—all the matter, all the forces, and even space itself—is the result of the vibrations of tiny strings moving in ten dimensions. There's a second question we also ought to consider: Do we have a prayer of verifying any of this—of gleaning any hints of extra dimensions, strings, branes, and the like?

The challenge facing string theorists remains what it has been since they first tried to re-create the Standard Model: Can we bring this marvelous theory into the real world—the goal being not just to make contact with our world, but also to show us something new, something we haven't seen before?

269

There's a huge chasm, at present, between theory and observation: The smallest things we can resolve with current technology are roughly sixteen orders of magnitude bigger than the Planck scale on which strings and the extra dimensions are thought to reside, and there appears to be no conceivable way of bridging that gap. With the "brute force" approach of direct observation apparently ruled out, it will take extraordinary cleverness, and some measure of luck, to test these ideas by indirect means. But that challenge must be met if string theorists are ever going to win over the doubters, as well as convince themselves that their ideas add up to more than just grandiose speculation on a very small scale.

So where do we start? Peer through our telescopes? Smash particles together at relativistic speeds and sift through the rubble for clues? The short answer is that we don't know which avenue, if any, will pay off. We still haven't found that one make-or-break experiment that's going to settle our questions once and for all. Until we do, we need to try all of the above and more, pursuing any lead that might furnish some tangible evidence. That is exactly what researchers are gearing up for right now, with *string phenomenology* becoming a growth area in theoretical physics.

A logical starting point is to look upward, to the skies, as Newton did in devising his theory of gravity and as astrophysicists have done to test Einstein's theory of gravity. A meticulous scan of the heavens may, for instance, shed light on one of string theory's more recent, and strangest, ideas—the notion that our universe is literally housed within a bubble, one of countless bubbles floating around within the cosmic landscape. While this may not seem like the most promising line of inquiry—being one of the more speculative ideas out there—it is pretty much where we left off in this narrative. And the example does illustrate some of the difficulties involved in translating these far-flung ideas into experiment.

When we discussed bubbles in the last chapter, it was in the context of decompactification, a process both extremely unlikely to be witnessed, taking possibly as long as $e^{(10^{120})}$ years to unfold, and probably not worth waiting for, anyway, as we wouldn't see a decompactification bubble coming until it literally hit us. And once it hit us, we wouldn't be "us" anymore, nor would we be capable of figuring out what the heck did us in. But there may be other bubbles out there. In fact, many cosmologists believe that right now, we're sitting in one that formed at the end of inflation, a fraction of a second after the Big Bang, when a tiny pocket of lower-energy material formed amid the higher-energy inflationary vacuum, and that has expanded since then to become the universe we presently know. It is widely believed, moreover, that inflation never fully ends— that once started, it will continue to spin off an endless number of bubble universes that differ in their vacuum energies and other physical attributes.

The hope held by proponents in the obscure (and sparsely populated) realm of bubble phenomenology is not to see our own bubble but rather signs of another bubble, filled with an entirely different vacuum state, that careened into ours at some time in the past. We might find evidence of such an encounter lurking, for instance, in the cosmic microwave background (CMB), the background radiation that bathes our universe. An aftereffect of the Big Bang, the CMB is remarkably homogeneous, uniform to 1 part in 100,000. From our point of view, the CMB is said to be isotropic, which means that no matter in what direction we look, the view is the same. A violent collision with another bubble, which deposits a huge amount of energy in one portion of the universe but not in another, would produce a localized departure from such uniformity called an *anisotropy*. This would impose a direction on our universe—an arrow that points straight toward the center of the other bubble, just before it slammed into us. Despite the hazards associated with our own universe's decompactifying, a collision with another universe inside another bubble would not necessarily be fatal. (Our bubble wall, believe it or not, would afford some protection.) Such a collision may, however, leave a discernible mark in the CMB that is not merely the result of random fluctuations.

That is just the sort of calling card that cosmologists look for, and a potential anisotropy, referred to as "the axis of evil" by its discoverers Joao Magueijo and Kate Land of Imperial College London, may have been uncovered within the CMB data. Magueijo and Land claim that hot spots and cold spots in the CMB appear to be aligned along a particular axis; if this observation is correct, it would suggest that the universe has a specific orientation, which would clash with hallowed cosmological principles avowing that all directions are equivalent. But for the moment, no one knows whether the putative axis is anything more than a statistical fluke.

If we could obtain firm evidence that another bubble crashed into us, what exactly would that prove? And would it have anything to do with string theory?

"If we didn't live in a bubble, there wouldn't be a collision, so we'd know, for starters, that we really are in a bubble," explains New York University physicist Matthew Kleban. Not only that, but we'd also know, by virtue of the collision, that there is at least one other bubble out there. "While that wouldn't prove that string theory is right, the theory makes a bunch of weird predictions, one being that we live in a bubble"—one of a vast number of such bubbles strewn across the string theory landscape. So at a minimum, says Kleban, "we'd be seeing something strange and unexpected that string theory also happens to predict."[1]

There's an important caveat, however, as Cornell's Henry Tye points out: Bubble collisions can also arise in quantum field theories that have nothing to do with strings. If traces of a collision were observed, Tye says, "I know of no good way of telling whether it came from string theory or quantum field theory."[2]

Then there's the question of whether we could ever see something like this, regardless of the source. The likelihood of such a detection, of course, depends on whether any other stray bubbles are within our path or "light cone." "It could go either way," says Ben Freivogel, a physicist at the University of California, Berkeley. "It's a question of probabilities, and our understanding is not good enough yet to determine those probabilities."[3] While no one can rate the exact odds of such a detection, most experts would probably rate it as very low.

If calculations eventually suggest that bubbles are not apt to be a fruitful avenue of investigation, many physicists still believe that cosmology offers the best chance of testing string theory, given that the almost-Planck-scale energies from which strings are thought to arise are so enormous, they could never be reproduced in the lab. Perhaps the best hope of ever seeing strings—presumed to be as small as 10^{-33} centimeters long—is if they were formed at the time of the Big Bang and have grown in step with the universe's expansion ever since. We're talking now about hypothetical entities called cosmic strings—an idea that originated before string theory took hold, but has gained renewed vigor through its association with that theory.

According to the traditional view (which is compatible with the string theory view), cosmic strings are slender, ultradense filaments formed during a "phase transition" within the first microsecond of cosmic history. Just as cracks inevitably appear in ice when water freezes, the universe in its earliest moments also went through phase transitions that were likely to have produced defects of various kinds. A phase transition would occur in different regions at the same time, and linear defects would form at junctures where these regions run into each other, leaving behind wispy filaments of unconverted material forever trapped in a primordial state.

Cosmic strings would emerge from this phase transition in a spaghetti-like tangle, with individual threads moving at speeds near the speed of light. They're either long and curvy, with a complex assortment of wiggles, or fragmented into smaller loops that resemble taut rubber bands. Far narrower than subatomic particles, cosmic strings are expected to be almost immeasurably thin, yet almost boundless in length, stretched by cosmic expansion to span the universe. These elongated filaments are physically characterized by their mass per unit length or tension, which provides a measure of their gravitational heft. Their linear density can reach incredibly high values—about 10^{22} grams per centimeter for strings formed at the so-called grand unified energy scale. "Even if we squeezed one billion neutron stars into the size of an electron, we would still hardly reach the matter-energy density characteristic of grand unified cosmic strings," says University of Buenos Aires astronomer Alejandro Gangui.[4]

These bizarre objects gained currency in the early 1980s among cosmologists who saw them as potential "seeds" for galaxy formation. However, a 1985 paper by Edward Witten argued that the presence of cosmic strings would create density inhomogeneities in the CMB far larger than observed, thus apparently ruling out their existence.[5]

Since then, cosmic strings have hit the comeback trail, owing much of their recent popularity to string theory, which has prompted many people to view these objects in a new light. Cosmic strings now seem to be common by-products of string-theory-based inflation models. The most recent versions of the theory show that so-called fundamental strings—the basic units of energy and matter in string theory—can reach astronomical sizes without suffering from the problems Witten identified in 1985. Tye and his colleagues explained how cosmic strings could be produced at the end of inflation and thus would not be diluted to oblivion during the brief period of runaway expansion, during which the universe doubled in size perhaps fifty to a hundred or more times. These strings,

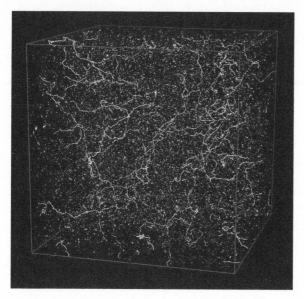

12.1—This image comes from a simulation that shows a network of cosmic strings when the universe was about ten thousand years old. (Courtesy of Bruce Allen, Carlos Martins, and Paul Shellard)

Tye demonstrated, would be less massive than the strings Witten and others contemplated in the 1980s, which means their influence on the universe would not be so pronounced as to have been already ruled out by observations. Meanwhile, Joe Polchinski of the University of California, Santa Barbara, showed how the newly conceived strings could be stable on cosmological timescales. The efforts of Tye, Polchinski, and others deftly addressed the objections Witten had raised two decades ago, leading to a resurgence of interest in cosmic strings.

Thanks to their postulated density, cosmic strings ought to exert a noticeable gravitational influence on their surroundings, and this ought to make them detectable. If a string ran between us and another galaxy, for example, light from that galaxy would go around the string symmetrically, producing two identical images close to each other in the sky. "Normally you'd expect three images, if lensing is due to a galaxy," explains Alexander Vilenkin, a cosmic string theorist at Tufts University.[6] Some light would pass straight through the lensing galaxy, and other rays would travel around on either side. But light can't go through a string, because the string's diameter is much smaller than the light's wavelength; thus strings, unlike galaxies, would produce just two images rather than three.

Hopes were stirred in 2003, when a Russian-Italian group led by Mikhail Sazhin of Moscow State University announced that they had taken double images of a galaxy in the Corvus constellation. The images were at the same distance or redshift and were spectrally identical at a 99.96 percent confidence level. Either two extremely similar galaxies were, by chance, closely aligned or it was the first case of lensing by a cosmic string. In 2008, a more detailed analysis drawing on Hubble Space Telescope data—which yielded sharper pictures than the ground-based telescope Sazhin and colleagues used—showed that what first appeared to be a lensed galaxy was, in fact, two different galaxies, thereby ruling out the cosmic string explanation.

A related approach called microlensing is based on the premise that a loop formed from the breakup of a cosmic string could lens individual stars in potentially detectable ways. Although it might be impossible to actually see two identical stars, astronomers might instead detect a star that periodically doubles in brightness, while remaining constant in color and temperature, which could signal the presence of a cosmic string loop oscillating in the foreground. Depending on where it is, how fast it's moving, its tension, and its precise oscillation mode, the loop would produce a double image at some times and not others—with stellar brightness changes occurring over the course of seconds, hours, or months. Such a signature might be secured by the Gaia Satellite, scheduled for launch in 2012, which is due to survey a billion stars in the Milky Way and its immediate vicinity. The Large Synoptic Survey Telescope (LSST) now under construction in Chile might also spot a similar signature. "Direct astronomical detection of superstring relics would constitute an experimental verification of some of the basic ingredients of string theory," claims Cornell astronomer David Chernoff, a member of the LSST Science Collaboration.[7]

Meanwhile, researchers continue to explore other means of detecting cosmic strings. Theorists believe, for instance, that cosmic strings could form cusps and kinks, in addition to loops, emitting gravitational waves as these irregularities straighten out or decay. The waves so produced might be at just the right frequency to be detected by the Laser Interferometer Space Antenna (LISA), the proposed orbital observatory now being developed for NASA. According to current plans, LISA will consist of three identical spacecraft separated from each other by 5 million kilometers in the configuration of an equilateral triangle. By closely monitoring changes in the distance between these spacecraft, LISA could sense the passage of gravitational waves. Vilenkin and Thibault Damour (of IHES in France) proposed that precise measurements of these waves could reveal the presence of cosmic strings. "Gravitational waves produced from a cosmic string source would have a specific waveform that would look very different from that produced by black hole collisions or other sources," Tye explains. "The signal would start at zero and then rapidly increase and decrease. The way it increases and decreases, which is what we mean by the 'waveform,' would be peculiar to cosmic strings alone."[8]

Another approach involves looking for distortions in the CMB produced by strings. A 2008 study along these lines, headed by Mark Hindmarsh of Sussex University, found that cosmic strings might account for the clumpy distribution of matter observed by the Wilkinson Microwave Anisotropy Probe (WMAP). This phenomenon of clumpiness is known as non-Gaussianity. While the data, according to Hindmarsh's team, suggests the presence of cosmic strings, many are skeptical, regarding the apparent correlation as a mere coincidence. This matter should be clarified as more sensitive CMB measurements become available. Investigating the potential non-Gaussianity of the matter distribution in the universe is, in fact, one of the goals of the Planck mission, launched by the European Space Agency in 2009.

"Cosmic strings may or may not exist," says Vilenkin. But the search for these entities is under way, and assuming they do exist, "their detection is very feasible within the next few decades."[9]

In some models of string inflation, the exponential growth of space occurs in a region of the Calabi-Yau manifold known as a *warped throat*. In the abstract realm of string cosmology, warped throats are considered both fundamental and generic features "that arise naturally from six-dimensional Calabi-Yau space," according to Princeton's Igor Klebanov.[10] While that doesn't guarantee that inflation takes place in such regions, warped throats offer a geometrical framework that can nevertheless help us understand inflation and other mysteries. For theorists, it's a setting rich in possibilities.

The throat, the most common defect seen in a Calabi-Yau, is a cone-shaped bump, or conifold, that juts out from the surface. The rest of space—often described as the *bulk*—can be thought of as a large scoop of ice cream sitting atop a slender and infinitely pointy cone, suggests Cornell physicist Liam McAllister. This throat becomes even more distended when the fields posited by string theory (technically called fluxes) are turned on. Cornell astronomer Rachel Bean argues that because a given Calabi-Yau space is likely to have more than one warped throat, a rubber glove makes a better analogy. "Our three-dimensional universe is like a dot moving down the finger of a glove," she explains. Inflation ends when the brane, or "dot," reaches the tip of the glove, where an antibrane, or a stack of antibranes, sits. Because the motion of the brane is constrained by the shape of the finger or throat, she says, "the specific features of inflation stem from the geometry of that throat."[11]

Regardless of the chosen analogy, different warped throat models lead to different predictions about the cosmic string *spectrum*—the full range of strings, of different tensions, expected to arise under inflation—which, in turn, could give us an indication of what Calabi-Yau geometry underlies our universe. "If we're lucky enough to see that [entire spectrum of cosmic strings]," Polchinski says, "we may be able to say that some specific picture of a warped throat looks right, whereas another does not."[12]

Even if we're not so lucky as to see a cosmic string—or, better yet, to see a whole network of them—we might still constrain the shape of Calabi-Yau space through cosmological observations that rule out some models of cosmic inflation while allowing others. At least that's the strategy being pursued by University of Wisconsin physicist Gary Shiu and his colleagues. "How are the extra dimensions of string theory curled up?" asks Shiu. "We argue that precise measurements of the cosmic microwave background (CMB) can give us clues."[13]

As Shiu suggests, the latest string-theory-based models of cosmic inflation are nearing the point where they can make detailed predictions about our universe. These predictions, which vary depending on the specific Calabi-Yau geometry wherein inflation is assumed to originate, can now be tested against CMB data.

The basic premise is that inflation is driven by the motions of branes. And the thing we call our universe actually sits on a brane (of three dimensions) as well. In this scenario, a brane and its counterpart, an antibrane, slowly move toward each other in the extra dimensions. (In a more refined version of the story, the brane motions take place in a warped throat region within those extra dimensions.) Because of the mutual attraction, the separation of these branes represents a source of potential energy that drives inflation. The fleeting process, during which our four-dimensional spacetime expands exponentially, continues until the branes smack into each other and annihilate, unleashing the heat of the Big Bang and making an indelible imprint on the CMB. "The fact that the branes have been moving allows us to learn much more about that space than if they were just sitting in the corner," Tye says. "Just like at a cocktail party: You don't pick up much if you stay in one corner. But if you move around, you're bound to learn more."[14]

Researchers like Tye are encouraged by the fact that data is getting precise enough that we can say that one Calabi-Yau space is consistent with experiment, whereas another space is not. In this way, cosmological measurements are starting to impose constraints on the kind of Calabi-Yau space we might live in. "You can take inflation models and divide them in half—those that are favored by observations and those that are not," says Perimeter Institute physicist Cliff Burgess. "The fact that we can now distinguish between inflation models means we can also distinguish between the geometric constructions that give rise to those models."[15]

Shiu and his former graduate student Bret Underwood (now at McGill University) have taken some additional steps in this direction. In a 2007 paper in *Physical Review Letters*, Shiu and Underwood showed that two different geometries for the hidden six dimensions—variations of familiar Calabi-Yau conifolds with warped throats—would lead to different patterns in the distribution of cosmic radiation. For their comparison, Shiu and Underwood picked two throat models—Klebanov-Strassler and Randall-Sundrum, whose geometries are reasonably well understood—and then looked at how inflation under these distinct conditions would affect the CMB. In particular, they focused on a standard CMB measurement, temperature fluctuations in the early universe. These fluctuations should appear roughly the same on small and large scales. The rate at which the fluctuations change as you go from small to large scales is called the spectral index. Shiu and Underwood found a 1 percent difference in the spectral index between the two scenarios, showing that the choice of geometry has a measurable effect.

Although that might seem inconsequential, a 1 percent difference can be significant in cosmology. And the recently launched Planck observatory should be able to make spectral index measurements to at least that level of sensitivity. In other words, the Klebanov-Strassler throat geometry might be allowed by Planck data, while the Randall-Sundrum throat might not, or vice versa. "Away from the tip of the throat, the two geometries look almost identical, and people used to think they could be used interchangeably," notes Underwood. "Shiu and I showed that the details do matter."[16]

However, going from the spectral index, which is a single number, to the geometry of the extra dimensions is a huge leap. This is the so-called inverse problem: If we see enough data in the CMB, can we determine what Calabi-Yau it is? Burgess doesn't think this will be possible in "our lifetime," or at least in the dozen or so years he has left before retirement. McAllister is also skeptical. "We'll be lucky in the next decade just to be able to say inflation did or did not occur," he says. "I don't think we'll get enough experimental data to flesh out the full shape of the Calabi-Yau space, though we might be able to learn what kind of throat it has or what sort of branes it contains."[17]

Shiu is more optimistic. While the inverse problem is much harder, he acknowledges, we still ought to give it our best shot. "If you can only measure the spectral index, it's hard to say something definitive about the geometry. But you'd get much more information if you could measure something like non-Gaussian features of the CMB." A clear indication of non-Gaussianity, he says, would impose "much more constraints on the underlying geometry. Instead of being one number like the spectral index, it's a whole function—a whole bunch of numbers that are all related to each other." A large degree of non-Gaussianity, Shiu adds, could point to a specific version of brane inflation—such as the Dirac-Born-Infeld (or DBI) model—occurring within a well-prescribed throat geometry. "Depending on the precision of the experiment, such a finding could, in fact, be definitive."[18]

Columbia physicist Sarah Shandera notes that a string-theory-motivated inflation model like DBI may, ironically, prove important even if we find out that string theory is not the ultimate description of nature. "That's because it predicts a kind of non-Gaussianity that cosmologists hadn't thought of before," Shandera says.[19] And in any experimental endeavor, knowing what questions to ask, how to frame them, and what to look for is a big part of the game.

Other clues for string inflation could come from gravitational waves emitted during the violent phase transition that spawned inflation. The largest of these primordial spacetime ripples cannot be observed directly, because their wavelengths would now span the entire visible universe. But they would leave a mark in the microwave background. While this signal would be hard to extract from

CMB temperature maps, say theorists, gravitational waves would create a distinctive pattern in maps of the polarization of the CMB's photons.

In some string inflation models, the gravitational wave imprint would be detectable; in others, it would not. Roughly speaking, if the brane moves a small distance on the Calabi-Yau during inflation, there'll be no appreciable gravitational wave signal. But if the brane travels a long way through the extra dimensions, says Tye, "tracing out small circles like the grooves on a record, the gravitational signal could be big." Getting the brane to move in this tightly circumscribed manner, he adds, "takes a special type of compactification and a special type of Calabi-Yau. If you see it, you know it must be that kind of manifold." The compactifications we're talking about here are ones in which the moduli are stabilized, implying the presence of warped geometry and warped throats in particular.[20]

Getting a handle on the shape of Calabi-Yau space, including its throaty appendages, will require precise measurements of the spectral index and, hopefully, sightings of non-Gaussianity, gravitational waves, and cosmic strings as well. Patience is also in order, suggests Shiu. "Although we now have confidence in the Standard Model of physics, that model did not materialize overnight. It came from a whole sequence of experiments over the course of many years. In this case, we'll need to bring a lot of measurements together to get an idea of whether extra dimensions exist or whether string theory seems to be behind it all."[21]

The overall goal here is not just to probe the geometry of the hidden dimensions. It's also to test string theory as a whole. McAllister, among others, believes that this approach may offer our best shot at an experimental test of the theory. "It's possible that string theory will predict a finite class of models, none of which are consistent with the observed properties of the early universe, in which case we could say the theory is excluded by observation. Some models have already been excluded, which is exciting because it means the cutting-edge data really does make a difference." While that kind of statement is not at all novel for physics, it is novel for string theory, which has yet to be experimentally verified. At the moment, he adds, warped throat inflation is one of the best models we've produced so far, "but in reality, inflation may not occur in warped throats even though the picture looks quite compelling."[22]

In the end, Bean agrees, "models of inflation in warped throats may not be the answer. But these models are based on geometries coming out of string theory for which we can make detailed predictions that we can then go out and test. In other words, it's a way of making a start."[23]

The good news is that there is more than one way of making a start. While some investigators are scouring the night (or day) sky for signs of extra dimensions, other eyes are trained on the Large Hadron Collider (LHC). Finding hints of extra dimensions may not be the top priority of the LHC, but it still ranks high on the list.

The most logical starting point for string theorists is to look for the supersymmetric partners of the particles we know. Supersymmetry is of interest to many physicists, string theory aside: The lowest-mass supersymmetric partner—which could be the neutralino, gravitino, or sneutrino, among others—is extremely important in cosmology because it's considered a leading candidate for dark matter. The presumed reason we haven't seen these particles yet—which is why they have remained invisible and hence "dark"—is that they are more massive than ordinary particles. Particle colliders up until now have not been powerful enough to produce these heavier "superpartners," whereas the LHC is on the cusp of being able to do so.

In the string-theory-inspired models of Harvard's Cumrun Vafa and Jonathan Heckman of the Institute for Advanced Study, the gravitino—which is the hypothetical superpartner of the graviton (the particle responsible for gravity)— is the lightest partner. Unlike heavier superpartners, the gravitino would be completely stable, because there is nothing lighter that it could decay into. The gravitino, in their model, accounts for the bulk of the universe's dark matter. Although it would be too weakly interacting to be observable at the LHC, Vafa and Heckman believe that another theorized supersymmetric particle—the *stau*, the superpartner of the so-called tau lepton—would be stable from anywhere from a second to an hour, more than long enough to leave a recognizable track in LHC's detectors.

Finding such particles would corroborate an important aspect of string theory. Calabi-Yau manifolds, as we've seen, were handpicked by string theorists as a suitable geometry for the extra dimensions partly because supersymmetry

12.2—Experiments at the Large Hadron Collider (LHC) at the CERN lab in Geneva could find hints of extra dimensions or the existence of supersymmetric particles. Apparatus from the LHC's ATLAS experiment is shown here. (Courtesy of CERN)

is automatically built into their internal structure. Discovering signs of supersymmetry at LHC would thus be encouraging news, to say the least, for string theory and the whole Calabi-Yau picture. For one thing, the attributes of the supersymmetric particles could tell us about the hidden dimensions themselves, explains Burt Ovrut, "because how you compactify the Calabi-Yau manifold affects the kind of supersymmetry you get and the degree of supersymmetry you get. You can find compactifications that preserve supersymmetry or break it completely."[24]

Confirmation of supersymmetry would not confirm string theory per se, but it would at least point in the same direction, showing that part of the story that string theory tells is correct. Not observing supersymmetric particles, on the other hand, would not bury string theory, either. It could mean that we've miscalculated and the particles are just beyond the reach of the LHC. (Vafa and Heckman, for instance, allow for the possibility that the LHC might generate a semistable and electrically neutral particle instead of the stau, which could not be seen directly.) If it turns out that the superpartners are slightly more massive than can be produced at this collider, it would take still higher energies to reveal them—and a long wait for the new machine that will eventually replace it.

Although it's a long shot, the LHC might turn up more direct, and less ambiguous, evidence of the extra dimensions predicted by string theory. In experiments already planned at the facility, researchers will look for particles bearing signs of the extra dimensions from where they came—so-called Kaluza-Klein particles. The idea here is that vibrations in higher dimensions would manifest themselves as particles in the four-dimensional realm we inhabit. We might either see remnants of the decay of these Kaluza-Klein particles or perhaps even hints of such particles (along with their energy) disappearing from our world and then crossing over to the higher-dimensional realm.

Unseen motion in the extra dimensions would confer momentum and kinetic energy to a particle, so Kaluza-Klein particles are expected to be heavier than their slower, four-dimensional counterparts. Kaluza-Klein gravitons are an example. They'd look like ordinary gravitons, which are the particles that transmit the gravitational force, only they'd be heavier by virtue of the extra momentum they carry. One way to pick out such gravitons amid the vast sea of particles produced at LHC is to look not only at the particle's mass but also at its spin. Fermions, such as electrons, have a certain amount of angular momentum that we classify as spin-½. Bosons, like photons and gluons, have somewhat more angular momentum and are classified as spin-1. Any particles with spin-2 detected at LHC are likely to be Kaluza-Klein gravitons.

Such a detection would be momentous indeed, for physicists not only would have caught the first glimpse of a long-sought particle, but also would have obtained strong evidence of extra dimensions themselves. Showing the existence of at least one extra dimension would be a breathtaking find in itself, but Shiu and his colleagues would like to go even further, gaining hints about the geometry of that extra-dimensional space. In a 2008 paper Shiu wrote with Underwood, Devin Walker of the University of California, Berkeley, and Kathryn Zurek of the University of Wisconsin, the team found that small changes in the shape of the extra dimensions would cause big changes—on the order of 50 to 100 percent—in both the mass and the interactivity of Kaluza-Klein gravitons. "When we changed the geometry a little bit, the numbers changed dramatically," Underwood notes.[25]

Though it's a far cry from being able to say anything conclusive about the shape of inner space or to specify the exact Calabi-Yau geometry, the analysis by Shiu et al. offers some hope of using experiments to "reduce the class of allowed shapes to a smaller range of possibilities. The power lies in cross-correlating between different types of experiments in both cosmology and high-energy physics," Shiu says.[26]

The mass of the particles, as determined at LHC, would also provide hints about the size of the extra dimensions. For particles that venture into the higher-dimensional domain, the smaller those dimensions are, the heavier the particles will be. As for why that's the case, you could ask how much energy it takes to stroll down a short hallway. Probably not much, you figure. But what if the hallway weren't short but was instead very narrow? Getting through that tunnel will involve a struggle every inch of the way—accompanied, no doubt, by a string of curses and dietary vows—while requiring a bigger expenditure of energy. That's roughly what's going on here, but in more technical terms, it really comes down to the Heisenberg uncertainty principle, which states that the momentum of a particle is inversely proportional to the accuracy of its position measurement. Phrased another way, if a wave or particle is confined to a tiny, tiny space, where its position is thus highly constrained, it will have tremendous momentum and a correspondingly high mass. Conversely, if the extra dimensions are large, the wave or particle will have more room to move in and correspondingly less momentum, and will therefore be lighter.

There is a catch, however: The LHC will only detect things like Kaluza-Klein gravitons if these particles are far, far lighter than we've traditionally expected, which is another way of saying that either the extra dimensions must be extremely warped or they must be much larger than the Planck-scale range traditionally assumed in string theory. In the Randall-Sundrum picture of warping, for instance, the extra-dimensional space is bounded by two branes, with the spacetime between them everywhere curved. On one brane, which exists at a high-energy scale, gravity is strong; on the other brane—the lower-energy one on which we live—gravity is feeble. An effect of that arrangement is that mass and energy values change drastically, depending on one's position with respect to the two branes. This means that the mass of fundamental particles, which we ordinarily held to be close to the Planck scale (on the order of 10^{28} electron volts), would instead be "rescaled" to something closer to 10^{12} electron volts, or 1 tera–electron volt (1 TeV), which might place them within the range of the LHC. The size of the extra dimensions in this picture could be as small as in conventional string theory models (though that is not required), whereas particles themselves would appear to be much lighter (and therefore lower-energy) than is typically assumed.

Another novel approach under consideration today was first proposed in 1998 by the physicists Nima Arkani-Hamed, Savas Dimopoulos, and Gia Dvali when they were all at Stanford. Challenging Oskar Klein's assertion that we can't see any extra dimensions because they are so small, Arkani-Hamed, Dimopoulos, and Dvali—a trio often abbreviated as ADD—claimed that the extra dimensions could be much larger than the Planck scale, at least 10^{-12} cm and possibly as big as 10^{-1} cm (a millimeter). This would be possible, they said, if our universe is stuck on a three-dimensional brane (with an additional dimension thrown in for time) and if that three-dimensional world is all we can see.

This might seem like an odd argument to make—after all, the idea that the extra dimensions are minuscule is the premise on which the vast majority of string theory models have been built. But it turns out that the size of Calabi-Yau space, which is often considered a given, "is still an open question," according to Polchinski. "For mathematicians, the size of a space is about the least interesting thing. If you double the size of something, mathematically that's trivial. But to a physicist, the size is anything but trivial because it tells us how much energy you need to see it."[27]

The ADD scenario does more than just expand the size of the extra dimensions; it lowers the energy scale at which gravity and the other forces become unified, and in the process lowers the Planck scale as well. If Arkani-Hamed and his colleagues are right, energy generated through particle collisions at the LHC may seep out into higher dimensions, showing up as an ostensible violation of energy conservation laws. And in their picture, even strings themselves, the basic unit of string theory, might become big enough to see—something that had never been deemed possible before. The ADD team was motivated, in part, to account for the apparent weakness of gravity in comparison with the other forces, given that a compelling explanation for that disparity had not yet been put forth. ADD theory suggests a novel answer: Gravity is not weaker than the other forces, but only appears to be weaker because, unlike the other forces, it "leaks" into the other dimensions so that we feel just a tiny fraction of its true strength. The same sort of thing happens when balls smack into each other on a pool table; some of the kinetic energy, which had been bound up in the moving balls and thus confined to the table's two-dimensional surface, escapes in the form of sound waves into the third dimension.

The details of how this might work suggest possible observational strategies: Gravity as we know it in four-dimensional spacetime obeys an inverse square law. The gravitational influence of a body drops off with the square of the distance from it. But if we added another dimension, gravity would drop off as the cube of distance. With ten dimensions, as is posited in string theory, gravity would drop off according to the eighth power of distance. In other words, the more extra dimensions, the weaker gravity would appear as measured from our four-dimensional perspective. (The electrostatic force, similarly, is inversely proportional to the square of the distance between two point charges in four-dimensional spacetime and inversely proportional to the eighth power of the distance in ten-dimensional spacetime.) In thinking about gravity over big distances, as we do in astronomy and cosmology, the inverse square law works perfectly well because those interactions take place in the space of three giant dimensions plus time. We wouldn't notice a gravitational pull in a strange new direction (corresponding to a hidden internal dimension) until we got down to a scale small enough to move around in those dimensions. And since we're physically barred from doing that, our best and perhaps only hope is to look for hints of the extra dimensions in the form of deviations from the inverse square law. That is precisely the effect that physicists at the University of Washington, the University of Colorado, Stanford University, and elsewhere are looking for through short-distance gravity measurements.

While these researchers have different experimental apparatus at their disposal, their goals are the same nevertheless: to measure the strength of gravity on a small scale at precisions never before dreamed of. Eric Adelberger's team at the University of Washington, for instance, performs "torsion balance" experiments that are similar in spirit to those conducted in 1798 by Henry Cavendish. The basic approach is to infer the strength of gravity by measuring

12.3—Minute rotations induced by gravitational attraction are measured at short distances, and with great precision, by the Mark VI pendulum designed and operated by the Eöt-Wash research group at the University of Washington. If gravity were observed to behave differently at close range and to deviate from the inverse square law well established in classical physics, it could signal the presence of the extra dimensions predicted by string theory. (University of Washington/Mary Levin)

the torque on a suspended pendulum. Adelberger's group employs a small metal pendulum dangling above a pair of metal disks, which exert a gravitational pull on the pendulum. The attractive forces from the two disks are balanced in such a way that if Newton's inverse square law holds, the pendulum will not twist at all.

In the experiments performed to date, the pendulum has shown no sign of twisting, as measured to an accuracy of a tenth of a millionth of a degree. By placing the pendulum ever closer to the disks, the researchers have ruled out the existence of dimensions larger than about 40 microns (or micrometers) in radius. In future experiments, Adelberger aims to test gravity at an even smaller scale, securing measurements down to a dimension size of about 20 microns. But that may be the limit, he says. In order to look at ever smaller scales, a different technological approach will probably be needed to test the large-dimension hypothesis.

Adelberger considers the idea of large extra dimensions revolutionary, but says that this doesn't make it right.[28] We not only need new tactics for probing the large-dimension question, but also need new tactics for dealing with more general questions about the existence of extra dimensions and the veracity of string theory.

So that's where things stand today, with various leads being chased down—only a handful of which have been discussed here—and no sensational results to speak of yet. Looking ahead, Shamit Kachru, for one, is hopeful that the range of experiments under way, planned, or yet to be devised will afford many opportunities to see new things. Nevertheless, he admits that a less rosy scenario is always possible, in the event that we live in a frustrating universe that offers little, if anything, in the way of empirical clues. "If we see nothing in cosmology, nothing in accelerator experiments, and nothing in lab experiments, then we're basically stuck," says Kachru. Although he considers such an outcome unlikely, he says this kind of situation is in no way peculiar to string theory or cosmology, as the dearth of data would affect other branches of science in the same way.[29]

What we do next, after coming up empty-handed in every avenue we set out, will be an even bigger test than looking for gravitational waves in the CMB or infinitesimal twists in torsion-balance measurements. For that would be a test of our intellectual mettle. When that happens, when every idea goes south and every road leads to a dead end, you either give up or try to think of other questions you can ask—questions for which there might be some answers.

Edward Witten, who, if anything, tends to be conservative in his pronouncements, is optimistic in the long run, feeling that string theory is too good not to be true. Though in the short run, he admits, it's going to be difficult to know exactly where we stand. "To test string theory, we will probably have to be lucky," he says. That might sound like a slender thread upon which to pin one's dreams for a theory of everything—almost as slender as a cosmic string itself. But fortunately, says Witten, "in physics, there are many ways of being lucky."[30]

I have no quarrel with that statement and, more often than not, tend to agree with Witten, as I've generally found this to be a wise policy. But if the physicists find their luck running dry, they might want to turn to their mathematical colleagues, who have enjoyed their fair share of that commodity as well.

Thirteen

TRUTH, BEAUTY, AND MATHEMATICS

How far can investigators, trying to survey the universe's hidden dimensions, proceed in the absence of any physical proof? The same question, of course, can also be put to string theorists trying to concoct a complete theory of nature without the benefit of empirical feedback. It's like exploring a vast, dark cavern—whose contours are largely unknown—with just a flickering candle, if that, to illuminate the path. Although proceeding under such circumstances may seem like sheer folly to some, the situation is by no means unprecedented in the history of science. In the early stages of theory building, periods of fumbling in the dark are rather common, especially when it comes to developing, and pushing through, ideas of great scope. At various junctures like this, when there is no experimental data to lean on, mathematical beauty may be all we have to guide us.

The British physicist Paul Dirac "cited mathematical beauty as the ultimate criterion for selecting the way forward in theoretical physics," writes the physicist Peter Goddard.[1] Sometimes this approach has paid off handsomely, as when Dirac predicted the existence of the positron (like an electron with positive charge), solely because mathematical reasoning led him to believe that such particles must exist. Sure enough, the positron was discovered a few years later, thereby affirming his faith in mathematics.

Indeed, one of the things we've found over and over again is that the ideas that hold up mathematically, and meet the criteria of simplicity and beauty, tend to be the ones that we eventually see realized in nature. Why that is the case is surely baffling. The physicist Eugene Wigner, for one, was perplexed by "the unreasonable effectiveness of mathematics in the natural sciences"—the mystery being how purely mathematical constructs, with no apparent connection to the natural world, can nevertheless describe that world so accurately.[2]

The physicist Chen Ning Yang was similarly astonished to find that the Yang-Mills equations, which describe the forces between particles, are rooted in gauge theories in physics that bear striking resemblances to ideas in bundle theory, which mathematicians began developing three decades earlier, as Yang put it, "without reference to the physical world." When he asked the geometer S. S. Chern how it was possible that "mathematicians dreamed up these concepts out of nowhere," Chern protested, "No, no. These concepts were not dreamed up. They were natural and real."[3]

There certainly is no shortage of abstract ideas that came to mathematicians seemingly out of thin air and that were later found to describe natural phenomena. Not all of these, by the way, were the products of modern mathematics. Conic sections, which are the curves—the circle, ellipse, parabola, and hyperbola—made by slicing a cone with a plane, were reportedly discovered by the Greek geometer Menaechmus around 300 B.C. and systematically explored a century later by Apollonius of Perga in his treatise *Conics*. These forms, however, did not find a major scientific application until the early 1600s, when Kepler discovered the elliptical orbits of planets in our solar system.

Similarly, "buckyballs," or buckminsterfullerenes—a novel form of carbon consisting of sixty carbon atoms arranged on a spherelike structure composed of pentagonal and hexagonal faces—were discovered by chemists in the 1980s. Yet the shape of these molecules had been described by Archimedes some two thousand years earlier.[4] Knot theory, a branch of pure mathematics that has evolved since the late 1800s, found applications more than a century later in string theory and in studies of DNA.

It's hard to say why ideas from mathematics keep popping up in nature. Richard Feynman found it equally hard to explain why "every one of our physical laws is a purely mathematical statement." The key to these puzzles, he felt,

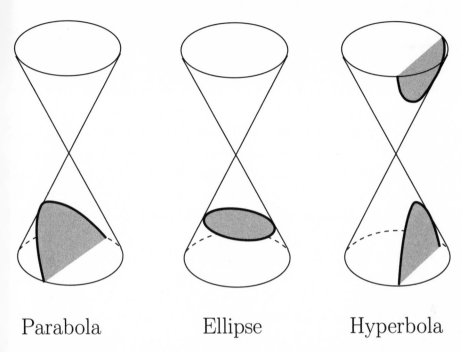

Parabola   Ellipse   Hyperbola

13.1—Conic sections are the three basic curves you get when you intersect a plane with a cone (or, actually, a kind of double cone attached at the pointy ends). Those curves are the parabola, ellipse (of which a circle is a special case), and hyperbola.

13.2—Whereas the regular icosahedron consists of twenty triangular faces, the *truncated icosahedron* (shown here) consists instead of twenty hexagonal faces and twelve pentagonal faces (in which no two pentagons have an edge in common). Unlike the regular icosahedron, which is classified as a Platonic solid, the truncated icosahedron is an Archimedean solid, named after the Greek mathematician who explored these shapes more than two thousand years ago. This shape resembles a soccer ball and one version of a so-called buckyball—the molecular structure of a form of carbon, consisting of sixty atoms, that was discovered in 1985 by the chemists Harold Kroto and Richard Smalley. The term *buckyball* is short for buckminsterfullerene—the name given to this class of molecules in honor of R. Buckminster Fuller, the inventor of the similarly shaped geodesic dome.

may lie somewhere in the connection between math, nature, and beauty. "To those who do not know mathematics," Feynman said, "it is difficult to get across a real feeling as to the beauty, the deepest beauty of nature."[5]

Of course, if beauty is to guide us in any way—even temporarily, until more tangible clues come in—that leaves the problem of trying to define it, a task that some feel might be best left to poets. While mathematicians and physicists may view this concept somewhat differently, in both disciplines the ideas we call beautiful tend to be those that can be stated clearly and concisely, yet have great power and broad reach. Even so, for a notion as subjective as beauty, personal taste inevitably comes into play as well. I'm reminded of a toast made at the wedding of a longtime bachelor who settled down relatively late in life after many years of playing the field. What kind of woman, people had wondered, would it take to get this guy to tie the knot? The bachelor himself was curious about that, too. "You'll know it when you see it," a friend repeatedly advised him prior to his finding "the one."

I know what he means. I felt that way when I met my wife in the Berkeley math library many decades ago, although I'd be hard-pressed to capture the exact feeling in words. And, with no offense intended to my wife, I felt something similar—a vague, tingly sense of euphoria—after proving the Calabi conjecture in the mid-1970s. With the proof complete after months of exertion and exhaustion—stretched out, of course, over the span of years—I was finally able to relax and admire the complex, multidimensional spaces I had found. You might say it was love at first sight, although after all that labor, I felt that I already knew these objects well, even upon first "viewing" them. Perhaps my confidence was misplaced, but I sensed then (as I still do) that somehow these spaces would play a role, and possibly an important one, in the physical world. Now it's up to string theorists—or perhaps researchers from some unrelated branch of science—to find out if that hunch is correct.

It ought to be reassuring to string theorists, mathematician Michael Atiyah argues, "that what they're playing with, even if we can't measure it experimentally, appears to have a very rich . . . mathematical structure, which not only is consistent but actually opens up new doors and gives new results and so on. They're onto something, obviously. Whether that something is what God's created for the universe remains to be seen. But if He didn't do it for the universe, it must have been for something."[6]

I don't know what that something is, but it strikes me as being far too much to be nothing. Yet I'm also fully aware—as is Atiyah—of the risk of being lulled by elegance onto shaky ground. "Beauty can be a slippery thing," cautions Jim Holt, a string theory skeptic writing in the *New Yorker*.[7] Or as Atiyah puts it, "the mathematical take-over of physics has its dangers, as it could tempt us into realms of thought which embody mathematical perfection but might be far re- moved, or even alien to, physical reality."[8]

There's no doubt that a blind adherence to mathematical beauty could lead us astray, and even when it does point us in the right direction, beauty alone can never carry us all the way to the goal line. Eventually, it has to be backed up by something else, something more substantial, or our theories will never go beyond the level of informed speculation, no matter how well motivated and plausible that speculation may be.

"Beauty cannot guarantee truth," asserted the physicist Robert Mills, the sec- ond half of the Yang-Mills duo. "Nor is there any logical reason why the truth must be beautiful, but our experience has repeatedly led us to expect beauty at the heart of things and to use this expectation as a guide in seeking deeper the- oretical understanding of the fundamental structures of nature." Conversely, Mills added, "if a proposed theory is inelegant, we have learned to be dubious."[9]

So where does that leave string theory and the mathematics behind it? Cornell physicist Henry Tye believes "string theory is too beautiful, rich, creative, and subtle not to be used by nature. That would be such a waste."[10] That in itself is not enough to make string theory right, and critical treatments of the subject, such as *The Trouble with Physics* and *Not Even Wrong*, have raised doubts in the public consciousness at a time when the theory is in a kind of doldrums, not having seen a major breakthrough in years. Even an enthusiast like Brian Greene, author of *The Elegant Universe*, acknowledges that a physical theory cannot be judged solely on the basis of elegance: "You judge it on the basis of whether it makes predictions that are going to be confirmed by experiment."[11]

While writing this book, I've had occasion to discuss its contents with a num- ber of laypeople—exactly the educated types who I hope might eventually want to read this sort of thing. When they heard it related to the mathematical foun- dations of string theory, the response often went like this: "Wait a minute. Isn't string theory supposed to be wrong?" Their questions suggested that writing a book about the mathematics of string theory was like writing a book about the fantastic blueprints that went into the making of the *Titanic*. A mathematics colleague of mine, who should have known better, even went on record as saying that because "the jury is still out on string theory," the jury is still out on the mathematics associated with string theory as well.

Such a claim implies a fundamental misconception about the nature of mathematics and its relation to the empirical sciences. Whereas the final proof in physics is an experiment, that is not the case in math. You can have a billion bits of evidence that something is true, yet on the billionth-and-first time, it fails. Until something is proven by pure logic, it remains a conjecture.

In physics and other empirical sciences, something thought to be true is always subject to revision. Newton's theory of gravitation held up well for more than two centuries until its limitations were finally appreciated and it was replaced by Einstein's version, whose own limitations may someday be addressed by a theory of quantum gravity like string theory. Nevertheless, the math that goes into Newtonian mechanics is 100 percent correct, and that will never change.

In fact, to formulate his theory of gravitation, Newton had to invent (or co-invent) calculus; when Newton's theory of gravity broke down at the limits that general relativity was designed to address, we didn't throw out calculus. We kept the math—which is not only sound but indispensable—realizing that Newtonian mechanics is a perfectly good tool in most situations, though it is not applicable in the most extreme cases.

Now for something a little more contemporary—and closer to my heart. Thirty-some years ago, I proved the existence of spaces we now call Calabi-Yau. Their existence, moreover, is not at all contingent on whether string theory turns out to be "the theory of nature." Admittedly, weak points can be uncovered in a proof, after the fact, and topple the argument like a house of cards. But in the case of the Calabi conjecture, the proof has been gone over so many times that the chances of finding a mistake are essentially nil. Not only are Calabi-Yau spaces here to stay, but the techniques I used to attack the problem have been applied with great success to many other mathematical problems, including those in algebraic geometry that have no obvious ties to the original conjecture.

Indeed, the utility of Calabi-Yau spaces in physics is, in some sense, irrelevant to the question of whether the mathematics is important. At the risk of sounding immodest, I might add that I was awarded the Fields Medal in 1982—one of the highest honors in mathematics—largely because of my proof of the Calabi conjecture. The award, you'll notice, was granted a couple of years before physicists knew about Calabi-Yau manifolds and before string theory itself was really on the map.

As for string theory, the mathematics underlying it or inspired by it can be absolutely correct, no matter what the jury finally decides regarding the theory itself. I'll go farther: If the mathematics associated with string theory is solid and has been rigorously proven, then it will stand regardless of whether we live in a ten-dimensional universe made of strings or branes.

So, what, if anything, can this tell us about the physics? As I said before, because I am a mathematician, it's not for me to judge the validity of string theory, but I will offer some opinions and observations. Granted, string theory remains not only unproven but untested. Nevertheless, one major tool for checking the work of physicists has been mathematical consistency, and so far, the theory has passed those exams with flying colors. Consistency in this case means there's no contradiction. It means that if what you put into the string theory equations is correct, what you get out of the equations is correct, too. It means that when you do a calculation, the numbers don't blow up and go to infinity. The functions remain well-spoken rather than spouting off gibberish. While that's not nearly enough to satisfy the strictures of science, it is an important starting point. And to me it suggests there must be some truth to this idea, even if nature doesn't follow the same script.

Edward Witten seems to share this opinion. Mathematical consistency has been, he claims, "one of the most reliable guides to physicists in the last century."[12]

Given how hard it is to devise an experiment that could access Planck-scale physics and how expensive such an experiment might be if we ever manage to come up with one, all we'll have in many cases will be these consistency checks, which, nevertheless, "can be very powerful," according to Berkeley mathematician Nicolai Reshetikhin. "That's why the high end of theoretical physics is becoming more and more mathematical. If your ideas are not mathematically consistent, you can rule them out right away."[13]

Beyond mathematical consistency, string theory also seems to be consistent with everything we've learned about particle physics, while offering new perspectives for grappling with issues of space and time—gravity, black holes, and various other conundrums. Not only does string theory agree with the established, well-tested physics of quantum field theories, but it appears to be inextricably tied to those theories as well. No one doubts, for example, that gauge theories—such as the Yang-Mills equations for describing the strong interaction—are a fundamental description of nature, argues Robbert Dijkgraaf, a physicist at the University of Amsterdam. "But gauge theories are fundamentally connected with strings." That's true because of all the dualities, which establish an equivalence between field theories and string theories, showing them to be different ways of looking at the same thing. "It isn't possible to argue whether string theory belongs in physics, since it's continuously connected to all the things we hold dear," Dijkgraaf adds. "So we can't get rid of string theory, regardless of whether our universe is described by it. It's just one more tool for thinking about the fundamental properties of physics."[14]

String theory was also the first theory to quantize gravity in a consistent way, which was the point all along. But it goes even further than that. "String theory has the remarkable property of predicting gravity," Witten proclaims. By that he means that string theory does more than just describe gravity. The phenomenon is embedded within the theory's very framework, and someone who knew nothing about gravity could discover it as a natural consequence of the theory itself.[15] In addition to quantizing gravity, string theory has gone far toward solving problems like the black hole entropy puzzle that had resisted solutions by other means. Viewed in that sense, string theory can already be considered a successful theory on some level, even if it doesn't turn out to be the ultimate theory of physics.

While that matter is being adjudicated, there's no denying that string theory has led to a treasure trove of new ideas, new tools, and new directions in mathematics. The discovery of mirror symmetry, for instance, has created a cottage industry in the fields of algebraic and enumerative geometry. Mirror symmetry—the notion that most Calabi-Yau spaces have a topologically distinct mirror partner (or partners) that gives rise to identical physics—was discovered in the context of string theory, and its validity was later confirmed by mathematicians. (That, as we have seen, is a typical pattern: Whereas string theory may provide concepts, hints, and indications, in most cases mathematics delivers the proof.)

One reason mirror symmetry has been so valuable for math is that a difficult calculation for one Calabi-Yau can be a much simpler calculation in its mirror partner. As a result, researchers were soon able to solve centuries-old problems in mathematics. Homological mirror symmetry and the theory of Strominger-Yau-Zaslow (SYZ), which have developed since the mid-1990s, have forged unexpected though fruitful connections between symplectic geometry and algebraic geometry—two branches of mathematics that had previously been considered separate. Although mirror symmetry was uncovered through string theory research, its mathematical basis does not depend on string theory for its truth. The phenomenon, notes Andrew Strominger, "can be described in a way that doesn't involve string theory at all, [but] it would have been a long time before we figured it out had it not been for string theory."[16]

To cite another example, a 1996 paper by my former postdoc Eric Zaslow and I used an idea from string theory to solve a classic problem in algebraic geometry related to counting the number of so-called rational curves on a four-dimensional K3 surface. (Please keep in mind that the term *K3* refers to a whole class of surfaces—not one but an infinite number of them.) The "curves" in this case are two-dimensional Riemann surfaces defined by algebraic equations that are the topological equivalent of spheres embedded on that surface. The counting of these curves, it turns out, depends only on the number of *nodes* a curve possesses—nodes being points where a curve crosses itself. A figure eight, for example, has just one node, whereas a circle has zero nodes.

Here's another way of thinking about nodes that relates to our previous discussion of conifold transitions (in Chapter 10): If you take a two-dimensional donut and shrink one of the circles that run through the hole down to a point, you'll get something that looks like a croissant whose two ends are attached. If you separate those two ends and blow up the surface, you'll have the topological equivalent of a sphere. So you can consider this pinched donut or "attached croissant" a sphere with one node (or crossing). Similarly, we could go to a higher genus and look at a double-holed donut: First we're going to pinch a circle down to a point in the "inner wall" between the two holes and do the same somewhere on the donut's "outer wall." The object with these two pinch points is actually a sphere with two nodes because if we were to separate those two points and inflate the surface, we'd have a sphere. The point is that if you start with a surface of higher genus (say two, three, or more holes), you can end up with a curve, or sphere, with more nodes, too.

Let me restate the problem in algebraic geometry that we were originally try-ing to solve: For a K3 surface, we'd like to know the number of rational curves with g nodes that can fit on that surface for any (positive integer) value of g. Using conventional techniques, mathematicians had devised a formula that worked for curves with six or fewer nodes but not beyond. Zaslow and I set out to tackle the more general situation of curves with an arbitrary number of nodes. Instead of the usual approach, we took a string theory perspective, view-ing the problem in terms of branes inside a Calabi-Yau.

String theory tells us there are branes associated with a K3 surface that con-sist of curves (or two-dimensional surfaces, as previously defined) plus a so-called flat line bundle attached to each curve. To get a sense of what this line bundle is all about, suppose a person walks around the equator holding a stick of any length—even an infinitely long one—that is perpendicular to the equator and tangent to the surface of the sphere. Eventually, the stick will trace out a cylinder, which is called a *trivial line bundle*. If the person rotated the stick 180 degrees once during the walk, it would trace a Möbius strip. Both of these line bundles, by the way, are "flat," meaning that they have zero curvature.

Zaslow and I observed that if you take the space of all the branes containing curves of a fixed genus g that are associated with a given K3 and then compute the Euler characteristic of that space, the number you get will be exactly equal to the number of rational curves with g nodes that fit in that K3 surface.

In this way, my colleague and I had converted the original problem into a different form, showing that it all came down to getting the Euler characteristic of the space of branes in question. We then used a string theory duality, devel-oped by Cumrun Vafa and Witten, to calculate the Euler characteristic. String theory had thus provided a new tool for attacking this problem, as well as hints about a new way of framing the problem. Algebraic geometers hadn't been able to solve the problem before, because they weren't thinking of branes: It never occurred to them to frame this problem in terms of the *moduli space*, which en-compasses the totality of all possible branes of this type.

Although Zaslow and I sketched out the general approach, the actual proof of this idea was completed by others—Jim Bryan of the University of British Columbia and Naichung Conan Leung of the University of Minnesota—a few

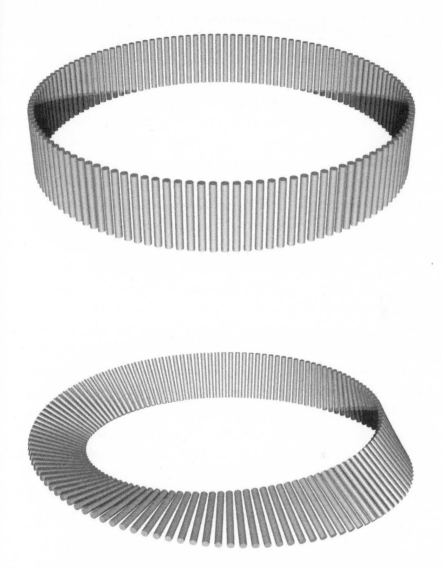

13.3—If you walked around the equator, all the while holding a pole parallel to the ground (and essentially tangent to the surface), you would sweep out a cylinder. If instead you gradually rotated the pole 180 degrees as you circumnavigated the globe, you would sweep out a more complicated surface—which has just one side rather than two—known as a Möbius strip.

years later. As a result, we now have a theorem in mathematics that should be true regardless of whether string theory is right or wrong.

In addition, the formula that Zaslow and I derived for counting rational curves on K3 surfaces gives you a function for generating all the numbers you get for rational curves with an arbitrary number of nodes. It turns out that this function essentially reproduces the famous *tau function*, which was introduced in 1916 by the Indian mathematician and self-taught genius Srinivasa Ramanujan.[17] This function and the conjectures that Ramanujan raised in conjunction with it have since led to many important advances in number theory. Our work, as far as I know, established the first solid link between enumerative geometry— the subject of counting curves—and the tau function.

This link has been strengthened by the recent contributions of Yu-Jong Tzeng, a young mathematician recently hired by Harvard and trained by my former student, Jun Li. Tzeng showed that not only are the rational curves on a K3 connected to the tau function but the counting of *any* curves of arbitrary genus on *any* algebraic surfaces is connected to the tau function. And Tzeng did this by proving a conjecture made by the German mathematician Lothar Goettsche, which generalized the so-called Yau-Zaslow formula for rational curves on K3 surfaces.[18] The new, generalized formula, whose validity Tzeng upheld, bears the name Goettsche-Yau-Zaslow. (A few years earlier, a former graduate student of mine, A. K. Liu, published a proof of the Goettsche-Yau-Zaslow formula.[19] But his proof, rooted in a highly technical, analytic approach, did not provide an explanation in the form that algebraic geometers were looking for. As such, Liu's paper was not regarded as the final confirmation of that formula. Tzeng's proof, which was based on algebraic geometry arguments, is more widely accepted.)

The broader point is that through a finding that originally stemmed from string theory, we've learned that the ties between enumerative geometry and Ramanujan's tau function are probably deeper than anyone imagined. We're always looking for connections like this between different branches of mathematics because these unexpected links can often lead us to new insights in both subjects. I suspect that, in time, more ties between enumerative geometry and the tau will be uncovered.

In the 1990s, in yet another, more celebrated example of how string theory has enriched mathematics, Witten and Nathan Seiberg of Rutgers University developed a set of equations called the Seiberg-Witten equations, which (as discussed in Chapter 3) have accelerated the study of four-dimensional spaces. The equations were much easier to use than existing methods, leading to an explosion in the things we can do in four dimensions—the main one being the attempt to categorize and classify all possible shapes. Although the Seiberg-Witten equations were initially discovered as a statement in field theory, it was soon shown that they could be derived from string theory as well. Putting this idea in the context of string theory, moreover, greatly enhanced our understanding of it. "On various occasions," says a publicity-shy colleague of mine, "Witten has basically said to mathematicians, 'Here, take these equations; they might be useful.' And, by golly, they were useful."

"String theory has been such a boon to mathematics, such a tremendous source of new ideas, that even if it turns out to be completely wrong as a theory of nature, it has done more for mathematics than just about any other human endeavor I can think of," claims my longtime collaborator Bong Lian of Brandeis University.[20] Although I'd phrase it more moderately than Lian, I agree that the payoff has been unexpectedly huge. On this point, we seem to be in accord with Atiyah, as well. String theory, he said, "has transformed and revitalized and revolutionized large parts of mathematics . . . in areas that seem far removed from physics." Many fields, "geometry, topology, and algebraic geometry and group theory, almost anything you want, seem to be thrown into the mixture—and in a way that seems to be very deeply connected with their central content, not just tangential contact, but into the heart of mathematics."[21]

While other areas of physics have informed math in the past, the influence of string theory has penetrated much deeper into the internal structure of mathematics, leading us to new conceptual breakthroughs. The advent of string theory, ironically, has led to harmonious collaborations within mathematics itself, because string theory has demanded a lot from mathematicians in areas that include differential geometry, algebraic geometry, Lie group theory, number theory, and others. In a funny way, our best hope so far for a unified theory of physics has helped promote the unification of mathematics as well.

Despite the beauty of string theory and its deeply felt impacts on mathematics, the question remains: How long must we wait for outside corroboration—for some connection, *any* connection, with the real world? Brian Greene believes that patience is in order, given that "we are trying to answer the most difficult, the most profound questions in the history of science. [Even] if we haven't gotten there in 50 or 100 years, that's a pursuit we should keep on with."[22] Sean Carroll, a physicist at the California Institute of Technology, agrees: "Profound ideas don't come with expiration dates."[23] Or, putting it in other terms, what's the big rush, anyway?

A historical precedent might be useful here. "In the nineteenth century, the question of why water boils at 100 degrees centigrade was hopelessly inaccessible," Witten notes. "If you told a 19th-century physicist that by the 20th century you would be able to calculate this, it would have seemed like a fairy tale."[24]

Neutron stars, black holes, and gravitational lenses—dense concentrations of matter that act like magnifying glasses in the sky—were similarly dismissed as sheer fantasy until they were actually seen by astronomers. "The history of science is littered with predictions that such and such an idea wasn't practical and would never be tested," Witten adds. But the history of physics also shows that "good ideas get tested."[25] New technology that could not even be guessed at a generation before can make ideas that seemed beyond the pale become science fact rather than science fiction.

"The more important the question is, the more patient one should be in the testing game," maintains MIT physicist Alan Guth, one of the architects of inflationary theory, which holds that our universe underwent a brief, explosive, expansionary burst during the earliest moments of the Big Bang. "When we were working on inflation in the early days, I never thought for a moment that it would be tested in my lifetime," Guth says. "We had to be lucky for inflation to be tested, and we were. Though it wasn't luck so much as the tremendous skill of the observers. The same thing could happen with string theory. And maybe we won't have to wait a hundred years."[26]

While string theory must be regarded as speculation, there's nothing necessarily wrong with that. The conjectures of mathematics, like Calabi's, are nothing more than speculation rooted in mathematical theory. They are absolutely essential to progress in my field. Nor could we get anywhere in physics and advance our understanding were it not for speculation of the learned, rather than idle, kind. Nevertheless, the word does imply some degree of doubt, and how you react to that is a matter of your temperament, as well as your personal investment in a problem. When it comes to string theory, some are in it for the long haul, hopeful that it will eventually pan out. Others, who can't get beyond the lingering questions, put the uncertainties front and center, waving metaphorical placards that read: "Stop! You're making a big mistake."

There was a time, not too many centuries ago, when people warned about sailing too far from one's shores, lest the ship and its passengers fall off the edge of the earth. But some intrepid travelers did set sail, nevertheless, and rather than falling off the edge of the world, they discovered the New World instead.

Perhaps that's where we are today. I'm of the camp of pushing forward because that's what mathematicians do. We carry on. And we can do so with or without any input from the external world—or the realm of experiment—while keeping productive as well.

Though, personally, I find it helpful to keep tabs with the physicists. Indeed, I've spent the bulk of my career working in the interstices between math and physics, partly owing to my conviction that interactions between the two fields are crucial for furthering our grasp of the universe. All told, these interactions have been mostly harmonious over the decades. Sometimes, the mathematics has developed before applications in physics were found—as occurred with the great works of Michael Atiyah, Elie Cartan, S. S. Chern, I. M. Singer, Hermann Weyl, and others—and sometimes the physics has outpaced the math, as occurred with the discovery of mirror symmetry. But perhaps I shouldn't characterize the current arrangement between mathematicians and physicists as entirely cozy. "There's a good deal of healthy and generally good-natured competition" between the two fields, according to Brian Greene, and I think that's a fair assessment.[27] Competition, however, is not always bad, as it usually helps move things along.

In different times in history, the divisions between the fields—or lack thereof—have changed significantly. People like Newton and Gauss were certainly comfortable moving freely between math, physics, and astronomy. Indeed, Gauss, who was one of the greatest mathematicians of all time, served as professor of astronomy at the Göttingen Observatory for almost fifty years, right up to the moment of his death.

The introduction of the Maxwell equations of electromagnetism, and subsequent developments in quantum mechanics, created a wedge between math and physics that persisted for the better part of a century. In the 1940s, 1950s, and 1960s, many mathematicians didn't think much of physicists and didn't interact with them. Many physicists, on the other hand, were arrogant as well, having little use for mathematicians. When the time came for math, they figured they could work it out for themselves.

MIT physicist Max Tegmark supports this interpretation, citing a "cultural gap" between the two fields. "Some mathematicians look down their nose at physicists for being sloppy—for doing calculations that lack rigor," he says. "Quantum electrodynamics is an example of an extremely successful theory that is not mathematically well defined." Some physicists, he adds, are scornful of mathematicians, thinking that "you guys take forever to derive things that we can get in minutes. And if you had our intuition, you'd see it's all unnecessary."[28]

Ever since string theory entered the picture—with theoretical physics relying to an increasing degree on advanced mathematics—that cultural gap has started to narrow. The mathematics that comes up in string theory is so complicated—and so integral to everything in the theory—that physicists not only needed help but also welcomed it. While mathematicians became interested in Calabi-Yau spaces before the physicists did, to pick one example, the physicists eventually got there, and when they did, they showed us a few tricks as well. We're now in a period of "reconvergence," as Atiyah puts it, and that's a good thing.

I can't say whether string theory will ever get past its most serious hurdle—coming up with a testable prediction and then showing that the theory actually gives us the right answer. (The math part of things, as I've said, is already on much firmer ground.) Nevertheless, I do believe the best chance for arriving at a successful theory lies in pooling the resources of mathematicians and physicists, combining the strengths of the two disciplines and their different ways of approaching the world. We can work on complementary tracks, sometimes crossing over to the other side for the benefit of both.

Cliff Taubes, a math colleague of mine at Harvard, summed up the differences between the fields well. Though the tools of math and physics may be the same, Taubes said, the aims are different. "Physics is the study of the world, while mathematics is the study of all possible worlds."[29]

That's one of the reasons I love mathematics. Physicists get to speculate about other worlds and other universes, just as we do. But at the end of the day,

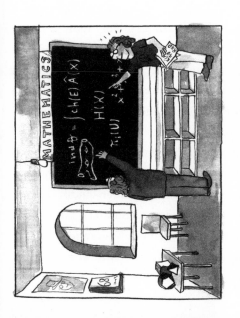
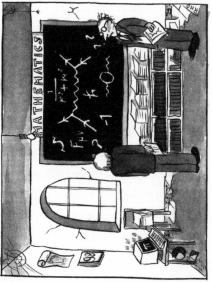

13.4 — This cartoon by the physicist Robbert Dijkgraaf shows the interplay between mathematicians and physicists. (Image courtesy of Robbert H. Dijkgraaf)

they eventually have to bring it back to our world and think about what's real. I get to think about what's possible—not only "all possible worlds," as Cliff put it, but the even broader category of all possible spaces. As I see it, that's our job. While physicists, by and large, tend to look at one space and see what it can tell us about nature, we mathematicians need to look at the totality of all spaces in order to find some general rules and guiding principles that apply to the cases of greatest concern.

Still, all spaces are not created equal, and some command my attention more than others, especially those spaces wherein the extra dimensions of nature are thought to reside. A critical challenge before us is to figure out the shape of that hidden realm, which, theory holds, dictates both the kinds of matter we see in the cosmos and the kinds of physics we see. That problem has been on my plate for a while, and it's not likely to be disposed of soon.

Although I take on a variety of projects from time to time, I keep coming back to this one. And, despite my forays into other areas of math and physics, I keep coming back to geometry. If peace comes through understanding, geometry is my way of trying to achieve some semblance of inner calm. Putting it more broadly, it's my way of trying to make sense of our universe and to fathom the mysterious spaces—named in part after me—that may lurk within.

Fourteen

THE END OF GEOMETRY?

Although geometry has served us well, there is a problem lurking just beneath the surface, and it may portend trouble for the future. To see this, one need travel no farther than to the nearest lake or pond. (And if there are no ponds in your neck of the woods, a backyard pool or bathtub might do.) A lake's surface may look perfectly smooth on a calm, windless day, but that is an illusion. When we examine the surface at higher resolution, it appears jagged rather than smooth. We see that it's actually composed of individual water molecules that are constantly jiggling around, moving within the pond itself and passing freely between the water and air. Viewed in this light, the surface is not a static, well-defined thing at all. In fact, it hardly qualifies as a surface, as we commonly use the term.

Classical geometry is like this, too, according to Harvard physicist Cumrun Vafa, in that it only provides an approximate description of nature rather than an exact, or a fundamental, one. To its credit, this approximate description holds up well and describes our universe almost flawlessly, except on the Planck scale (10^{-33} centimeters)—a realm at which standard geometry gets swamped by quantum effects, rendering simple measurements impossible.

The chief difficulty in resolving things at very fine scales stems from the Heisenberg uncertainty principle, which makes it impossible to localize a single point or to secure a precise fix on the distance between two points. Instead of standing still, everything at the Planck scale fluctuates—including points, lengths, and curvature. Whereas classical geometry tells us that two planes intersect in a line and three planes intersect in a point, the quantum perspective tells us that we might instead imagine three planes intersecting in a sphere that encompasses a range of probable positions for that point.

To probe the universe at the level of the hidden dimensions or individual strings, we're going to need a new kind of geometry—sometimes referred to as quantum geometry—that is capable of operating on both the largest and smallest scales imaginable. A geometry of this sort would have to be consistent with general relativity on large scales and quantum mechanics on small scales and consistent with both in places where the two theories converge. For the most part, quantum geometry does not yet exist. It is as speculative as it is important, a hope rather than a reality, a name in search of a well-defined mathematical theory. "We have no idea what such a theory will look like, or what it should be called," Vafa says. "It's not obvious to me that it should be called geometry."[1] Regardless of its name, geometry as we know it will undoubtedly come to an end, only to be replaced by something more powerful—geometry as we don't know it. This is the way of all science, as it should be, since stagnation means death.

"We're always looking for the places where science breaks down," explains University of Amsterdam physicist Robbert Dijkgraaf. "Geometry is closely tied to Einstein's theory, and when Einstein's theory becomes stressed, geometry is stressed, too. Ultimately, Einstein's equations must be replaced in the same way that Newton's equations were replaced, and geometry will go along with it."[2]

Not to pass the buck, but the problem has more to do with physics than with math. For one thing, the Planck scale where all this trouble starts is not a mathematical concept at all. It's a *physical* scale of length, mass, and time. Even the fact that classical geometry breaks down at the Planck scale doesn't mean there's anything wrong with the math per se. The techniques of differential calculus that underlie Riemannian geometry, which in turn provides the basis for general relativity, do not suddenly stop working at a critical length scale. Differential geometry is designed by its very premise to operate on infinitesimally small lengths that can get as close to zero as you want. "There's no reason that extrapolating general relativity to the smallest distance scales would be a problem from a mathematics standpoint," says David Morrison, a mathematician at the University of California, Santa Barbara. "There's no real problem from a physics standpoint, either, except that we know it's wrong."[3]

In general relativity, the metric or length function tells you the curvature at every point. At very small length scales, the metric coefficients fluctuate wildly, which means that lengths and curvature will fluctuate wildly as well. The geometry, in other words, would be undergoing shifts so violent it hardly makes sense to call it geometry. It would be like a rail system where the tracks shrink, lengthen, and curve at will—a system that would never deliver you to the right destination and, even worse, would get you there at the wrong time. That's no way to run a railroad, as they say, and it's no way to do geometry, either.

As with many problems we've grappled with in this book, this geometric weirdness springs from the fundamental incompatibility of quantum mechanics and general relativity. Quantum geometry might be thought of as the language of quantum gravity—the mathematical formalism needed to fix the compatibility problem—whatever that theory turns out to be. There's another way to consider this problem, which also happens to be the way that many physicists think about this: Geometry, as it appears in physics, might be a phenomenon that's "emergent" rather than fundamental. If this view is correct, it might explain why traditional geometric descriptions of the world appear to falter at the realm of the very small and the very energetic.

Emergence can be seen, for instance, in the lake or pond we discussed earlier in this chapter. If you look at a sizable body of water, it makes sense to think of the water as a fluid that can flow and form waves—as something that has bulk properties like viscosity, temperature, and thermal gradients. But if you were to examine a tiny droplet of water under an extremely powerful microscope, it would not look anything like a fluid. Water, as everyone knows, is made out of molecules that on a small scale behave more like billiard balls than a fluid. "You cannot look at waves on the surface of a lake and, from that, deduce anything about the molecular structure or molecular dynamics of H_2O," explains MIT physicist Allan Adams. "That's because the fluid description is not the most fundamental way of thinking about water. On the other hand, if you know where all the molecules are and how they're moving, you can in principle deduce everything about the body of water and its surface features. The microscopic description, in other words, contains the macroscopic information."[4] That's why we consider the microscopic description to be more fundamental, and the macroscopic properties emerge from it.

So what does this have to do with geometry? We've learned through general relativity that gravity is a consequence of the curvature of spacetime, but as we've seen, this long-distance (low-energy) description of gravity—what we're now calling classical geometry—breaks down at the Planck scale. From this, a number of physicists have concluded that our current theory of gravity, Einstein's theory, is merely a low-energy approximation of what's really going on. Just as waves on the surface of a lake emerge from underlying molecular processes we cannot see, these scholars believe that gravity and its equivalent formulation as geometry also emerge from underlying, ultramicroscopic processes that we assume must be there, even if we don't know exactly what they are. That is what people mean when they say gravity or geometry is "emergent" from the sought-after Planck scale description of quantum geometry and quantum gravity.

Vafa's concerns regarding the possible "end of geometry" are legitimate, but such an outcome may not be a tragedy—Greek or otherwise. The downfall of classical geometry should be celebrated rather than dreaded, assuming we can replace it with something even better. The field of geometry has constantly changed over the millennia. If the ancient Greek mathematicians, including the great Euclid himself, were to sit in on a geometry seminar today, they'd have no idea what we were talking about. And before long, my contemporaries and I will be in the same boat with respect to the geometry of future generations. Although I don't know what geometry will ultimately look like, I fully believe it will be alive and well and better than ever—more useful in more situations than it presently is.

On this point, Santa Barbara physicist Joe Polchinski appears to agree. He doesn't think the breakdown of conventional geometry at the Planck length signals "the end of the road" for my favorite discipline. "Usually when we learn something new, the old things that took us there are not thrown out but are instead reinterpreted and enlarged," Polchinski says. Paraphrasing Mark Twain, he considers reports of the death of geometry to be greatly exaggerated. For a brief period in the late 1980s, he adds, geometry became "old hat" in physics. Passé. "But then it came back stronger than ever. Given that geometry has played such a central role in the discoveries to date, I have to believe it is part of something bigger and better, rather than something that will ultimately be discarded."[5] That's why I argue that quantum geometry, or whatever you call it, has to be an "enlargement" of geometry, as Polchinski put it, since we need something that can do all the great things geometry already does for us, while also providing reliable physical descriptions on the ultra-tiny scale.

Edward Witten seems to be in accord. "What we now call 'classical geometry' is much broader than what geometry was understood to be just a century ago," he says. "I believe that the phenomenon at the Planck scale very likely involves a new kind of generalization of geometry or a broadening of the concept."[6]

Generalizations of this sort—which involve taking a theory that is valid in a certain regime and extending its scope and applicability into an even larger milieu—have been introduced to geometry repeatedly. Consider the invention of non-Euclidean geometry. "If you asked Nikolai Lobachevsky what geometry was when he was young," early in the nineteenth century, "he probably would have listed the five postulates of Euclid," says Adams. "If you asked him later in his career, he might have said there were five postulates, but maybe we don't need them all."[7] In particular, he would have singled out Euclid's fifth postulate—which holds that parallel lines never intersect—as one we could let go. It was Lobachevsky, after all, who realized that excluding the parallel postulate made a whole new geometry—which we call hyperbolic geometry—possible. For while parallel lines clearly do not intersect on a plane—the domain in which Euclid's plane geometry operates—this is certainly not the case on the surface of a sphere. We know, for instance, that all longitudinal lines on a globe converge at the north and south poles. Similarly, while the angles of triangles drawn on a plane must always add up to 180 degrees, on the surface of a sphere those angles always add up to more than 180 degrees, and on the surface of a saddle they add up to less than 180 degrees.

Lobachevsky published his controversial ideas on non-Euclidean geometry in 1829, although they were buried in an obscure Russian journal called the *Kazan Messenger*. A few years later, the Hungarian mathematician János Bolyai published his own treatise on non-Euclidean geometry, though the work, unfortunately, was relegated to the appendix of a book written by his father, the mathematician Farkas (Wolfgang) Bolyai. At roughly the same time, Gauss had been developing similar ideas on curved geometry. He immediately recognized that these new notions of curved spaces and "intrinsic geometry" were inter-twined with physics. "Geometry should be ranked not with arithmetic, which is purely *a priori*, but with mechanics," Gauss claimed.[8] I believe he meant that geometry, unlike arithmetic, must draw on empirical science—namely, physics (which was called mechanics at the time)—for its descriptions to carry weight. Gauss's intrinsic geometry laid the groundwork for Riemannian geometry, which, in turn, led to Einstein's dazzling insights on spacetime.

In this way, pioneers like Lobachevsky, Bolyai, and Gauss did not throw out all that came before but merely opened the door to new possibilities. And the breakthrough they helped usher in created a more expansive geometry, as its tenets were no longer confined to a plane and could instead apply to all manner of curved surfaces and spaces. Yet the elements of Euclid were still retained in this enlarged, more general picture. For if you take a small patch of Earth's surface— say, a several-block area in Manhattan—the streets and avenues really are par-allel for all practical purposes. The Euclidean description is valid in that limited domain, where the effects of curvature are negligible, but does not hold when you are looking at the planet as a whole. You might also consider a triangle drawn on a spherical balloon. When the balloon is relatively small, the angles will add up to more than 180 degrees. But if we keep inflating the balloon, the radius of curvature (r) will get bigger and bigger, and the curvature itself (which scales with $1/r^2$) will get smaller and smaller. If we let r go to infinity, the cur-vature will go to zero and the angles of the triangle, at this limit, will be exactly 180 degrees. As Adams puts it, "there's this one situation, on a flat plane, where Euclidean geometry works like a champ. It works pretty well on a slightly curved sphere, but as you inflate the balloon and the sphere gets flatter and flatter, the agreement gets better and better. So we can see that Euclidean geometry is really just a special case of a more general story—the case in which the radius of cur-vature is infinite, the angles of the triangle add up to 180 degrees, and all the postulates of Euclidean geometry are recovered."[9]

Similarly, Newton's theory of gravity was an extremely practical theory, in the sense that it gave us a simple way of computing the gravitational force ex-erted on any object in a system. Specifically, it worked well so long as the objects in question were not moving too fast or in situations where the gravitational potential is not too large. Then along came Einstein with his new theory, in which gravity was seen as a consequence of curved spacetime rather than a force propagated between objects, and we realized that Newton's gravity was just a special case of this broader picture that was only valid in the limit when objects were moving slowly and gravity was weak. In this way, we can see that general relativity, as the name suggests, truly was general: It was not only a generalization of Einstein's special theory of relativity—by taking a theory that did not have gravity and expanding the setting to include gravity—but also a generalization of Newtonian gravity.

In the same way, quantum mechanics is a generalization of Newtonian mechanics, but we don't need to invoke quantum mechanics to play baseball or tiddlywinks. Newton's laws of motion work fine for large objects like baseballs (and even for smaller objects like winks), where the corrections imposed by quantum theory would be immeasurably small and can thus be safely overlooked. But the macroscopic domain in which baseballs and rockets fly about, and Newton's laws prevail, is just a special case of the broader, more general domain of quantum theory, which also holds for objects that are considerably smaller. Using quantum mechanics, we can accurately predict the trajectories of relativistic electrons at a high-energy collider, whereas in that domain, Newtonian mechanics will let us down.

Now we're approaching this very same juncture with respect to geometry. Classical Riemannian geometry is not capable of describing physics at the quantum level. Instead we seek a new geometry, a more general description, that applies equally well to a Rubik's Cube as it does to a Planck-length string. The question is how to proceed. To some extent, we're groping in the dark, as Isaac Newton might have been when trying to write his own theory of gravity.

Newton had to invent new techniques to achieve that end, and out of that, calculus was born. Just as Newton's mathematics was motivated by physics, so must ours be today. We can't create quantum geometry without some input from physics. While we can always conjure up some new interpretation of geometry, if it is to be truly fruitful, geometry must describe nature at some basic level. And for that, as Gauss wisely acknowledged, we need some guidance from the outside.

The relevant physics gives us the technical demands that our math must satisfy. If classical geometry is used, physics at the Planck scale would appear to involve discrete changes and sudden discontinuities. The hope is that quantum

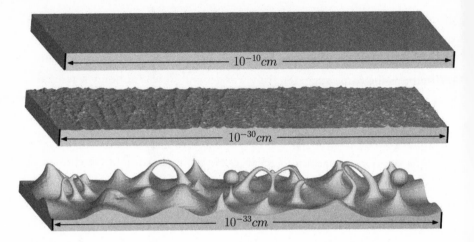

14.1—The physicist John Wheeler's concept of *quantum foam*. The top panel looks completely smooth. But if that surface is blown up by twenty orders of magnitude (middle panel), irregularities become evident. When the surface is blown up another thousandfold, all the little bumps become mountains, and the surface is now the antithesis of smooth.

geometry will eliminate those discontinuities, creating a smoother picture that's simpler to grasp and easier to work with.

String theory, almost by definition, is supposed to deal with problems of that sort. Because "the fundamental building block of . . . string theory is not a point but rather a one-dimensional loop, it is natural to suspect that classical geometry may not be the correct language for describing string physics," Brian Greene explains. "The power of geometry, however, is not lost. Rather, string theory appears to be described by a modified form of classical geometry . . . with the modifications disappearing as the typical size in a given system becomes large relative to the string scale—a length scale which is expected to be within a few orders of magnitude of the Planck scale."[10]

Previous theories of fundamental physics regarded their basic building block, the particle, as an infinitely small, zero-dimensional point—an object that the mathematics of the time was ill-equipped to handle (and that the mathematics of today is still ill-equipped to handle). Strings are not infinitely small particles, so the quantum fluctuations, which had been so troublesome for classical geometry at ultramicroscopic scales, are spread out over a substantially larger area, thereby diminishing their strength and making the fluctuations more

14.2a—This photograph is called "The Blue Marble" to suggest that when viewed from a distance, Earth's surface looks as smooth and unblemished as a marble. (NASA Goddard Space Flight Center)

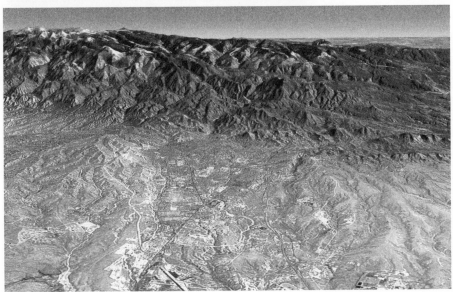

14.2b—A close-up photograph of Santa Fe, New Mexico (which lies near the center of the Blue Marble image), as taken by the Landsat 7 Earth-observing satellite, showing that the surface is anything but smooth. Together, these two photographs illustrate the notion of quantum foam—the idea that what might appear to be a smooth, featureless object from a distance can look extremely irregular from up close. (Visualization created by Jesse Allen, Earth Observatory; data obtained, coregistered, and color balanced by Laura Rocchio, Landsat Project Science Office)

manageable. In this way, the vexing problem of singularities in physics, where the curvature and density of spacetime blow up to infinity, is deftly bypassed. "You never get to the point where the disasters happen," says Nathan Seiberg of the Institute of Advanced Study. "String theory prevents it."[11]

Even if outright catastrophe is averted, it is still instructive to look at the close calls and near misses. "If you want to study situations where geometry breaks down, you want to pick cases where it only breaks down a little bit," says Andrew Strominger. "One of the best ways of doing that is studying how Calabi-Yau spaces break down, because in those spaces, we can isolate regions where space-time breaks down, while the rest of the region stays nice."[12]

The hope to which my colleague speaks is that we might be able to gain some insights on quantum geometry, and what it entails, by doing string theory in the controlled setting of Calabi-Yau spaces—a theme, of course, that runs throughout this book. One promising avenue has been to look for situations in string theory where geometry behaves differently than it does in classical geometry. A prime example of this is a topology-changing transition that can sometimes proceed smoothly in string theory but not in conventional physics. "If you're restricted to standard geometric techniques—and by that I mean always keeping a Riemannian metric in place—the topology cannot change," says Morrison.[13] The reason topological change is considered a big deal is that you can't transform one space into another without ripping it in some way—just as you can't scramble eggs without breaking some shells. Or turn a sphere into a donut without making a hole.

But piercing a hole in an otherwise smooth space creates a singularity. That, in turn, poses problems for general relativists, who now have to contend with infinite curvature coefficients and the like. String theory, however, might be able to sidestep this problem. In 1987, for example, Gang Tian (my graduate student at the time) and I demonstrated a technique known as a flop transition, which yielded many examples of closely related but topologically distinct Calabi-Yau manifolds.

Conifold transitions, which were discussed in Chapter 10, are another, even more dramatic example of topological change involving Calabi-Yau spaces. Think of a two-dimensional surface such as a football standing on end inside a Calabi-Yau space, as illustrated in Figure 14.3. We can shrink the football down to an increasingly narrow strand that eventually disappears, leaving behind a tear—a vertical slit—in the fabric of spacetime. Then we tilt the slit by pushing the "fabric" above it and that below it toward each other. In this way, the vertical slit is gradually transformed into a horizontal slit, into which we can insert and then reinflate another football. The football has now "flopped" over from its original configuration. If you do this procedure in a mathematically precise way—tearing space at a certain point, opening it up, reorienting the tear, and inserting a new two-dimensional surface with a shifted orientation back into the six-dimensional space—you can produce a Calabi-Yau space that is topologically distinct, and thus has a different overall shape, from the one you started with.

The flop transition was of interest mathematically because it showed how you could start with one Calabi-Yau, whose topology was already familiar, and end up with other Calabi-Yau spaces we'd never seen before. As a result, we mathematicians could use this approach to generate more Calabi-Yau spaces to study or otherwise play around with. But I also had a hunch that the flop transition might have some physical significance as well. Looking back with the benefit of hindsight, one might think I was particularly prescient, though that's not necessarily the case. I feel that any general mathematical operation we can do to a Calabi-Yau should have an application in physics, too. I encouraged Brian Greene, who was my postdoc at the time, to look into this, as well as mentioning the idea to a few other physicists who I thought might be receptive.

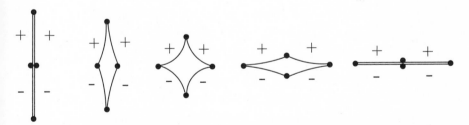

14.3—One way to think about the *flop transition* is to make a vertical slit in a two-dimensional fabric. Now push on the fabric from both the top and the bottom so that the vertical slit becomes wider and wider and eventually becomes a horizontal slit. So in this sense, the slit or tear, which was once standing upright, has now "flopped" over onto its side. Calabi-Yau manifolds can undergo flop transitions, too, when internal structures are toppled over in similar fashion (often after an initial tear), resulting in a manifold that's topologically distinct from the one you started with. What makes the flop transition especially interesting is that the four-dimensional physics associated with these manifolds is identical, despite the difference in topology.

Greene ignored my advice for several years, but finally started working on the problem in 1992 in concert with Paul Aspinwall and Morrison. In light of what they came up with, it was well worth the wait.

Aspinwall, Greene, and Morrison wanted to know whether something like a flop transition could occur in nature and whether space itself would rip apart, despite general relativity's picture of a smoothly curved spacetime that is not prone to rupture. Not only did the trio want to determine whether this type of transition could occur in nature, they also wanted to know whether it could occur in string theory.

To find out, they took a Calabi-Yau manifold with a sphere (rather than football) sitting inside it, put it through a flop transition, and used the resultant (topologically altered) manifold to compactify six of spacetime's ten dimensions to see what kind of four-dimensional physics it yielded. In particular, they wanted to predict the mass of a certain particle, which, in fact, they were able to compute. They then repeated the same process, this time using the mirror partner of the original Calabi-Yau space. In the mirror case, however, the sphere did not shrink down to zero volume as it went through the flop transition. In other words, there was no tearing of space, no singularity; the string physics, as Greene puts it, was "perfectly well behaved."[14] Next, they computed the mass of that same particle—this time associated with the mirror manifold—and compared the results. If the predictions agreed, that would mean the tearing of space and the singularity that came up in the first case were not a problem; string theory and the geometry on which it rests could handle the situation seamlessly. According to their calculations, the numbers matched almost perfectly, meaning that tears of this sort could arise in string theory without dire consequences.

One question left unanswered by their analysis was how this could be true— how, for instance, could a sphere shrink down to zero volume (the size of a point in traditional geometry) when the smallest allowable size was that of an individual string? Possible answers were contained in a Witten paper that came out at the same time. Witten showed how a loop of string could somehow encircle the spatial tear, thereby protecting the universe from the calamitous effects that might otherwise ensue.

"What we learned is that when the classical geometry of the Calabi-Yau appears to be singular, the four-dimensional physics looks smooth," explains Aspinwall. "The masses of particles do not go to infinity, and nothing bad happens." So the quantum geometry of string theory must somehow have a "smoothing effect," taking something that classically looks singular and making it nonsingular.[15]

The flop transition can shed light on what quantum geometry might look like, by showing us those situations that classical geometry cannot handle. Classical geometry can describe the situation at the beginning and end of the flop with no trouble, but fails at the center, where the width of the football (or basketball) shrinks to zero. By seeing exactly what string theory does differently, in this case and in as many other cases as we can find, we can figure out how classical geometry has to be modified—what kind of quantum corrections are needed, in other words.

The next question to ask, says Morrison, "is whether the quantum modifications we have to make to geometry are sufficiently geometric-looking to still be called geometry or whether they are so radically different that we would have to give up the notion of geometry altogether." The quantum corrections we know about so far, through examples like the flop transition, "can still be described geometrically, even though they may not be easily calculated," he says. But we still don't know whether this is generally true.[16]

I personally am betting that geometry will prevail in the end. And I believe the term *geometry* will remain in currency, not simply for the sake of nostalgia but rather because the field itself will continue to provide useful descriptions of the universe, as it has always done in the past.

Looking to the future, it's clear that creating a theory of quantum geometry, or whatever you care to call it, surely stands as one of the greatest challenges facing the field of geometry, if not all of mathematics. It's likely to be a decades-long ordeal that will require close collaboration between physicists and mathematicians. While the task certainly demands the mathematical rigor we always try to apply, it may benefit just as much from the intuition of physicists that never ceases to amaze us mathematicians.

At this stage of my career, having already been in the game for about forty years, I certainly have no illusions of solving this problem on my own. In contrast to a more narrowly defined proof that one person might be able to solve single-handedly, this is going to take a multidisciplinary effort that goes beyond the labors of a lone practitioner. But given that Calabi-Yau spaces have been central in some of our early attempts to gain footholds on quantum geometry, I'm hoping to make some contributions to this grand endeavor as part of my long-standing quest to divine the shape of inner space.

Ronnie Chan, a businessman who generously supported the Morningside Institute of Mathematics in Beijing (one of four math institutes I've helped start in China, Hong Kong, and Taiwan), once said, "I've never seen any person as persistent on one subject as Yau. He only cares about math." Chan is right about my persistence and devotion to math, though I'm sure that if he were to look hard enough, he'd find plenty of people as persistent and devoted to their own pursuits as I am to mine. On the other hand, the question I have thrown myself at—trying to understand the geometry of the universe's internal dimensions—is an undeniably big one, even though the dimensions themselves may be small. Without some degree of persistence and patience, my colleagues and I would never have gotten as far as we have. That said, we still have a long way to go.

I read somewhere, perhaps in a fortune cookie, that life is in the getting there—the time spent negotiating a path between point A and point B. That is also true of math, especially geometry, which is all about getting from A to B. As for the journey so far, all I can say is that it's been quite a ride.

Epilogue

ANOTHER DAY, ANOTHER DONUT

Recently, one of us—the less mathematically inclined of the two—stood in the halls of the Jefferson Laboratory's theory group at Harvard, waiting to speak with Andrew Strominger, who was ensconced in an animated conversation with a colleague. Some minutes later, Cumrun Vafa burst out of the office, and Strominger apologized for the delay, saying that "Cumrun had a new idea related to Calabi-Yau spaces that couldn't wait." After a brief pause, he added, "It seems I hear a new idea about Calabi-Yau's just about every day."[1]

Upon further reflection, Strominger downgraded that statement to "about every week." Over the past several years, consistent with Strominger's remark, scholarly articles with the term "Calabi-Yau" in the title are being published at the rate of more than one per week—in the English language alone. These manifolds are not just relics of the first string revolution or mathematical curiosities of only historical import. They are alive and well and, if not living in Paris, are at least still prominent fixtures on the mathematics and theoretical physics archives.

That's not bad considering that in the late 1980s, many physicists thought that Calabi-Yau spaces were dinosaurs and that the whole subject was doomed. Calabi-Yau enthusiasts like myself—the more mathematically inclined of our duo—have often been told that we were talking nonsense. In that era, Philip Candelas, among others, got a bad grant review, which reduced his funding drastically. The cutbacks came for the simple reason that he was still investigating Calabi-Yau spaces, which were considered "the language of the past." A physicist who was then teaching at Harvard criticized the whole approach in even stronger terms: "Why are you idiots still working on this stupid theory?" Although I was taken aback by the question at the time, I have, after two decades of sustained cogitation, formulated what seems to be an appropriate response: "Well, maybe it's not so stupid after all."

Strominger, for one, doesn't think so, but then again, Calabi-Yau spaces have figured prominently in his career. In fact, it's possible he's done more than just about any other person to establish the importance of this class of spaces in physics. "It's surprising after all this time how central the role of Calabi-Yau's has remained," he says. "They keep popping up again and again—the black hole story being one example."[2] Another example involves a new strategy for realizing the Standard Model in *eight*-dimensional (rather than six-dimensional) Calabi-Yau manifolds—where the shape of the two extra dimensions is determined by string coupling constants—as discussed in recent papers by Vafa, Chris Beasley, and Jonathan Heckman and separately by Ron Donagi and Martijn Wijnholt.[3]

"It's not often an idea holds center stage for so long," adds Strominger, referring to the enduring reign of Calabi-Yau manifolds in string theory. "And it's not as if the idea has just stuck around for old-time's sake. It's not just a bunch of fogies from the eighties remembering the good old days. The idea has continued to branch off and spawn new buds."[4]

Vafa, his collaborator from across the hall, concurs: "If you're interested in four-dimensional gauge theories, you might think they have nothing to do with Calabi-Yau manifolds. Not only do they have something to do with Calabi-Yau manifolds, they have something to do with Calabi-Yau threefolds, which are of greatest interest to string theory. Similarly, you might think the theory of Riemann surfaces has nothing to do with Calabi-Yau threefolds, but studying them in the context of these threefolds turns out to be the key to understanding them."[5]

And then there's Edward Witten, the physicist sometimes called Einstein's successor (and if string theory is ever proven right, that might be a fair comparison). Witten has had what might reasonably be called an intimate relationship with Calabi-Yau manifolds, as well as with string theory as a whole, where he's contributed mightily to the first two string "revolutions." And he is likely to have a hand (or foot) in the third, if and when it ever comes to pass. For as Brian Greene once said, "Everything I've ever worked on, if I trace its intellectual roots, I find they end at Witten's feet."[6]

During a meeting with Strominger at Princeton, Witten recently mused, "Who would have thought, twenty-some years ago, that doing string theory on Calabi-Yau manifolds would turn out to be so interesting?" He went on to say: "The deeper we dig, the more we learn because Calabi-Yau's are such a rich and central construction." Almost every time we learn a new way of looking at string theory, he adds, these manifolds have helped us by providing basic examples.[7]

Indeed, almost all of the major calculations in string theory have been done in a Calabi-Yau setting simply because that is the space where we know how to do the calculations. Thanks to the "Calabi-Yau theorem" that emerged from the Calabi conjecture proof, says mathematician David Morrison of the University of California, Santa Barbara, "we have these techniques from algebraic geometry that enable us, in principle, to study and analyze all Calabi-Yau manifolds. We don't have techniques of similar strength for dealing with non-Kähler manifolds or the seven-dimensional G_2 manifolds that are important in M-theory. As a result, a lot of the progress that's been made has come from Calabi-Yau manifolds because we have the tools for studying them we don't have for other kinds of solutions."[7] In that sense, Calabi-Yau manifolds have provided a kind of laboratory for experiments, or at least thought experiments, that can teach us about string theory and hopefully about our universe as well.

"It's a testament to the human mind that we began thinking about Calabi-Yau's strictly as mathematical objects, before there was any obvious role for them in physics," notes Stanford mathematician Ravi Vakil. "We're not forcing Calabi-Yau's on nature, but nature seems to be forcing them upon us."[9]

That does not mean, however, that Calabi-Yau spaces are necessarily the last word or that we even *live* in such a space. The study of these manifolds has enabled physicists and mathematicians to learn many interesting and unexpected things, but these spaces can't explain everything, nor can they take us everywhere we might conceivably want to go. Although Calabi-Yau spaces may not be the ultimate destination, they may well be "stepping-stones to the next level of understanding," Strominger says.[10]

Speaking as a mathematician, which I suppose is the only way I can speak (with any authority, that is), I can say that a complete understanding of Calabi-Yau spaces is not there yet. And I have my doubts as to whether we can ever know everything there is to know about such spaces. One reason I'm skeptical stems from the fact that the one-dimensional Calabi-Yau is called an elliptic curve, and these curves—solutions to cubic equations in which at least some of the terms are taken to the third power—are enigmatic objects in mathematics. Cubic equations have fascinated mathematicians for centuries. Although the equations assume a simple form (such as $y^2 = x^3 + ax + b$), which might even look familiar to someone in a high school algebra class, their solutions hold many deep mysteries that can transport practitioners to the farthest reaches of mathematics. Andrew Wiles's famous proof of Fermat's Last Theorem, for example, revolved around understanding elliptic curves. Yet despite Wiles's brilliant work, there are many unsolved problems associated with such curves—and, equivalently, with one-dimensional Calabi-Yau manifolds—for which there appears to be no resolution in sight.

There is reason to believe that generalizations of elliptic curves to higher dimensions, of which Calabi-Yau threefolds offer one possibility, might be used to address similarly deep puzzles in mathematics, for we often learn something new by taking a special case—such as the elliptic curve—into a more general and higher-dimensional (or arbitrary dimensional) setting. On this front, the study of Calabi-Yau spaces of two complex dimensions, K3 surfaces, has already helped answer some questions in number theory.

But this work is just beginning, and we have no idea where it might take us. At this stage, it's fair to say that we've barely scratched the surface—regardless of whether that surface is the K3 or another variety of Calabi-Yau. Which is why I believe a thorough understanding of these spaces may not be possible until we understand a large chunk of mathematics that cuts across geometry, number theory, and analysis.

Some might take that as bad news, but I see it as a good thing. It means that Calabi-Yau manifolds, like math itself, are developing stories on a road that undoubtedly holds many twists and turns. It means there's always more to be learned, always more to be done. And for those of us who worry about keeping employed, keeping engaged, and keeping amused, it means there should be plenty of challenges, as well as fun, in the years ahead.

Postlude

ENTERING THE SANCTUM

Let us end where we began, looking to the past in the hopes of gleaning hints about the road ahead. The year was 387 B.C. or thereabouts, when, in an olive grove in the northern suburbs of Athens, Plato established his Academy, which is sometimes referred to as the world's first major university. (The Academy he founded remained in use for more than 900 years, until the Roman emperor Justinian shut it down in A.D. 526—a record that makes the 370-plus-year lifetime of my school, Harvard, seem paltry by comparison.) As legend has it, Plato placed an inscription above the entryway to the school that read: *Let no one ignorant of geometry enter here.*

The exact wording of this is in doubt, as I've seen it rendered several ways. Some experts deny whether it said that at all. Piers Bursill-Hall, a scholar of Greek mathematics at the University of Cambridge, suggests the inscription might just as easily have been "No parking in front of these gates."[1] For many, however, there's little reason to doubt the claim. "It was asserted by one of the ancient authorities to be true, and there's nothing to suggest that it's apocryphal," argues Donald Zeyl, a Plato expert at the University of Rhode Island. "It makes sense to me, given that Plato considered geometry an essential prerequisite to the study of philosophy."[2]

I, of course, am neither a historian nor a classicist and am thus in no position to arbitrate such a dispute. However, given the little I know about Plato, and the good deal more that I know about geometry, I'm inclined to side with Zeyl on this matter, if for no other reason than that despite the 2,400 or so years that separate Plato from me, we do see eye to eye on the importance of geometry. Plato regarded the truths of geometry as eternal and unchanging, whereas he regarded the knowledge derived from empirical science as more ephemeral in nature, unavoidably subject to revision. On these points, I heartily agree: Geometry can carry us far toward explaining the universe on scales both big and small (though perhaps not all the way down to the Planck scale), and when we prove something through rigorous mathematics, we can be sure it will stand the test of time. Geometrical proofs, like the diamonds advertised on TV, are forever.

Even though the particulars of Plato's "theory of everything," as outlined in *Timaeus*, strike the modern sensibility as absurd (if not borderline psychotic), there are many parallels between his picture of the universe and that embodied in string theory. Geometrization—the idea that the physics we observe springs directly from geometry—stands at the core of both approaches. Plato used the solids named after him to pursue this end (unsuccessfully, I might add) in much the same way that string theory relies on Calabi-Yau manifolds, though we're hoping the results will be better this time around.

The Platonic solids are literally built on symmetry, as are contemporary theories of physics. In the end, the search for a single, all-encompassing theory of nature amounts, in essence, to the search for the symmetry of the universe. The individual components of this overarching theory have their own symmetries as well, such as in the inherently symmetrical gauge fields, which we've considered at length, that provide our best current descriptions of the electromagnetic, strong, and weak interactions. What's more, the symmetry groups these fields inhabit are actually related to the symmetries of the Platonic solids, although not in the ways the ancient Greeks imagined it.

Today's physics is awash in dualities—the idea that the same physical reality can be described in two mathematically distinct ways. These dualities associate four-dimensional quantum field theories with ten-dimensional string theories, link ten-dimensional string theory with eleven-dimensional M-theory, and even reveal the physical equivalence between two Calabi-Yau manifolds that, on the face of it, have little in common. Likewise, the Platonic solids have their own dualities. The cube and octahedron, for instance, form a dual pair: Each can be rotated twenty-four different ways and still remain unchanged. The icosahedron and dodecahedron belong to a larger symmetry group, remaining invariant after sixty twists and turns. The tetrahedron, meanwhile, is its own dual. Curiously, when my mathematics colleague Peter Kronheimer (whose office is just a few doors down the hall from mine) tried to classify a group of four-dimensional Calabi-Yau manifolds by symmetry, he discovered that they followed the same classification scheme as the Platonic solids.

I'm by no means trying to suggest that Plato, promulgating his ideas in the early days of mathematics, got it all right. On the contrary, his notions about the origins of the elements are flat-out wrong. Similarly, the astronomer Johannes Kepler's attempts to explain the orbits of our solar system's planets by means of nested Platonic solids, lying within concentric spheres, were also doomed to failure. The details in these scenarios don't add up, and they don't even come close. But in terms of the big picture, Plato was on the right track in many ways, identifying some of the key pieces of the puzzle—such as symmetry, duality, and the general principle of geometrization—that we now believe any workable attempt to explain it all ought to include.

In view of that, it seems plausible to me that Plato would have made some reference to geometry at the entryway to his famed Academy. Much as I share his respect for the discipline that I chose so many years later, were I to mount a sign above the doorway to my decidedly uncelebrated Harvard office, I would amend the wording to this: *Let no one ignorant of geometry leave here.* The same words, I trust, would apply to readers now "leaving" the pages of this slim volume and, I hope, viewing the world somewhat differently.

A FLASH IN THE MIDDLE
OF A LONG NIGHT

Once I stood on the edge of a bridge,
strolled along the lakeshore,
striving to catch from afar
a glimpse of your matchless beauty.

Once I climbed atop a pavilion,
seeking a road at land's end,
yearning for a vision
of your fair, ineffable presence.

Ten thousand yards of silk
your lightness cannot contain.
The moon hovers miles away,
its shadow broad, expansive.

The boundless sky, the neverending stream,
waves breaking without surcease.
All excite the unfathomable depth
of which we call eternity.

The road is long, the view obscured,
with thousands of threads entangled.
Pursuing the truth, you toil endlessly.
In dreams you've surely been there.

Yet inspiration has struck, time and again,
lifting us onto the shoulders of giants.
From Euclid to Descartes, Newton to Gauss,
and Riemann to Poincaré.

O! The phantom of all things,
So hard to divine in daylight's glare.
Then suddenly, when the radiance dims,
she reveals a glimpse of her unseen form.

A chance encounter, unrivalled splendor,
from genius through the ages.
Let us celebrate the poetry of the universe
and the geometry through which it sings.

SHING-TUNG YAU
CAMBRIDGE, 2006

NOTES

PRELUDE

1. Plato, *Timaeus*, trans. Donald J. Zeyl (Indianapolis: Hackett, 2000), p. 12.
2. Ibid., pp. 46–47.
3. Ibid., p. 44.

CHAPTER 1

1. Max Tegmark, interview with author, May 16, 2005. (Note: All interviews were conducted by Steve Nadis unless otherwise noted.)
2. Aristotle, *On the Heavens*, at Ancient Greek Online Library, http://greektexts .com/library/Aristotle/On_The_Heavens/eng/print/1043.html.
3. Michio Kaku, *Hyperspace* (New York: Anchor Books, 1995), p. 34.
4. H. G. Wells, *The Time Machine* (1898), available at http://www. bartleby.com/ 1000/1.html.
5. Abraham Pais, *Subtle Is the Lord* (New York: Oxford University Press, 1982), p. 152.
6. Oskar Klein, "From My Life of Physics," in *The Oskar Klein Memorial Lectures*, ed. Gosta Ekspong (Singapore: World Scientific, 1991), p. 110.
7. Leonard Mlodinow, *Euclid's Window* (New York: Simon & Schuster, 2002), p. 231.
8. Andrew Strominger, "Black Holes and the Fundamental Laws of Nature," lecture, Harvard University, Cambridge, MA, April 4, 2007.
9. Ibid.

CHAPTER 2

1. Georg Friedrich Bernhard Riemann, "On the Hypotheses Which Lie at the Foundations of Geometry," lecture, Göttingen Observatory, June 10, 1854.
2. E. T. Bell, *Men of Mathematics* (New York: Simon & Schuster, 1965), p. 21.
3. Leonard Mlodinow, *Euclid's Window* (New York: Simon & Schuster, 2002), p. xi.

4. Edna St. Vincent Millay, "Euclid Alone Has Looked on Beauty Bare," quoted in Robert Osserman, *Poetry of the Universe* (New York: Anchor Books, 1995), p. 6.

5. Andre Nikolaevich Kolmogorov, *Mathematics of the 19th Century* (Boston, Birkhauser, 1998).

6. Deane Yang (Polytechnic Institute of New York University), e-mail letter to author, April 20, 2009.

7. Mlodinow, *Euclid's Window*, p. 205.

8. Brian Greene, *The Elegant Universe* (New York: Vintage Books, 2000), p. 231.

9. C. N. Yang, "Albert Einstein: Opportunity and Perception," speech, 22nd International Conference on the History of Science, Beijing, China, 2005.

10. Chen Ning Yang, "Einstein's Impact on Theoretical Physics in the 21st Century," *AAPPS Bulletin* 15 (February 2005).

11. Greene, *The Elegant Universe*, p. 72.

CHAPTER 3

1. Robert Greene (UCLA), interview with author, March 13, 2008.

2. Lizhen Ji and Kefeng Liu, "Shing-Tung Yau: A Manifold Man of Mathematics," Proceedings of Geometric Analysis: Present and Future Conference, Harvard University, August 27–September 1, 2008.

3. Leon Simon (Stanford University), interview with author, February 6, 2008.

4. Greene, interview with author, March 13, 2008.

5. Cameron Gordon (University of Texas), interview with author, March 14, 2008.

6. Robert Geroch (University of Chicago), interview with author, February 28, 2008.

7. Edward Witten (Institute for Advanced Study), interview with author, March 31, 2008.

8. Edward Witten, "A New Proof of the Positive Energy Theorem," *Communications in Mathematical Physics* 80 (1981): 381–402.

9. Roger Penrose, "Gravitational Collapse: The Role of General Relativity," 1969, reprinted in *Mathematical Intelligencer* 30 (2008): 27–36.

10. Richard Schoen (Stanford University), interview with author, January 31, 2008.

11. Demetrios Christodoulou, *The Formation of Black Holes in General Relativity* (Zurich: European Mathematical Society, 2009).

12. John D. S. Jones, "Mysteries of Four Dimensions," *Nature* 332 (April 7, 1998): 488–489.

13. Simon Donaldson (Imperial College), interview with author, April 3, 2008.

14. Faye Flam, "Getting Comfortable in Four Dimensions," *Science* 266 (December 9, 1994): 1640.

15. Ibid.

16. Mathematical Institute at the University of Oxford, "Chart the Realm of the 4th Dimension," http://www2.maths.ox.ac.uk/~dusautoy/2soft/4D.htm.

17. Grisha Perelman, "The Entropy Formula for the Ricci Flow and Its Geometric Applications," November 11, 2002, http://arxiv.org/abs/math/0211159v1.

CHAPTER 4

1. Eugenio Calabi (University of Pennsylvania), interview with author, October 18, 2007.

2. Robert Greene (UCLA), interview with author, October 18, 2008.

3. Calabi, interview with author, October 18, 2008.

4. Greene, interview with author, June 24, 2008.

5. Calabi, interview with author, October 18, 2007.

CHAPTER 5

1. Robert Greene (UCLA), interview with author, January 29, 2008.

2. Eugenio Calabi (University of Pennsylvania), interview with author, May 14, 2008.

3. Ibid.

4. Erwin Lutwak (Polytechnic Institute of NYU), interview with author, May 15, 2008.

5. Calabi, interview with author, May 14, 2008.

6. Calabi, interview with author, June 16, 2008.

7. Ibid.

8. Calabi, interview with author, October 18, 2007.

CHAPTER 6

1. Cumrun Vafa (Harvard University), interview with author, January 19, 2007.

2. John Schwarz (California Institute of Technology), interview with author, August 13, 2008.

3. Michael Green (University of Cambridge), e-mail letter to author, August 15, 2008.

4. Schwarz, interview with author, August 13, 2008.

5. Andrew Strominger (Harvard University), interview with author, February 7, 2007.

6. Strominger, interview with author, November 1, 2007.

7. Raman Sundrum (Johns Hopkins University), interview with author, January 25, 2007.

8. Strominger, interview with author, February 7, 2007.

9. Dennis Overbye, "One Cosmic Question, Too Many Answers," *New York Times*, September 2, 2003.

10. Juan Maldacena (Princeton University), interview with author, September 9, 2007.

11. Dan Freed (University of Texas), interview with author, June 24, 2008.

12. Tristan Hubsch (Howard University), interview with author, August 30, 2008.

13. Gary Horowitz (University of California, Santa Barbara), interview with author, February 15, 2007.

14. Eugenio Calabi (University of Pennsylvania), interview with author, October 18, 2007.

15. Woody Allen, "Strung Out," *New Yorker,* July 28, 2003.

16. Liam McAllister (Cornell University), e-mail letter to author, April 24, 2009.

17. Allan Adams (MIT), interview with author, August 10, 2007.

18. Joe Polchinski (University of California, Santa Barbara), interview with author, January 29, 2007.

19. Brian Greene, *The Fabric of the Cosmos* (New York: Alfred A. Knopf, 2004), p. 372.

20. P. Candelas, G. Horowitz, A. Strominger, and E. Witten, "Vacuum Configurations for Superstrings," *Nuclear Physics B* 258 (1985): 46–74.

21. Edward Witten (IAS), e-mail letter to author, July 24, 2008.

22. Volker Braun, Philip Candelas, and Rhys Davies, "A Three-Generation Calabi-Yau Manifold with Small Hodge Numbers," October 28, 2009, http://arxiv.org/PS_cache/arxiv/pdf/0910/0910.5464v1.pdf.

23. Dennis Overbye, "One Cosmic Question, Too Many Answers," *New York Times,* September 2, 2003.

24. Dale Glabach and Juan Maldacena, "Who's Counting?" *Astronomy,* (May 2006), 72.

25. Andrew Strominger, "String Theory, Black Holes, and the Fundamental Laws of Nature," lecture, Harvard University, Cambridge, MA, April 4, 2007.

26. Witten, e-mail letter to author, July 21, 2008.

27. Petr Horava (University of California, Berkeley), interview with author, July 6, 2007.

28. Ibid.

29. Leonard Susskind (Stanford University), interview with author, May 25, 2007.

CHAPTER 7

1. Ronen Plesser (Duke University), interview with author, September 3, 2008.

2. Ibid.

3. Marcus Grisaru (McGill University), interview with author, August 18, 2008.

4. Plesser, interview with author, September 3, 2008.

5. Shamit Kachru (Stanford University), interview with author, August 19, 2008.

6. Ashoke Sen (Harish-Chandra Research Institute), interview with author, August 22, 2008.

7. Jacques Distler and Brian Greene, "Some Exact Results on the Superpotential from Calabi-Yau Compactifications," *Nuclear Physics B* 309 (1988): 295–316.

8. Doron Gepner, "Yukawa Couplings for Calabi-Yau String Compactification," *Nuclear Physics B* 311 (1988): 191–204.

9. Kachru, interview with author, August 19, 2008.

10. Paul Aspinwall (Duke University), interview with author, August 14, 2008.

11. Wolfgang Lerche, Cumrun Vafa, and Nicholas Warner, "Chiral Rings in $N = 2$ Superconformal Theories," *Nuclear Physics B* 324 (1989): 427–474.

12. B. R. Greene, C. Vafa, and N. P. Warner, "Calabi-Yau Manifolds and Renormalization Group Flows," *Nuclear Physics B* 324 (1989): 371–390.

13. Brian Greene (Columbia University), interview with author, March 11, 2010.

14. Ibid.

15. Doron Gepner, interview with author, August 19, 2008.

16. B. R. Greene and M. R. Plesser, "Duality in Calabi-Yau Moduli Space," *Nuclear Physics B* 338 (1990): 15–37.

17. Brian Greene, *The Elegant Universe* (New York: Vintage Books, 2000), p. 258.

18. Plesser, interview with author, September 19, 2008.

19. Greene, interview with author, March 11, 2010.

20. Greene, *The Elegant Universe*, p. 259.

21. Cumrun Vafa (Harvard University), interview with author, September 19, 2008.

22. Greene, interview with author, March 13, 2010.

23. Mark Gross (UCSD), interview with author, October 31, 2008.

24. Andreas Gathmann (University of Kaiserslautern), interview with author, August 25, 2008.

25. David Hilbert, "Mathematical Problems," lecture, International Congress of Mathematicians, Paris, 1900, http://aleph0.clarku.edu/~djoyce/hilbert/problems.html (html version prepared by David Joyce, Mathematics Department, Clark University, Worcester, MA).

26. Andreas Gathmann, "Mirror Principle I," *Mathematical Reviews*, MR1621573, 1999.

27. David Cox (Amherst College), interview with author, June 13, 2008.

28. Andrew Strominger (Harvard University), interview with author, February 7, 2007.

29. Gross, interview with author, September 19, 2008.

30. Ibid.

31. Gross, interview with author, September 24, 2008.

32. Eric Zaslow (Northwestern University), interview with author, June 26, 2008.

33. A. Strominger, S. T. Yau, and E. Zaslow, "Mirror Symmetry Is T Duality," *Nuclear Physics* 479 (1996): 243–259.

34. Gross, interview with author, September 24, 2008.

35. Mark Gross, e-mail letter to author, September 29, 2008.

36. Ibid.

37. Strominger, interview with author, August 1, 2007.

38. Zaslow, interview with author, June 26, 2008.

39. Gross, interview with author, September 19, 2008.

40. Yan Soibelman (Kansas State University), interview with author, September 26, 2008.

41. Aspinwall, interview with author, June 23, 2008.

42. Michael Douglas (Stony Brook University), interview with author, August 20, 2008.

43. Aspinwall, interview with author, June 23, 2008.

44. Gross, interview with author, September 24, 2008.

CHAPTER 8

1. Avi Loeb (Harvard University), interview with author, September 25, 2008.

2. American Mathematical Society, "Interview with Heisuke Hironaka," *Notices of the AMS* 52, no. 9 (October 2005): 1015.

3. Steve Nadis, "Cosmic Inflation Comes of Age," *Astronomy* (April 2002).

4. Andrew Strominger, "String Theory, Black Holes, and the Fundamental Laws of Nature," lecture, Harvard University, Cambridge, MA, April 4, 2007.

5. Ibid.

6. Hirosi Ooguri (California Institute of Technology), interview with author, October 8, 2008.

7. Strominger, lecture.

8. Andrew Strominger and Cumrun Vafa, "Microscopic Origin of the Bekenstein-Hawking Entropy," *Physics Letters B* 379 (June 27, 1996): 99–104.

9. Andrew Strominger, quoted in Gary Taubes, "Black Holes and Beyond," *Science Watch,* May/June 1999, http://archive.sciencewatch.com/may-june99/sw_may-june 99_page3.htm.

10. Ooguri, interview with author, October 8, 2008.

11. Strominger, quoted in Taubes, "Black Holes and Beyond."

12. Xi Yin (Harvard University), interview with author, October 14, 2008.

13. Ibid.

14. Yin, interview with author, October 22, 2008.

15. Frederik Denef (Harvard University), interview with author, August 26, 2008.

16. Yin, interview with author, October 14, 2008.

17. Aaron Simons, interview with author, February 9, 2007.

18. Ooguri, interview with author, October 8, 2008.

19. Simons, interview with author, February 9, 2007.

20. J. M. Maldacena, A. Strominger, and E. Witten, "Black Hole Entropy in M-Theory," *Journal of High Energy Physics* 9712 (1997), http://arxiv.org/PS_cache/hep-th/pdf/ 9711/9711053v1.pdf.

21. Juan Maldacena (IAS), interview with author, September 4, 2008.

22. Hirosi Ooguri, Andrew Strominger, and Cumrun Vafa, "Black Hole Attractors and the Topological String," *Physical Review D* 70 (2004).

23. Cumrun Vafa (Harvard University), interview with author, September 26, 2008.

24. James Sparks (Harvard University), interview with author, February 6, 2007.

25. Amanda Gefter, "The Elephant and the Event Horizon," *New Scientist* (October 26, 2006): 36–39.

26. John Preskill, "On Hawking's Concession," July 24, 2004, http://www.theory .caltech.edu/~preskill/jp_24jul04.html.

27. Andrew Strominger (Harvard University), interview with author, February 7, 2007.

28. Juan Maldacena, "The Illusion of Gravity," *Scientific American*, November 2005, pp. 57–58, 61.

29. Davide Castelvecchi, "Shadow World," *Science News* 172 (November 17, 2007).

30. Taubes, "Black Holes and Beyond."

CHAPTER 9

1. L. Frank Baum, *The Wizard of Oz* (Whitefish, MT: Kessinger, 2004), p. 111.

2. Volker Braun (Dublin Institute for Advanced Studies), interview with author, November 4, 2008.

3. Philip Candelas (Oxford University), interview with author, December 1, 2008.

4. Ibid.

5. Andrew Strominger (Harvard University), interview with author, February 7, 2007.

6. Cumrun Vafa, "The Geometry of Grand Unified Theories," lecture, Harvard University, Cambridge, MA, August 29, 2008.

7. Chris Beasley (Stony Brook University), interview with author, November 13, 2008.

8. Burt Ovrut (University of Pennsylvania), interview with author, July 20, 2008.

9. Ovrut, interview with author, February 2, 2007.

10. Ron Donagi (University of Pennsylvania), interview with author, November 14, 2008.

11. Donagi, interview with author, November 19, 2008.

12. Candelas, interview with author, December 1, 2008.

13. Donagi, interview with author, May 3, 2008

14. Ovrut, interview with author, November 20, 2008.

15. Donagi, interview with author, November 20, 2008.

16. Ovrut, interview with author, November 20, 2008.

17. Shamit Kachru (Stanford University), interview with author, November 4, 2008.

18. Michael Douglas (Stony Brook University), interview with author, August 20, 2008.

19. Candelas, interview with author, December 1, 2008.

20. Simon Donaldson (Imperial College), interview with author, November 29, 2008.

21. Ovrut, interview with author, November 19, 2008.

22. Candelas, interview with author, December 1, 2008.

23. Strominger, interview with author, February 7, 2007.

24. Adrian Cho, "String Theory Gets Real—Sort Of," *Science* 306 (November 26, 2004): 1461.

25. Candelas, interview with author, December 1, 2008.

26. Allan Adams (MIT), interview with author, November 15, 2008.

CHAPTER 10

1. Gary Shiu, quoted in Adrian Cho, "String Theory Gets Real—Sort Of," *Science* 306 (November 26, 2004): 1461.

2. Shamit Kachru (Stanford University), e-mail letter to author, December 6, 2008.

3. Shamit Kachru, Renata Kallosh, Andrei Linde, and Sandip Trivedi, "De Sitter Vacua in String Theory," *Physical Review D* 68 (2003).

4. Raman Sundrum (Johns Hopkins University), interview with author, February 22, 2007.

5. Liam McAllister (Cornell University), interview with author, November 12, 2008.

6. Kachru, interview with author, September 8, 2007.

7. McAllister (Princeton University), interview with author, February 20, 2007.

8. Joe Polchinski (University of California, Santa Barbara), interview with author, February 6, 2006.

9. David Gross, quoted in Dennis Overbye, "Zillions of Universes? Or Did Ours Get Lucky?" *New York Times,* October 28, 2003.

10. Burton Richter, "Randall and Susskind," letter to editor, *New York Times,* January 29, 2006.

11. Leonard Susskind, *The Cosmic Landscape* (New York: Little, Brown, 2006), pp. 354–355.

12. Tristan Hubsch (Howard University), interview with author, November 7, 2008.

13. Mark Gross (UCSD), interview with author, October 31, 2008.

14. Gross, interview with author, September 19, 2008.

15. Miles Reid (University of Warwick), interview with author, August 12, 2007.

16. Allan Adams (MIT), interview with author, October 31, 2008.

17. Gross, interview with author, October 31, 2008.

18. Adams, interview with author, October 31, 2008.

19. Tristan Hubsch, e-mail letter to author, December 15, 2008.

20. Melanie Becker (Texas A&M University), interview with author, February 1, 2007.

21. Andrew Strominger (Harvard University), interview with author, February 7, 2007.

22. Li-Sheng Tseng (Harvard University), interview with author, December 17, 2008.

23. Becker, interview with author, February 1, 2007.

24. Polchinski, interview with author, January 29, 2007.

25. Strominger, interview with author, August 1, 2007.

26. Burt Ovrut (University of Pennsylvania), interview with author, February 2, 2007.

CHAPTER 11

1. Geoffrey Landis, "Vacuum States," *Asimov's Science Fiction* 12 (July 1988): 73–79.

2. Andrew R. Frey, Matthew Lippert, and Brook Williams, "The Fall of Stringy de Sitter," *Physical Review D* 68 (2003).

3. Sidney Coleman, "Fate of the False Vacuum: Semiclassical Theory," *Physical Review D* 15 (May 15, 1977): 2929–2936.

4. Steve Giddings (University of California, Santa Barbara), interview with author, September 24, 2007.

5. Matthew Kleban (New York University), interview with author, January 17, 2008.

6. Dennis Overbye, "One Cosmic Question, Too Many Answers," *New York Times*, September 2, 2003.

7. Andrei Linde (Stanford University), interview with author, December 27, 2007.

8. Giddings, interview with author, October 17, 2007.

9. Shamit Kachru (Stanford University), interview with author, September 18, 2007.

10. Linde, interview with author, December 27, 2007.

11. Henry Tye (Cornell University), interview with author, September 12, 2007.

12. Linde, interview with author, January 10, 2008.

13. S. W. Hawking, "The Cosmological Constant," *Philosophical Transactions of the Royal Society A* 310 (1983): 303–310.

14. Kleban, interview with author, January 17, 2008.

15. Steven B. Giddings, "The Fate of Four Dimensions," *Physical Review D* 68 (2003).

CHAPTER 12

1. Matthew Kleban (New York University), interview with author, March 4, 2008.

2. Henry Tye (Cornell University), e-mail letter to author, May 15, 2008.

3. Ben Freivogel (University of California, Berkeley), interview with author, February 4, 2008.

4. Alejandro Gangui, "Superconducting Cosmic Strings," *American Scientist* 88 (May 2000).

5. Edward Witten, "Cosmic Superstrings," *Physics Letters B* 153 (1985): 243–246.

6. Alexander Vilenkin (Tufts University), interview with author, November 23, 2004.

7. David F. Chernoff (Cornell University), personal communication, February 8, 2010.

8. Tye, interview with author, February 8, 2010.

9. Alexander Vilenkin, "Alexander Vilenkin Forecasts the Future," *New Scientist*, November 18, 2006.

10. Igor Klebanov (Princeton University), interview with author, April 26, 2007.

11. Rachel Bean (Cornell University), interview with author, April 25, 2007.

12. Joe Polchinski (University of California, Santa Barbara), interview with author, April 23, 2007.

13. Gary Shiu (University of Wisconsin), e-mail letter to author, May 19, 2007.

14. Tye, interview with author, January 22, 2007.

15. Cliff Burgess (Perimeter Institute), interview with author, April 13, 2007.

16. Bret Underwood (University of Wisconsin), interview with author, April 13, 2007.

17. Liam McAllister (Princeton University), interview with author, January 26, 2007.

18. Shiu, interview with author, May 19, 2007.

19. Sarah Shandera (Columbia University), interview with author, May 15, 2007.

20. Tye, interview with author, January 28, 2008.

21. Shiu, interview with author, May 19, 2007.

22. McAllister, interview with author, January 26, 2007.

23. Bean, interview with author, April 25, 2007.

24. Burt Ovrut (University of Pennsylvania), interview with author, February 2, 2007.

25. Underwood, interview with author, July 20, 2007.

26. Shiu, interview with author, July 20, 2007.

27. Polchinski, interview with author, August 31, 2007.

28. The Eöt-Washington Group, Laboratory Tests of Gravitational and Sub-Gravitational Physics, "Short-Range Tests of Newton's Inverse-Square Law," 2008–2009, http://www.npl.washington.edu/eotwash/experiments/shortRange/sr.html.

29. Shamit Kachru (Stanford University), interview with author, February 1, 2007.

30. K. C. Cole, "A Theory of Everything," *New York Times Magazine,* October 18, 1987.

CHAPTER 13

1. Peter Goddard, ed., *Paul Dirac: The Man and His Work* (New York: Cambridge University Press, 1998).

2. Eugene Wigner, "The Unreasonable Effectiveness of Mathematics in the Natural Sciences," *Communications in Pure and Applied Mathematics* 13 (February 1960).

3. Chen Ning Yang, *S. S. Chern: A Great Geometer of the 20th Century* (Boston: International Press, 1998), p. 66.

4. Robert Osserman, *Poetry of the Universe* (New York: Anchor Books, 1996), pp. 142–143.

5. Richard P. Feynman, *The Character of Physical Law* (New York: Modern Library, 1994), p. 50.

6. Michael Atiyah, quoted in Patricia Schwarz, "Sir Michael Atiyah on Math, Physics and Fun," The Official String Theory Web site, http://www.superstringtheory.com/people/atiyah.html.

7. Jim Holt, "Unstrung," *New Yorker*, October 2, 2006, p. 86.

8. Michael Atiyah, "Pulling the Strings," *Nature* 438 (December 22–29, 2005): 1081.

9. Robert Mills, "Beauty and Truth," in *Chen Ning Yang: A Great Physicist of the 20th Century,* ed. Shing-Tung Yau and C. S. Liu (Boston: International Press, 1995), p. 199.

10. Henry Tye (Cornell University), e-mail letter to author, December 19, 2008.

11. Brian Greene, interview by Ira Flatow, "Big Questions in Cosmology," *Science Friday*, NPR, April 3, 2009.

12. K. C. Cole, "A Theory of Everything," *New York Times Magazine*, October 18, 1987.

13. Nicolai Reshetikhin (University of California, Berkeley), interview with author, June 5, 2008.

14. Robbert Dijkgraaf (University of Amsterdam), interview with author, February 8, 2007.

15. Brian Greene, *The Elegant Universe* (New York: Vintage Books, 2000), p. 210.

16. Andrew Strominger (Harvard University), interview with author, August 1, 2007.

17. S. Ramanujan, "On Certain Arithmetic Functions," *Transactions of the Cambridge Philosophical Society* 22 (1916): 159–184.

18. Lothar Goettsche, "A Conjectural Generating Function for Numbers of Curves on Surfaces," November 11, 1997, arXiv.org, Cornell University archives, http://arxiv .org/PS_cache/alg-geom/pdf/9711/9711012v1.pdf.

19. Ai-Ko Liu, "Family Blowup Formula, Admissible Graphs and the Enumeration of Singular Curves, I," *Journal of Differential Geometry* 56 (2000): 381–579.

20. Bong Lian (Brandeis University), interview with author, December 12, 2007.

21. Michael Atiyah, "Pulling the Strings," *Nature* 438 (December 22–29, 2005): 1082.

22. Glennda Chui, "Wisecracks Fly When Brian Greene and Lawrence Krauss Tangle over String Theory," *Symmetry* 4 (May 2007): 17–21.

23. Sean Carroll, "String Theory: Not Dead Yet," Cosmic Variance blog, *Discover* online magazine, May 24, 2007, http://cosmicvariance.com.

24. Edward Witten, quoted in K. C. Cole, "A Theory of Everything," *New York Times Magazine*, October 18, 1987.

25. Ibid.

26. Alan Guth (MIT), interview with author, September 13, 2007.

27. Greene, *The Elegant Universe* (New York: Vintage Books, 2000), p. 261.

28. Max Tegmark (MIT), interview with author, October 23, 2007.

29. Faye Flam, "Getting Comfortable in Four Dimensions," *Science* 266 (December 9, 1994): 1640.

CHAPTER 14

1. Cumrun Vafa (Harvard University), interview with author, January 19, 2007.

2. Robbert Dijkgraaf (University of Amsterdam), interview with author, February 8, 2007.

3. David Morrison (University of California, Santa Barbara), interview with author, May 27, 2008.

4. Allan Adams (MIT), interview with author, May 23, 2008.

5. Joe Polchinski (University of California, Santa Barbara), interview with author, August 31, 2007.

6. Edward Witten (Institute for Advanced Study), e-mail letter to author, January 30, 2007.

7. Adams, interview with author, May 23, 2008.

8. Turnbull WWW Server, "Quotations by Gauss," School of Mathematical Sciences, University of St. Andrews, St. Andrews, Fife, Scotland, February 2006, http://www-groups .dcs.st-and.ac.uk/~history/Quotations/Gauss.html.

9. Adams, interview with author, May 23, 2008.

10. Brian Greene, "String Theory on Calabi-Yau Manifolds," lectures given at Theoretical Advanced Study Institute, 1996 session (TASI-96), Boulder, CO, June 1996.

11. K. C. Cole, "Time, Space Obsolete in New View of Universe," *Los Angeles Times*, November 16, 1999.

12. Andrew Strominger (Harvard University), interview with author, August 1, 2007.

13. Morrison, interview with author, May 29, 2008.

14. Brian Greene, *Elegant Universe* (New York: Vintage Books, 2000), pp. 268, 273.

15. Paul Aspinwall (Duke University), interview with author, June 6, 2008.

16. Morrison, interview with author, May 27, 2008.

EPILOGUE

1. Andrew Strominger (Harvard University), interview with author, February 7, 2007.

2. Ibid.

3. Chris Beasley, Jonathan Heckman, and Cumrun Vafa, "GUTs and Exceptional Branes in F-Theory—I," November 18, 2008, http://arxiv.org/abs/0802.3391; Chris Beasley, Jonathan Heckman, and Cumrun Vafa, "GUTs and Exceptional Branes in F-Theory—II: Experimental Predictions," June 12, 2008, http://arxiv.org/abs/arxiv:0806.0102; Ron Donagi and Martijn Wijnholt, "Model Building with F-Theory," March 3, 2008, http://lanl.arxiv.org/pdf/0802.2969v2; and Ron Donagi and Martijn Wijnholt, "Breaking GUT Groups in F-Theory," August 17, 2008, http://lanl.arxiv.org/pdf/0808.2223v1.

4. Strominger, interview with author, July 23, 2007.

5. Cumrun Vafa (Harvard University), interview with author, November 2, 2007.

6. Leonard Mlodinow, *Euclid's Mirror* (New York: Simon & Schuster, 2002), p. 255.

7. Edward Witten (Institute for Advanced Study), e-mail letter to author, February 12, 2007.

8. David Morrison (University of California, Santa Barbara), interview with author, May 27, 2008.

9. Ravi Vakil (Stanford University), interview with author, May 28, 2008.

10. Strominger, interview with author, August 1, 2007.

POSTLUDE

1. Piers Bursill-Hall, "Why Do We Study Geometry? Answers Through the Ages," lecture, Faulkes Institute for Geometry, University of Cambridge, May 1, 2002.

2. Donald Zeyl (University of Rhode Island), interview with author, October 18, 2007.

GLOSSARY

algebraic geometry: a branch of mathematics that applies algebraic techniques—particularly those involving polynomial equations—to problems in geometry.

anisotropy: a property that varies in magnitude depending on the direction of measurement. Astronomers, for example, have detected variations in temperature—hot spots and cold spots—at different points in the sky, which are indications of temperature (and density) of anisotropies.

anomaly: a symmetry violation that is not apparent in classical theory but becomes evident when quantum effects are taken into consideration.

anthropic principle: the notion that the observed laws of nature must be consistent with the presence of intelligent life and, specifically, the presence of intelligent observers like us. Put in other terms, the universe looks the way it does because if conditions were even slightly different, life would not have formed and humans would not be around to observe it.

Big Bang: a theory that holds that our universe started from a state of extremely high temperature and density and has been expanding ever since.

black hole: an object so dense that nothing, not even light, can escape its intense gravitational field.

boson: one of two kinds of particles found in quantum theory, the other being fermions. Bosons include "messenger particles" that are carriers of the fundamental forces. (See *fermion.*)

brane: the basic object of string theory and M-theory, which can assume the form of a one-dimensional string and higher-dimensional objects, including a two-dimensional sheet or "membrane" (from which the term originated). When string theorists speak of branes, they're generally referring to these higher-dimensional objects rather than to strings.

bundle (or vector bundle or fiber bundle): a topological space that is constructed from, or attached to, a manifold. To picture it, let's assume the manifold is something familiar like a sphere. Then assume, to pick the simplest example, that a particular array of vectors (or a vector space) is attached to every point on the surface of the

345

sphere. The bundle consists of the entire manifold—the sphere, in this case—plus all the arrays of vectors attached to it at every point. (See *tangent bundle*.)

Calabi conjecture: a mathematical hypothesis, put forth in the early 1950s by the geometer Eugenio Calabi, stating that spaces that satisfy certain topological requirements can also satisfy a stringent geometric (curvature) condition known as "Ricci flatness." The conjecture also covered more general cases, where the Ricci curvature was not zero.

Calabi-Yau manifold: a broad class of geometric spaces with zero Ricci curvature that were shown to be mathematically possible in the proof of the Calabi conjecture. These spaces, or shapes, are "complex," meaning that they must be of even dimension. The six-dimensional case is of special interest to string theory, where it serves as a candidate for the geometry of the theory's six hidden, or "extra," dimensions.

calculus: a set of tools—involving derivatives, integrals, limits, and infinite series—that were introduced to "modern" mathematics by Isaac Newton and Gottfried Leibniz.

Cartesian product: a way of combining two distinct geometric objects to create a new shape. The product of a circle and a line, for instance, is a cylinder. The product of two circles is a two-dimensional torus, or donut.

Chern class: a set of fixed properties, or invariants, that are used for characterizing the topology of complex manifolds. The number of Chern classes for a particular manifold equals the number of complex dimensions, with the last (or "top") Chern class being equal to the Euler characteristic. Chern classes are named after the geometer S. S. Chern, who introduced the concept in the 1940s.

classical physics: physical laws, mostly developed before the twentieth century, that do not incorporate the principles of quantum mechanics.

compact space: a space that is bounded and finite in extent. A sphere is compact, whereas a plane is not.

compactification: the rolling up of a space so that it is "compact" or finite in extent. In string theory, different ways of wrapping, or compactifying, the extra dimensions lead to different physics.

complex manifold: a manifold that can be described in a mathematically consistent way by complex coordinates—its ordinary or real dimension being twice the complex dimension. All complex manifolds are also real manifolds of even dimension. However, not all real manifolds of even dimension are complex manifolds, because, in some cases, it is not possible to describe the entire manifold consistently with complex numbers. (See *manifold*.)

complex numbers: numbers of the form $a + bi$, where a and b are real numbers and i is the square root of -1. Complex numbers can be broken down into two components, with a called the real part and b the imaginary part.

conformal field theory: a quantum field theory that retains scale invariance and conformal invariance. Whereas in an ordinary quantum field theory, the strong force that binds quarks changes with distance, in conformal field theory, that force remains the same at any distance.

conformal invariance: transformations that preserve angles. The notion of conformal invariance includes scale invariance, as changes in scale—such as uniformly blowing up or shrinking a space—also leave angles intact. (See *scale invariance*.)

conifold: a cone-shaped singularity. Singularities of this sort are commonly found on a Calabi-Yau manifold.

conifold transition: a process during which space tears in the vicinity of a conifold singularity on a Calabi-Yau manifold and is then repaired in a way that changes the topology of the original manifold. Topologically distinct Calabi-Yau manifolds can thus be linked through a conifold transition.

conjecture: a mathematical hypothesis that is initially proposed without a complete proof.

convex: an object, such as a sphere, that curves or bulges "outward" such that a line segment connecting every two points within that object also lies within that object.

coordinates: numbers that specify the position of a point in space or in spacetime. Cartesian coordinates, for example, are the standard coordinates on a plane in which each point is specified by two numbers, one being the distance from the origin in the x direction and the other being the distance from the origin in the y direction. This coordinate system is named after the French mathematician (and philosopher) René Descartes. More coordinates are required, of course, to localize a point in a higher-dimensional space.

cosmic microwave background: electromagnetic (microwave) radiation that is left over from the Big Bang and that has since cooled and diffused and now permeates the universe.

cosmic strings: one-dimensional objects—which can assume the form of long, extremely thin, and extremely massive filaments—that are predicted by some quantum field theory models to have formed during a phase transition in the early universe. Cosmic strings also arise naturally in some versions of string theory, corresponding to the fundamental strings of those theories.

cosmological constant: a term that counters the effects of gravity in the famous Einstein equation; the constant corresponds to the energy locked up in spacetime itself. The cosmological constant is basically the vacuum energy—a form of energy thought to pervade all of spacetime, thereby offering a possible explanation for the phenomenon of dark energy. (See *dark energy* and *vacuum energy*.)

coupling constant: a number that determines the strength of a physical interaction. The string coupling constant, for instance, governs the interactions of strings, determining how likely it is for one string to split into two, or for two strings to come together to make one.

cubic equation: an equation whose highest term is third order, as in $ax^3 + bxy^2 + cy + d = 0$.

curvature: a quantitative way of measuring the extent to which a surface or space deviates from flatness. For example, the curvature of a circle is given by the inverse of its radius: The smaller the curvature of a circle, the larger its radius. In more than one dimension, curvature not only is given by a number but also takes into account the different directions along which a manifold can curve. While two-dimensional surfaces can be completely described by one kind of curvature, different kinds of curvature are possible in higher dimensions.

dark energy: a mysterious form of energy, constituting more than 70 percent of the universe's total energy, according to recent measurements. Dark energy may be the measured value of the vacuum energy. Cosmologists believe it is causing the universe to expand in an accelerated fashion.

dark matter: nonluminous matter of unknown form whose presence has been inferred but not directly detected. Dark matter is thought to comprise the bulk of the universe's matter, accounting for about 25 percent of the universe's total energy.

D-brane: a brane or multidimensional surface in string theory upon which *open strings* (those that are not closed loops) can end.

decompactification: the process during which curled up, compact dimensions "unwind" and become infinitely large.

derivative: the measure of how a function, or quantity, changes with respect to a particular variable or variables. For a given input (or number), a function yields a specific output (or number). The derivative measures how the output changes as the input itself deviates slightly from the original value. If one were to graph a function on, say, the *x*-*y* plane, the derivative of that function at a particular point equals the slope of the tangent line at that point.

differentiable: a term that applies to "smooth" functions whose derivative can be taken at every point. A function is called infinitely differentiable if a limitless number of derivatives can be taken.

differential equation: an equation involving derivatives that shows how something changes with respect to one or more variables. Ordinary differential equations involve just one variable, whereas partial differential equations involve two or more independent variables. When processes in the physical and natural world are described mathematically, it is usually through differential equations.

differential geometry: the branch of geometry involving calculus (as opposed to algebra) that studies how the property of a space, such as its curvature, changes as you move about the space.

dimension: an independent direction, or "degree of freedom," in which one can move in space or time. We can also think of the dimensionality of a space as the minimum number of coordinates needed to specify the position of a point in space. We call a plane "two-dimensional" because just two numbers—an *x* and a *y* coordinate—are needed to specify a position. Our everyday world has three spatial dimensions (left-right, forward-backward, up-down), whereas the spacetime we're thought to inhabit has four dimensions—three of space and one of time. In addition, string theory (among other theories) holds that spacetime has additional spatial dimensions that are small, curled up, and concealed from view.

Dirac equation: a set of four interconnected equations, formulated by the British physicist Paul Dirac, that describe the behavior and dynamics of freely moving (and hence noninteracting) "spin-½" particles such as electrons.

duality: two theories that, at least superficially, appear to be different yet give rise to identical physics.

Einstein equations: the equations of general relativity that describe gravity, taking into account the theory of special relativity. Expressed in other terms, these equations can be used to determine the curvature of spacetime due to the presence of mass and energy.

electromagnetic force: one of four known forces in nature, this force combines electricity and magnetism.

elementary particle: a particle that is not known to have any substructure. Quarks, leptons, and gauge bosons are the elementary particles of the Standard Model—particles that we believe to be indivisible and fundamental.

entropy: a measure of the disorder of a physical system, with disordered systems having large entropy and ordered systems having low entropy. The entropy can also be thought of as the number of ways of arranging a system's ingredients without changing the system's overall properties (such as its volume, temperature, or pressure).

Euclidean geometry: the mathematical system attributed to the Greek mathematician Euclid in which the Pythagorean theorem always holds; the angles of a triangle always add up to 180 degrees; and for a straight line and a point not on that straight line, only one line can be drawn through that point (in the same plane) that does not intersect the given line. (This is the so-called parallel postulate.) Other versions of geometry have subsequently been developed, falling under the rubric of "non-Euclidean," where these principles do not always hold.

Euler characteristic (or Euler number): an integer that helps characterize a topological space in a very general sense. The Euler characteristic, the simplest and the oldest known "topological invariant" of a space, was first introduced by Leonhard Euler for polyhedra and has since been generalized to other spaces. The Euler characteristic of a polyhedron, for example, is given by the number of vertices minus the number of edges plus the number of faces.

event horizon: the surface surrounding a black hole beyond which nothing, not even light, can escape.

family (of particles): See *generation.*

fermion: a particle of half-integer spin. This class of particles includes quarks and leptons, the so-called matter particles of the Standard Model.

field: a physical concept, introduced by the nineteenth-century physicist Michael Faraday, that assigns a specific value, such as a number or vector, to each point in spacetime. While a field can describe the force exerted on a particle at a given point in space, it can also describe the particle itself.

field theory: a theory in which both particles and forces are described by fields.

flux: lines of force, like the familiar electric and magnetic fields, that correspond to the special fields of string theory.

function: a mathematical expression of the form, for example, of $f(x) = 3x^2$, where every input value of x leads to a single output value for the function $f(x)$.

fundamental group: a way of classifying spaces in topology. In spaces with a *trivial* fundamental group, every loop you can draw in that space can be shrunk down to a point without tearing a hole in the space. Spaces with a *nontrivial* fundamental group have noncontractible loops—that is, loops that cannot be shrunk down to a point owing to the presence of some obstruction (such as a hole).

gauge theory: a field theory, such as the Standard Model, in which symmetries are "gauged." If a particular symmetry is gauged (which we then refer to as a gauge symmetry), that symmetry can be applied differently to a field at different points in spacetime, and yet the physics doesn't change. Special fields called gauge fields must be added to the theory so that the physics remains invariant when symmetries are gauged.

Gaussian: a random probability distribution that is sometimes called a bell curve. This probability distribution is named after the geometer Carl Friedrich Gauss, who used it in his astronomical analyses, among other applications.

general relativity: Albert Einstein's theory that unites Newtonian gravity with his own theory of special relativity. General relativity describes the gravitational potential as a metric and the gravitational force as the curvature of four-dimensional spacetime.

generation (or family): the organization of matter particles into three groups, each consisting of two quarks and two leptons. The particles in these groups would be identical, except that the masses increase with each generation.

genus: simply put, the number of holes in a two-dimensional surface or space. An ordinary donut, for instance, is of genus 1, whereas a sphere, which lacks a hole, is of genus 0.

geodesic: a path that is generally the shortest distance between two points on a given surface. On a two-dimensional plane, this path is a line segment. On a two-dimensional sphere, the geodesic lies along a so-called great circle that passes through the two points and has its center in the middle of the sphere. Depending on which way one travels on this great circle, the geodesic can be either the shortest path between those two points or the shortest path between those points compared with any path nearby.

geometric analysis: a mathematical approach that applies the techniques of differential calculus to geometric problems.

geometry: the branch of mathematics that concerns the size, shape, and curvature of a given space.

gravitational waves: disturbances or fluctuations of the gravitational field due to the presence of massive objects or localized sources of energy. These waves travel at the speed of light, as predicted by Einstein's theory of gravity. Although there have been no direct detections of gravitational waves, there has been indirect evidence that they exist.

gravity: the weakest of the four forces of nature on the basis of current measurements. Newton viewed gravity as the mutual attraction of two massive objects, whereas Einstein showed that the force can be thought of in terms of the curvature of spacetime.

Heisenberg uncertainty principle: See *uncertainty principle.*

heterotic string theory: a class that includes two of the five string theories—the $E_8 \times E_8$ and the $SO(32)$ theories—which differ in terms of their choice of (gauge) symmetry groups. Both heterotic string theories involve only "closed" strings or loops rather than open strings.

Higgs field: a hypothetical field—a component of which consists of the Higgs boson or Higgs particle—that is responsible for endowing particles in the Standard Model with mass. The Higgs field is expected to be observed for the first time at the Large Hadron Collider.

Hodge diamond: a matrix or an array of Hodge numbers that provides detailed topological information about a Kähler manifold from which one can determine the Euler characteristic and other topological features. The Hodge diamond for a six-dimensional Calabi-Yau manifold consists of a four-by-four array. (Arrays of different sizes correspond to spaces of other even dimensions.) Hodge numbers, which are named after the Scottish geometer W. V. D. Hodge, offer clues into a space's internal structure.

holonomy: a concept in differential geometry, related to curvature, that involves moving vectors around a closed loop in a parallel manner. The holonomy of a surface (or manifold), loosely speaking, is a measure of the degree to which tangent vectors are transformed as one moves around a loop on that surface.

inflation: a postulated exponential growth spurt during the universe's first fraction of a second. The idea, first suggested by the physicist Alan Guth in 1979, simultaneously solves many cosmological puzzles, while also helping to explain the origin of matter and our expanding universe. Inflation is consistent with observations in astronomy and cosmology but has not been proven.

integral: one of the principal tools of calculus, taking an integral (or integration) offers a way of finding the area bounded by a curve. The calculation breaks up the bounded region into infinitesimally thin rectangles and adds up the areas of all the rectangles contained therein.

invariant (or topological invariant): a number or another fixed property of a space that does not change under the allowed transformations of a given mathematical theory. A topological invariant, for example, does not change under the continuous deformation (such as stretching, shrinking, or bending) of the original space from one shape to another. In Euclidean geometry, an invariant does not change under translations and rotations. In a conformal theory, an invariant does not change under conformal transformations that preserve angles.

K3 surface: a Calabi-Yau manifold of four real dimensions—or, equivalently, of two complex dimensions—named after the geometers Ernst Kummer, Erich Kähler, and Kunihiko Kodaira, the three K's of K3. The name of these surfaces, or manifolds, also refers to the famous Himalayan mountain K2.

Kähler manifold: a complex manifold named after the geometer Erich Kähler and endowed with a special kind of holonomy that preserves the manifold's complex structure under the operation of parallel transport.

Kaluza-Klein theory: Originally an attempt to unify general relativity and electromagnetism by introducing an extra (fifth) dimension, *Kaluza-Klein* is sometimes used as shorthand for the general approach of unifying the forces of nature by postulating the existence of an additional, unseen dimension (or dimensions).

landscape: in string theory, this is the range of possible shapes, or geometries, that the unseen dimensions could assume, which also depends on the number of ways that fluxes can be placed in that internal space. Put in other terms, the landscape consists of the range of possible vacuum states, or vacua, that are allowed by string theory.

lemma: a proven statement in mathematics that, rather than being considered an endpoint in itself, is normally regarded as a stepping-stone toward the proof of a broader, more powerful statement. But lemmas can be useful in themselves, too—sometimes more than initially realized.

lepton: the class of elementary particles that includes electrons and neutrinos. Unlike quarks, which are the other fermions, leptons do not experience the strong force and hence do not get trapped in atomic nuclei.

linear equation: an equation (in the case of two variables) of the general form $ax + by + c = 0$. An equation of this sort has no higher-order terms (such as x^2, y^2, or xy) and maps out a straight line. Another key feature of a linear equation is that a change in one variable, x, leads to a proportional change in the other variable, y, and vice versa. However, linear equations need not have only two variables, x and y, and can instead have any number of variables.

manifold: a topological space that is locally Euclidean, meaning that every point lies in a neighborhood that resembles flat space.

matrix: a two-dimensional (rectangular or square) array of numbers or more complicated algebraic entities. Two matrices can be added, subtracted, multiplied, and divided according to a relatively simple set of rules. A matrix can be expressed in the abbreviated form, a_{ij} where i is the row number and j the column number.

metric: a mathematical object (technically called a tensor) used to measure distances on a space or manifold. On a curved space, the metric determines the extent to which the actual distance deviates from the number given by the Pythagorean formula. Knowing a space's metric is equivalent to knowing the geometry of that space.

minimal surface: a surface whose area is "locally minimized," meaning that the surface area cannot be reduced by replacing small patches with any other possible nearby surface in the same ambient (or background) space.

mirror symmetry: a correspondence between two topologically distinct Calabi-Yau manifolds that give rise to the exact same physical theory.

moduli space: For a given topological object such as a Calabi-Yau manifold, the moduli space consists of the set of all possible *geometric* structures—the continuous set of manifolds encompassing all possible shape and size settings.

M-theory: a theory that unites the five separate string theories into a single, all-embracing theory with eleven spacetime dimensions. The principal ingredients of M-theory are branes, especially the two-dimensional (M2) and five-dimensional (M5) branes. In

M-theory, strings are considered one-dimensional manifestations of branes. M-theory was introduced by Edward Witten—and largely conceived by him as well—during the "second string revolution" of 1995.

neutron star: a dense star, composed almost entirely of neutrons, that forms as a remnant following the gravitational collapse of a massive star that has exhausted its nuclear fuel.

Newton's constant: the coefficient G, which determines the strength of gravity according to Newton's law. Although Newton's law has, of course, been supplanted by Einstein's general relativity, it still remains a good approximation in many cases.

non-Euclidean geometry: the geometry that applies to spaces that are not flat, such as a sphere, where parallel lines can intersect, contrary to Euclid's fifth postulate. In a non-Euclidean space, the sum of the angles of a triangle may be more or less than 180 degrees.

non-Kähler manifold: a class of complex manifolds that includes Kähler manifolds but also includes manifolds that cannot support a Kähler metric.

nonlinear equation: an equation that is not linear, meaning that changing one variable can lead to a disproportionate change in another variable.

orthogonal: perpendicular.

parallel transport: a way of moving vectors along a path on a surface or manifold that keeps the lengths of those vectors constant while keeping the angles between any two vectors constant as well. Parallel transport is easy to visualize on a flat, two-dimensional plane, but in more complicated, curved spaces, we may need to solve differential equations to determine the precise way of moving vectors around.

phase transition: the sudden change of matter, or a system, from one state to another. Boiling, freezing, and melting are familiar examples of phase transitions.

Planck scale: a scale of length (about 10^{-33} centimeters), time (about 10^{-43} seconds), energy (about 10^{28} electron-volts), and mass (about 10^{-8} kilograms) at which the quantum mechanical effects on gravity must be taken into account.

Platonic solids: the five "regular" (convex) polyhedrons—the tetrahedron, hexahedron (cube), octahedron, dodecahedron, and icosahedron—that satisfy the following properties: Their faces are made up of congruent polygons with every edge being the same length, and the same number of faces meet at every vertex. The Greek philosopher Plato theorized that the elements of the universe were composed of these solids, which were subsequently named after him.

Poincaré conjecture (in three dimensions): a famous conjecture posed by Henri Poincaré more than a century ago, which holds that if any loop drawn in a three-dimensional space can be shrunk to a point, without ripping the space or the loop itself, then that space is equivalent, topologically speaking, to a sphere.

polygon: a closed path in geometry composed of line segments such as a triangle, square, and pentagon.

polyhedron: a geometric object consisting of flat faces that meet at straight edges. Three-dimensional polyhedrons consist of polygonal faces that meet at edges, and the edges, in turn, meet at vertices. The tetrahedron and cube are familiar examples.

polynomial: functions that involve addition, subtraction, and multiplication, and non-negative, whole-number exponents. Although polynomial equations may look simple at first glance, they are often very difficult (and sometimes impossible) to solve.

positive mass theorem (or positive energy theorem): the statement that in any isolated physical system, the total mass or energy must be positive.

product: the result of multiplying two or more numbers (or other quantities).

Pythagorean theorem: a formula dictating that the sum of the squares of the two sides of a right triangle equals the square of the hypotenuse ($a^2 + b^2 = c^2$).

quadratic equation: an equation with a second-order (or "squared") term, as in $ax^2 + bx + c = 0$.

quantum field theory: a mathematical framework that combines quantum mechanics and field theory. Quantum field theories serve as the principal formalism underlying particle physics today.

quantum fluctuations: random variations on submicroscopic scales due to quantum effects such as the uncertainty principle.

quantum geometry: a form of geometry thought to be necessary to provide realistic physical descriptions on ultramicroscopic scales, where quantum effects become important.

quantum gravity: a sought-after theory that would unite quantum mechanics and general relativity and provide a microscopic or quantum description of gravity that's comparable to the descriptions we have for the three other forces. String theory offers one attempt at creating a theory of quantum gravity.

quantum mechanics: a set of laws dictating the behavior of the universe on atomic scales. Quantum mechanics holds, among other things, that a particle can be equivalently expressed as a wave, and vice versa. Another central notion is that in some situations, physical quantities like energy, momentum, and charge only come in discrete amounts called *quanta*, rather than assuming any possible value.

quark: the class of elementary, subatomic particles—of which there are six varieties in all—that make up protons and neutrons. Quarks experience the strong force, unlike the other fermions, leptons, which do not.

relativistic: a term that applies to particles or other objects traveling at velocities that approach the speed of light.

Ricci curvature: a kind of curvature that is related to the flow of matter in spacetime, according to Einstein's equations of general relativity.

Riemann surface: a one-dimensional complex manifold or, equivalently, a two-dimensional real surface. In string theory, the surface swept out by a string moving through spacetime is considered a Riemann surface.

Riemannian geometry: a mathematical framework for studying the curvature of spaces of arbitrary dimension. This form of geometry, introduced by Georg Friedrich Bernhard Riemann, lies at the heart of general relativity.

scalar field: a field that can be completely described by a single number at each point in space. A number corresponding to the temperature at each point in space is one example of a scalar field.

scale invariance: a phenomenon that is true, regardless of physical scale. In a scale-invariant system, the physics remains unchanged if the size of a system (or the notion of distance in the system) is uniformly expanded or shrunk.

singularity: a point in spacetime where the curvature and other physical quantities such as density become infinite, and conventional laws of physics break down. The center of a black hole and the moment of the Big Bang are both thought to be singularities.

slope: a term denoting the steepness or gradient of a curve—a measure of how much the steepness changes vertically compared with changes in the horizontal direction.

smooth: infinitely differentiable. A *smooth manifold* is a manifold that is everywhere differentiable, arbitrarily often, which means that a derivative can be taken at any point on the manifold, as many times as one cares to.

spacetime: In the four-dimensional version, spacetime is the union of the three dimensions of space with the single dimension of time to create a single, combined entity. This notion was introduced at the turn of the twentieth century by Albert Einstein and Hermann Minkowski. The concept of spacetime, however, is not restricted to four dimensions. String theory is based on a ten-dimensional spacetime, and M-theory, to which it is related, is based on an eleven-dimensional spacetime.

special relativity: a theory devised by Einstein that unifies space and time, stating that the laws of physics should be the same for all observers moving at constant velocity, regardless of their speed. The speed of light (c) is the same for all observers, according to special relativity. Einstein also showed that for a particle at rest, energy (E) and mass (m) are related by the formula $E = mc^2$.

sphere: as employed by geometers, the term typically refers to the two-dimensional surface of a ball, rather than the three-dimensional object itself. The concept of the sphere, however, is not limited to two dimensions and can instead apply to objects of any dimension, from zero on up.

Standard Model: a theory of particle physics that describes the known elementary particles and the interactions (strong, weak, and electromagnetic) between them. Gravity is not included in the Standard Model.

string theory: a physical theory, incorporating both quantum mechanics and general relativity, that is widely regarded as the leading candidate for a theory of quantum gravity. String theory posits that the fundamental building blocks of nature are not pointlike particles but are instead tiny, one-dimensional strands called strings, which come in either open or closed (loop) forms. There are five varieties of string theories—Type I, Type IIA, Type IIB, Heterotic $E_8 \times E_8$, Heterotic $SO(32)$—all of which are related to each other. The term *superstring theory* is sometimes used in place of *string theory* to explicitly show that the theory incorporates supersymmetry.

strong force: the force responsible for binding quarks inside protons and neutrons and for keeping protons and neutrons together to form atomic nuclei.

submanifold: a space of lower dimension sitting inside a higher-dimensional space. One can think of a donut, for example, as a continuous ring of circles, with each of those circles being a submanifold within the bigger structure or manifold—that being the donut itself.

superpartners: pairs of particles, consisting of a fermion and a boson, that are related to each other through supersymmetry.

supersymmetry: a mathematical symmetry that relates fermions to bosons. It is important to note that the bosons inferred by supersymmetry, which would be "superpartners" to the known quarks and leptons, have yet to be observed. Although supersymmetry is an important feature of most string theories, finding firm evidence of it would not necessarily prove that string theory is "right."

symmetry: an action on an object, physical system, or equation in a way that leaves it unchanged. A circle, for example, remains unchanged under rotations about its center. A square and an equilateral triangle, similarly, remain unchanged under rotations about the center of 90 and 120 degrees, respectively. A square is not symmetric, however, under rotations of 45 degrees, as its appearance would change to what is sometimes called a square diamond that's tipped on one corner.

symmetry breaking: a process that reduces the amount of observed symmetry in a system. Bear in mind that after symmetry is "broken" in this manner, it can still exist, though remaining hidden rather than visible.

symmetry group: a specific set of operations—such as rotations, reflections, and translations—that leaves an object invariant.

tangent: the best linear approximation to a curve at a particular point on that curve. (The same definition holds for higher-dimensional curves and their tangents.)

tangent bundle: a particular type of bundle made by attaching a tangent space to every point on the manifold. The tangent space encompasses all the vectors that are tangent to the manifold at that point. If the manifold is a two-dimensional sphere, for instance, then the tangent space is a two-dimensional plane that contains all the tangent vectors. If the manifold is a three-dimensional object, then the tangent space will be three-dimensional as well. (See *bundle.*)

tension: a quantity that measures a string's resistance to being stretched or vibrated. A string's tension is similar to its linear energy density.

theorem: a statement or proposition proven through formal, mathematical reasoning.

topology: a general way of characterizing a geometrical space. Topology concerns itself only with the gross features of that space rather than its exact size or shape. In topology, shapes are classified into groups that can be deformed into each other by stretching or compressing, without tearing their structure or changing the number of interior holes.

torus (plural, tori): a class of topological shapes that include the two-dimensional surface of a donut, plus higher-dimensional generalizations thereof.

tunneling (or quantum tunneling): a phenomenon, such as a particle crossing through a barrier into a different region, that is forbidden according to classical physics but is allowed (or has a nonzero probability) in quantum physics.

uncertainty principle (also known as the Heisenberg uncertainty principle): a tenet of quantum mechanics that holds that both the position and the momentum of an object cannot be known with absolute precision. The more precisely one of those variables is known, the greater the uncertainty that is attached to the other.

unified field theory: an attempt to account for all the forces of nature within a single, overarching framework. Albert Einstein devoted the last thirty years of his life to this goal, which has not yet been fully realized.

vacuum: a state, essentially devoid of matter, that represents the lowest possible energy density, or *ground state*, of a given system.

vacuum energy: the energy associated with "empty" space. The energy carried by the vacuum is not zero, however, because in quantum theory, space is never quite empty. Particles are continually popping into existence for a fleeting moment and then disappearing into nothingness. (See *cosmological constant*.)

vector: a geometric object (a line segment in one dimension) that has both length (or magnitude) and direction.

vertex: the point at which two or more edges of a shape meet.

warping (related to "warp factor" and "warped product"): the idea that the geometry of the four-dimensional spacetime we inhabit is not independent of the hidden extra dimensions but is instead influenced by the internal dimensions of string theory.

weak force: this force, one of the four known forces of nature, is responsible for radioactive decay, among other processes.

world sheet: the surface traced out by a string moving in spacetime.

Yang-Mills equations: generalizations of Maxwell's equations that describe electromagnetism. The Yang-Mills equations are now used by physicists to describe the strong and weak forces, as well as the electroweak force that combines the electromagnetic and weak interactions. The equations are part of Yang-Mills theory or *gauge theory*, which was developed in the 1950s by the physicists Chen Ning Yang and Robert Mills.

Yukawa coupling constant: a number that determines the strength of coupling, or interaction, between a scalar field and a fermion—a noted example being the interactions of quarks or leptons with the Higgs field. Since the mass of particles stems from their interactions with the Higgs field, the Yukawa coupling constant is closely related to particle mass.

INDEX

Adams, Allan
  Calabi-Yau manifolds, 242, 243
  on geometry, 309–310, 312
  hidden dimensions, 138
  Standard Model, 224
Adelberger, Eric, 286–287
Ahlfors, Lars/theorem, 116
Algebraic geometry
  about, 180–181
  B-branes, 180
  enumerative geometry, 165–166,
    167 (fig.), 169, 173, 296, 300
  mirror symmetry and, 165–166, 169,
    173, 296
Allen, Woody, 137
Ampère, André-Marie, 106
  See also Monge-Ampère equations
Anisotropy, 271
Anomaly cancellation, 125–126, 212,
  213, 214
Anti-deSitter Space/Conformal Field
    Theory (AdS/CFT), 196–198
Antibrane, 233, 277–278
Apollonius/problem, 166, 168 (fig.),
  290
A priori estimates, 111–113
Archimedes/work, 290, 291 (fig.)
Aristotle, 3, 4

Arkani-Hamed, Nima, 285
Aspinwall, Paul, 151, 157, 181, 317–318
Associative law, 41
Atiyah, Michael, 292, 293, 301, 303, 304
Aubin, Thierry, 116, 122

Bass, Hyman, 55
Bean, Rachel, 277, 281
Beasley, Chris, 322
Becker, Katrin, 173–174, 251
Becker, Melanie, 173–174, 244, 250,
  251, 252
Bekenstein, Jacob, 187, 188
Bekenstein-Hawking formula, 188, 189,
  193, 195, 263
Bell, E. T., 22
Bergmann, Peter, 13
Beta function, 153
Betti numbers, 161–162, 161 (fig.)
Big Bang
  about, 14, 184, 256, 267, 272, 278
  cosmic inflation and, 184, 231–232,
    270–271, 272, 302
Big Crunch, 256
Black holes
  about/description, 60, 184–185, 185
    (photo)
  curvature, 60, 62, 185

Black holes (*continued*)
    D-branes association, 189–193, 191
        (fig.), 195–196
    entropy, 187–188
    escape velocity, 184
    event horizon, 63, 64, 187, 190, 193,
        195, 263
    general relativity and, 184–185
    general relativity/quantum
        mechanics dispute, 185–198
    information destruction/preservation
        and, 186–187, 197
    Penrose process, 63–64
    as singularities, 19, 60, 184, 195
    supersymmetry and, 192–193,
        194–195
    trapped surface and, 60, 62
"Blue Marble, The" Earth, 315 (fig.)
Bogomolov-Miyaoka-Yau inequality, 122
    See also Severi conjecture
Boltzmann, Ludwig, 188–189
Bolyai, Farkas, 311
Bolyai, János, 29, 311, 312
Bonnet, Pierre, 94
    Gauss-Bonnet formula, 94–95
Bosons, 133, 141–142, 283
Bouchard, Vincent, 217, 218
Boundary (topology) defined, 94
Boundary value problems, 114
Bourguignon, Jean-Pierre, 222–223
Bourguignon-Li-Yau work, 222–223
Braun, Volker, 145, 199, 217, 223
Bryan, Jim, 298, 300
Bubble phenomenology, 271–272
"Buckyballs" (buckminsterfullerenes),
    290, 291 (fig.)
Bundle theory, 290
Bundles
    about, 205
    tangent bundle, 207, 208, 208 (fig.),
        210, 211–212, 213–215, 218
    See also Gauge fields

Burgess, Cliff, 278, 279
Bursill-Hall, Piers, 17, 325

Calabi, Eugenio
    about, 77, 78 (photo), 136 (photo)
    conjecture proof, 118
    Dirichlet problem, 114, 115, 116
    Monge-Ampère equations, 106, 107
    on physics/conjecture, 137
    a priori estimates, 112
    Yau's "disproving" conjecture and,
        106
Calabi conjecture
    "best" metric for, 100–101
    complex manifolds and, 80–81, 85
    description, 77–80
    Einstein equation and, 108
    general relativity and, 78–79
    gravity in vacuum, 78
    Kähler manifolds and, 89, 90, 93–94,
        100
    nonlinear equations and, 105
    Ricci curvature, 78, 79, 98, 99, 100,
        101
    skepticism on, 103–106
    summary of, 104
    zero Ricci curvature and, 100, 101
Calabi conjecture/proving
    beginnings, 106
    checking proof, 118–119
    Einstein equations and, 119, 123
    Monge-Ampère equations and,
        106–108, 113, 114, 116, 117
    nonlinear partial differential
        equations approach, 109–113
    requirements summary, 106
    significance, 79–80, 119–120
    zeroth-order estimate and, 116–118
Calabi-Yau equation, 119
Calabi-Yau manifolds or spaces
    compactness, 137–139, 138 (fig.),
        181, 228

conformal field theory, 156–157

continuing significance of, 321, 322–323

description, 15 (fig.), 19, 136 (fig.), 137–139, 138 (fig.)

doubts on, 151–154, 321–322

Einstein's differential equations and, 123

Euler characteristic and, 143–145

flop transition, 316–319, 317 (fig.)

holonomy and, 129, 131

hope for unique solution, 236

mirror symmetry, 157–158, 160–164, 160 (fig.), 163 (fig.), 165

need for better understanding, 238, 324

non-Kähler manifolds and, 242–244

as non-simply connected, 139–140

number of families, 238–240

numbers of, 146–147

numbers of particle families, 143–153

particles/masses derivation, 139–142

popular culture and, 137

possible configurations, 233–234, 236–237

quintic threefold example, 146

relatedness question, 240, 241 (fig.), 242–244

resurrection of, 154–157

Ricci-flat metric precision and, 153, 154–156

significance of, 19–20, 139, 294–295, 321, 322–323

spacetime and, 316

string theory early connection, 129, 131–132, 134–135, 137–139, 138 (fig.)

SU(3) holonomy group, 131

supersymmetry and, 129, 131–132, 134, 141–142, 190

warped throats, 276–277, 280

Calabi-Yau theorem

immediate applications overview, 121–123

physics fit possibilities, 123–124

Severi conjecture and, 122

string theory and, 127, 128

Calculus invention, 24, 45, 294, 313

Candelas, Philip

bundles, 218

Calabi-Yau manifolds, 321–322

mirror symmetry, 163, 164

models of universe, 224

Schubert problem, 165–166, 168, 169, 170, 171, 172–173

string theory, 128, 131, 134–135, 137, 145, 147, 246

string theory paper (1985), 135, 137, 142–143, 146, 200, 214–215

"Canonical" way, 47

Cardano, Girolamo, 82

Carley-Salmon theorem, 167 (fig.)

Carroll, Sean, 302

Cartan, Elie, 303

Cartesian coordinate system, 8, 23–24

Casson, Andrew, 68

Cavendish, Henry, 286

Cayley, Arthur, 167 (fig.)

Chan, Ronnie, 320

Chaos theory and butterfly effect, 45

Cheeger, Jeff, 74

Cheng, S. Y.

about, 72, 109 (photo)

Dirichlet problem, 113–114, 115–116

Minkowski problem, 108, 109

Monge-Ampère equations, 108

Chern, S. S.

mathematics and, 290, 303

as Yau's mentor, 35, 35 (photo), 36–37

See also Chern classes

Chern classes
  about, 95–96, 97 (fig.), 98, 121–122,
    212–213, 214–215
  Calabi conjecture/proof and, 79,
    100, 122
  characterizing manifolds and, 95–96
  Euler characteristic and, 95, 97 (fig.),
    98
Chernoff, David, 275
Chiral fermions, 200
Christodolou, Demetrios, 63
Classical geometry. *See* Geometry
Clemens, Herb, 240, 242, 249
Closed group, 41
Cohomology, 164–165, 217
Coleman, Sidney, 254, 256
Compact objects defined, 12, 64
Complex manifolds, 80, 81
Complex numbers, 81–83
Complex space, 77
Conformal field theories, 156–157
Conformal invariance investigations,
    152–154
Confucius, 36
Conic sections, 290, 291 (fig.)
Conics (Apollonius), 290
Conifold singularity, 240
Conifold transitions, 240, 241 (fig.),
    242, 297–298
Convex bodies/space, 53
Convex regular polyhedra
  about, xvii–xviii
  See also Platonic solids
Cosmic inflation
  Big Bang, 184, 231–232, 270–271,
    272, 302
  branes and, 277–278, 279, 280
  warped throat models, 276–280, 281
Cosmic Landscape, The (Susskind), 236
Cosmic microwave background (CMB),
    271
Cosmic strings, 272–276, 274 (fig.)

Cosmological constant, 123, 127–128,
    235–236, 237, 254, 256,
    262–264, 266
Cox, David, 172
Cubic equations, 324
Curve shortening flow, 48 (fig.)
Cycles
  branes/black holes and, 191, 192,
    196
  mathematicians and, 165
  moduli and, 227
  physicists and, 165, 190–191

D-branes
  about, 148, 174, 180, 189
  black hole association, 189–193, 191
    (fig.), 195–196
  Calabi-Yau manifolds and, 190–192,
    193–194, 196–197
  SYZ conjecture and, 174–175
Damour, Thibault, 275–276
Dark energy
  cosmological constant and, 123
  description, 1, 232
  extra dimensions and, 258
  KKLT scenario and, 232–233
Dark matter, 1, 201, 281
Davies, Rhys, 145
De Giorgi, Ennio, 46
De Sitter, Willem, 262
De Sitter space, 262–263, 266
Decompactification
  definition, 260 (fig.)
  energy states and, 258–262, 260
    (fig.), 261 (fig.)
  four vs. ten dimensions and,
    257–262, 266–267
  landscapes and, 259, 261 (fig.)
  vacuum decay, 257–262, 260 (fig.),
    261 (fig.), 264–265, 265 (fig.),
    270
Dehn's lemma, 53–54, 54 (fig.), 55

Derivatives
  description, 24, 25
  use of (overview), 24–25
  velocity curve and, 44 (fig.)
Descartes, René, 7–8, 23, 24
Determinism, 186–187
Diffeomorphic manifolds, 66–67, 68
Differential equations
  scope, 43
  See also Partial differential equations;
    specific applications
Differential geometry, 24–25
Dijkgraaf, Robbert, 296, 305 (fig.), 308
Dimensions
  "degree of freedom" and, 3–4
  three/four dimensions
    classifications, 65
  See also Hidden dimensions; Higher-
    dimensional systems; String
    theory
Dimopoulos, Savas, 285
Dirac, Paul, 289
Dirac-Born-Infield (DBI) model, 279
Dirac equation, 141, 142, 217
Dirichlet, Lejeune/problem, 113–116
Discretization, 220, 221 (fig.)
Distler, Jacques, 156
Dixon, Lance, 157
Dodecahedron
  description, xvii, xviii (fig.), xix, 327
  See also Platonic solids
Donagi, Ron, 209, 214, 215, 217, 218,
    322
Donaldson, Simon, 66 (photo)
  DUY theorem, 209, 210
  embedding, 223
  four-dimensional space, 65–66, 68,
    69
  Severi conjecture, 122
  See also DUY theorem
Douglas, Jesse, 51–52
Douglas, Michael, 180, 218, 223

Down quark, 218–219
DUY theorem
  about, 209–212, 216
  non-Kähler manifolds and, 249
  tangent bundle and, 211, 212
Dvali, Gia, 285

Einstein, Albert
  black holes and, 184
  general relativity/spacetime, 9, 10,
    31–32, 312–313
  geometrization of physics, xix
  Kaluza's ideas and, 10–11, 13
  mass/energy relationship, 56
  Riemann's geometry and, 8–9,
    31–32, 33
  special theory of relativity, 9, 31, 313
Elegant Universe, The (Brian Greene),
    293
Elements, The (Euclid), xviii, 23
Ellingsrud, Geir, 170
Ellipses, 246–247, 247 (fig.)
Ellipsoid, 51
Elliptic curves, 324
Embedding, 220–222, 222 (fig.), 223
Enneper, Alfred, 50 (fig.)
Enneper surface, 50 (fig.)
Entropy
  about, 187, 263–264
  black holes, 187–188
  microstates, 188, 189, 192
  Strominger/Vafa work, 189,
    192–196
Enumerative geometry, 165–166, 167
    (fig.), 169, 173, 296
Euclid, xviii, 23, 310
Euclidean geometry
  about, 311, 312
  curved space and, 25, 26, 27, 27
    (fig.)
  flat space, 25–26
  limits of, 25, 26, 27, 27 (fig.), 30

Euler, Leonhard, 24, 45
Euler characteristic
    Chern classes and, 95, 97 (fig.), 98
    computing, 95 (fig.)
    description, 94, 95 (fig.), 300
    Lagrangian submanifolds/SYZ
        conjecture, 178–179
    mirror symmetry and, 160–161, 164
    of Riemann surfaces, 84 (fig.), 95
Exotic spaces, 66, 67

"Fate of Four Dimensions, The"
    (Giddings), 266
Fermat, Pierre de, 49
Fermat's Last Theorem, 324
Fermions, 133, 141–142, 200, 283
Feynman, Richard, 290, 292
Field theory, 152–153
Finster, Felix, 63
Flop transition, 316–319, 317 (fig.)
Fluxes
    Calabi-Yau manifolds and, 229–230
    magnetic/electric field comparison,
        229, 229 (fig.), 230 (fig.)
    moduli problem and, 228–231
    as quantized/effects, 230, 233
Four-dimensional topology, 64–69
Fourfold symmetry, 92, 93 (fig.)
Freed, Dan, 132
Freeman, Michael, 68
Freivogel, Ben, 272
Freud, Sigmund, 183
Frey, Andrew, 255
Friedman, Robert, 240
Fu, Ji-Xian, 249, 250, 251
Fuller, R. Buckminster, 291 (fig.)
Function defined, 24
Fundamental group, 41, 69, 144, 212
Fundamental strings, 124, 273

G_2 manifolds, 149–150, 323
Gaia Satellite, 275

Gangui, Alejandro, 273
Gathman, Andreas, 170, 172
Gauge fields
    Calabi-Yau manifolds and, 205–206
    physicists/mathematicians
        terminology, 205
    Standard Model, 204, 205, 206–212
Gauge theories
    truth of, 296
    See also specific theories
Gauss, Carl Friedrich
    about, 303, 311–312, 313
    abstract space and, 29
    curved space and, 25, 26, 27–28, 27
        (fig.), 28 (fig.), 94
    differential geometry and, 24
Gauss-Bonnet formula, 94–95
Gauss curvature, 27–28, 28 (fig.), 98, 99
General relativity
    about, 9, 10, 13, 14, 31–32, 312–313
    ground state of, 59
    Riemann's geometry and, 31–32, 33
Geodesic defined, 49, 50 (fig.), 51
Geometric analysis
    about, 46
    examples, 47–48, 48 (fig.)
    four-dimensional topology, 64–69
    goal/scope of, 39, 64
    history, 45, 46
    Poincaré conjecture, 69–75
    See also Calabi conjecture; Minimal
        surfaces; *specific applications*
Geometric flow, 48
    See also Ricci flow
Geometrization of physics, xix, 65, 71,
    74, 326, 327
Geometry
    conjectures defined, 20
    as dynamic, 20, 39, 310–313
    Einstein's theory and, 308–309, 310
    as "emergent," 309–310
    general relativity and, 308–309

history of contributions (overview),
xvii–xviii, xviii (fig.), xix, 20–30,
311–313
intrinsic geometry, 27–28, 28 (fig.)
Planck scale and, 307–308, 310
quantum mechanics/general
relativity incompatibility and,
309
reducing dimensions, 144–145, 145
(fig.)
scope of, 17–20
space and, 18–19
term derivation, 18
topology vs., 5, 6 (fig.), 64–65
See also Mathematics; Quantum
geometry; *specific branches/types*
Gepner, Doron, 156–157
Gepner model, 156–158
Geroch, Robert, 56, 58, 105
Giddings, Steve, 232, 256, 259, 266
Givental, Alexander, 172
Global symmetry, 90
Goddard, Peter, 289
Goettsche, Lothar, 300
Gordon, Cameron, 55
Göttingen Observatory, 303
Gravitino, 281
Gravitons, 281, 283
Gravity
force description, 19–20, 32, 202,
232, 246, 285
geometry and, 32
hidden dimensions verification and,
285–287, 287 (photo)
inverse square law and, 286–287, 287
(photo)
as nonlinear/effects, 59
Randall-Sundrum model, 284
structure in universe and, 32–33
tangent bundle and, 213–214
"torsion balance" experiments,
286–287, 287 (photo)

Green, Michael
canceling anomalies, 125–126, 212,
213, 214
string theory, 125–126, 127, 212
Green, Paul
Calabi-Yau manifolds, 243
Schubert problem, 165–166, 168,
169, 170, 171, 172–173
Greene, Brian, 159 (photo)
Calabi-Yau manifolds, 139, 145–146,
151, 156
flop transitions, 317–318
geometry's future, 314
on gravity, 31
mirror symmetry, 152, 157–158,
160, 163, 164, 165, 168–170
on physical theory, 293
on physics/mathematics
relationship, 303
Standard Model, 200–201
string theory verification, 301–302
Witten and, 323
Greene, Robert, 43, 51, 101, 104
Grisaru, Marcus, 154
Gromoll, Detlef, 74
Gromov, Misha, 56–57
Gross, David, 154–155, 236, 237 (photo)
Gross, Mark
Calabi-Yau manifolds, 239–240, 243,
252
mathematics/physics relationship, 169
SYZ conjecture, 175, 177, 181
Group theory, 41
Guth, Alan, 184, 302

Hamilton, Richard, 70–72, 73–74, 75
Hawking, Stephen, 61 (photo)
Bekenstein-Hawking formula, 188,
189, 193, 195, 263
black holes, 60, 62, 63, 64, 186, 187,
188, 189, 197
string theory, 196

Headrick, Matt, 220
Heckman, Jonathan, 281, 322
Heisenberg, Werner/uncertainty
  principle, 255, 284, 307–308
Hermitian manifolds, 88–89
Hermitian Yang-Mills equations, 208,
  209, 210, 211, 212, 247
Hermitian Yang-Mills theory, 249
Heterotic string theory
  fluxes and, 230–231
  Standard Model and, 204, 205–212
Hexahedron
  description, xvii, xviii (fig.)
  See also Platonic solids
Hidden dimensions
  large extra dimension hypothesis,
    285–287
  size, 2, 138–139, 231, 269, 285
  See also Decompactification; String
    Theory; String theory verification
Higgs field, 218–219
Higgs particle/boson, 217, 218, 246
Higher-dimensional systems
  early thinking on, 7–8
  mathematicians and, 11
  shape possibilities and, 7
  vantage point of, 3, 4
  See also String theory
Hilbert, David, 171
Hindmarsh, Mark, 276
Hironaka, Heisuke, 184
Hitchin, Nigel, 177
Hodge, W. V. D., 162
Hodge diamond
  about, 142, 162, 163 (fig.)
  Calabi-Yau manifolds and, 142,
    162–163, 163 (fig.), 238–239
Hodge numbers, 162, 163, 164
Holonomy definition/description,
  129–131, 130 (fig.)
Holt, Jim, 293

Homeomorphic manifolds, 66–67, 68
Homological mirror symmetry
  D-branes and, 180
  effects on mathematics, 297
  mirror symmetry explanation and,
    173, 179–181
  SYZ conjecture, 179–180, 297
Homology described, 214, 217
Hooft, Gerard 't, 186
Horava, Petr, 150
Horowitz, Gary
  string theory, 128, 135, 137
  string theory paper (1985), 135,
    137, 142–143, 146, 200,
    214–215, 246
Hubble Space Telescope data, 275
Hubble volume, 1
Hubsch, Tristan, 133, 134, 239, 241
  (fig.), 243
Hurricane Fran, 97 (photo)
Hyperbolic geometry, 311
Hypersurface, 144

Icosahedron
  description, xvii, xviii (fig.), xix, 327
  See also Platonic solids
Integration, 25, 26 (fig.), 219
Internal symmetry, 90, 92–93
Intrinsic geometry, 27–28, 28 (fig.)

J transformation/operation, 90–91, 92,
  92 (fig.), 93, 93 (fig.)
Jang, P.S., 58
Jones, John D.S., 65
Joyce, Dominic, 149

K3 surfaces
  about, 127, 128, 157
  rational curves on, 297, 298, 300
  as string theory models, 128
  SYZ conjecture, 175, 176 (fig.)

Kachru, Shamit
  Calabi-Yau manifolds, 155, 157
  decompactification, 260
  KKLT scenario/paper, 231, 233, 257
  moduli problem, 23, 229–230, 231,
    233
  Standard Model, 218
  string theory/gravity, 232
  string theory verification, 288
  See also KKLT scenario/paper
Kähler, Erich, 85, 128
Kähler-Einstein manifolds, 122–123
Kähler geometry, 79, 85
Kähler manifolds
  Calabi conjecture and, 89, 90, 93–94,
    100
  description, 88–91, 92
  J operation under parallel transport,
    91, 92
  supersymmetry, 94
  symmetry of, 89–90, 92, 93–94
  See also Calabi-Yau manifolds
Kallosh, Renata, 231, 257
  See also KKLT scenario/paper
Kaluza, Theodor, 10–11, 12
Kaluza-Klein gravitons, 283
Kaluza-Klein particles, 283
Kaluza-Klein theory
  description, 12–13, 12 (fig.), 15
    (fig.), 126, 244
  string theory and, 124, 126–127
Kamran, Niky, 63
Katz, Sheldon, 166, 172
Kazan Messenger, 311
Kazdan, Jerry, 118
Kepler, Johannes, 290, 327
Kirklin, Kelley, 145–146
KKLT scenario/paper, 231, 232–233,
  237, 257
Kleban, Matthew, 257, 272
Klebanov, Igor, 196, 276

Klebanov-Strassler throat model, 278
Klein, Oskar
  extra dimensions, 285
  Kaluza's ideas and, 11, 12
  See also Kaluza-Klein theory
Knot theory/applications, 290
Kodaira, Kunihiko, 128
Kohn, J. J., 116
Kontsevich, Maxim, 171, 179–180
Kronheimer, Peter, 327
Kroto, Harold, 291 (fig.)

Lagrange, Joseph, 24
Lagrangian submanifolds, 174–175, 176
  (fig.), 178
Land, Kate, 271
Landis, Geoffrey, 253, 254
Laplace equation, 142
Large Hadron Collider (LHC)
  about, 201, 282 (photo)
  string theory verification and, 134,
    281–285, 282 (fig.)
Large Synoptic Survey Telescope
  (LSST), 275
Laser Interferometer Space Antennae
  (LISA), 275–276
Lawson, Blaine, 36, 42
Leibniz, Gottfried, 24
Lemma defined, 53
Lerche, Wolfgang, 157
Leung, Naichung Conan, 177, 298, 300
Li, Jun, 248, 249, 300
Li, Peter, 72, 222, 223
Li-Yau inequality, 72–74
Lian, Bong, 301
Linde, Andrei
  decompactification, 257–258, 261,
    263
  KKLT scenario/paper, 231, 257
  See also KKLT scenario/paper
Linear systems/theory overview, 45

Liu, A. K., 300
Liu, Melissa, 60
Lobachevsky, Nikolai, 29, 311, 312
Loeb, Avi, 183
Loewner, Charles, 114–115
Loop quantum gravity, 15
Loop theorem, 54–55
Luo, Wei, 223
Lutwak, Erwin, 109

M-theory
    about, 125, 147–150, 148 (fig.), 189,
        215, 323
    branes (membranes), 148, 150, 174,
        193
    dimension numbers, 148–150
    See also D-branes; String theory
Magueijo, Joao, 271
Maldacena, Juan
    AdS/CFT, 196, 197, 198
    black holes, 195, 196, 197, 198
    hidden dimensions, 148
    supersymmetry, 132
Mamet, David, 262
Manifolds
    comparisons, 66–67, 68
    definition/description, 64, 81
    Euclidean space and, 85–89
    See also specific types
Mass, "local" mass defined, 59–60
Massless particles, 141, 142
Mathematical Sciences Research
        Institute (MSRI) conference
        (1991), 169–171
Mathematical theorem
    about, 20–21, 22, 23
    See also specific theorems
Mathematics
    beauty of, 289–290, 292, 293
    enrichment by string theory
        (examples), 165–166, 169, 173,
        296–298, 300–301

physics relationship (overview), 43,
        128–129, 131–132, 289–290,
        291, 302–304, 305 (fig.), 306
    proof and, 294
    See also specific branches
Maximum principle, 72
Maxwell equations, 303–304
McAllister, Liam
    Calabi-Yau manifolds, 137, 232, 233,
        276
    verifying string theory, 276, 279,
        280
Meeks, William, 53 (photo)
    minimal surfaces work, 53, 54–55,
        54 (fig.)
Menaechmus, 290
Mercator projection maps, 26, 83
Metric coefficients and distance
        calculations, 86–88
Metric tensor
    curved space and, 29–30, 32
    description, 10
    gravity and, 10, 32
Microlensing, 275
Millay, Edna St. Vincent, 23
Mills, Robert, 205 (photo)
    gauge theory, 203–204
    truth/beauty, 293
    Yang-Mills equations, 65, 69
    See also Hermitian Yang-Mills
        equations; Yang-Mills equations;
        Yang-Mills theory
Milnor, John, 39, 40, 54, 67
Minimal Supersymmetric Standard
        Model, 201–202
    See also Standard Model
Minimal surfaces
    description/definition, 49
    geometric analysis and, 48
    Plateau problem, 50 (fig.), 51–52, 52
        (fig.), 53, 54, 114
    singularities and, 52–53

Minimization concept
in geometry/physics, 49
See also specific applications
Minkowski, Hermann/problem, 9,
108–109
Miron, Paul, 145–146
Mirror conjecture proof/controversy,
171–173
Mirror symmetry
Betti numbers and, 161–162, 161
(fig.)
Calabi-Yau manifolds, 157–158,
160–164, 160 (fig.), 163 (fig.),
165
as duality, 158
effects on algebraic/enumerative
geometry, 165–166, 169, 173,
296
Euler characteristic and, 160–161,
164
example, 160 (fig.)
Schubert problem and, 165–166,
168–173
SYZ conjecture and, 173–175, 176
(fig.), 177–180
See also Homological mirror
symmetry
Mirror Symmetry and Algebraic Geometry
(Katz and Cox), 172
Möbius strip, 6 (fig.), 7, 83, 178, 298,
299 (fig.)
Moduli
Calabi-Yau manifolds and, 228–229,
237
cycles and, 227
description, 227
with torus, 227–228
Moduli problem
fluxes and, 228–231
KKLT scenario/paper, 231,
232–233, 237, 257
significance, 227, 228

Moduli space, 298
Monge, Gaspard, 24, 106
Monge-Ampère equations
about, 106–108, 250
Calabi conjecture/proving, 106–108,
113, 114, 116, 117
elliptic equations, 106, 107
hyperbolic (nonlinear) equations,
106–107
parabolic equations, 107
Mori, Shigefumi, 242
Morrey, Charles, 42 (photo)
Berkeley and, 36, 42–43
Calabi-Yau manifolds, 323
partial differential equations, 46–47,
48
Morrison, David
flop transition, 318–319
mirror symmetry, 170–171
small scale general relativity, 308–309
topological change, 316
Morse Theory (Milnor), 39
Multiple universes. *See* String theory
landscape
Mumford, David, 121, 122

Nash, John, 46, 221
Nash embedding theorem, 221–222
Negative curvature, 27, 27 (fig.)
Nemeschansky, Dennis, 155
Neutrinos
mass and, 200, 201
Standard Model of particle physics,
143, 200
New York Times, 132, 236
New Yorker, 137, 293
Newton, Isaac, 24, 36, 294, 312–313
Newton's method (nonlinear partial
differential equations), 110, 111,
111 (fig.)
Nirenberg, Louis, 115 (photo)
Calabi conjecture, 118

Nirenberg, Louis (*continued*)
  Dirichlet problem, 114–115, 116
  higher-dimensional theories, 46
  a priori estimates, 112
Nodes, 297–298
Non-Euclidean geometry invention,
  311–312
Non-Gaussianity phenomenon, 276,
  279, 280
Non-Kähler manifolds
  Calabi-Yau manifolds and, 242–244
  compactification and, 244, 250, 251
  Standard Model, 244–252
  warped products and, 244–246, 245
    (fig.)
Non-simply connected manifolds,
  139–140
Nonlinear equations, 45–46
Nonlinear partial differential equations
  approaches, 109–113
  Newton's method, 110, 111, 111
    (fig.)
  a priori estimates and, 111–113
Nonorientable surfaces, 7
Nontrivial loops, 40
Number theory, 104–105

Octahedron
  description, xvii, xviii (fig.), 326–327
  See also Platonic solids
Olbers, Heinrich Wilhelm Matthäus, 29
On the Heavens (Aristotle), 3
Ooguri, Hirosi, 186, 189, 194, 195–196
Orientable surfaces, 5, 6 (fig.), 7, 64
Ossa, Xenia de la, 165–166, 168, 169,
  170, 171, 172–173
Osserman, Robert, 23, 52–53
Ovrut, Burt
  bundles, 218
  hidden dimensions, 282
  particle masses, 223–224

Standard Model, 216–217, 252
string theory/M-theory, 150,
  208–209

P-branes, 148
Pantev, Tony, 217
Papakyriakopoulos, Christos, 54
Parallel translation (transport), 90–91,
  92 (fig.)
Parity violation, 125–126, 127
Parks, Linda, 165–166, 168, 169, 170,
  171, 172–173
Partial differential equations, 43, 45
Particle physics
  families of particles, 143–147
  handedness of particles, 200
  as quantum field theory, 203
  See also Minimal Supersymmetric
    Standard Model; Standard Model
Penrose, Roger, 61 (photo)
  black holes, 60, 62, 63, 64
Perelman, Grisha, 72, 74–75
Physical Review D, 255
Physical Review Letters, 278
Physics
  dualities, 326–327
  geometrization of, xix, 65, 71, 74,
    326, 327
  local laws and, 43–44
  mathematics relationship
    (overview), 43, 128–129,
    131–132, 289–290, 291,
    302–304, 305 (fig.), 306
Planck length, 12
Planck telescope, 1
Plateau, Joseph/problem, 50 (fig.),
  51–52, 52 (fig.), 53, 54, 114
Plato, xvii–xix, 325–326, 327
Platonic solids
  about, xvii–xix, xviii (fig.), 291 (fig.),
    326–327

cosmology and, xviii–xix
properties, xviii, xix
Plesser, Ronen, 159 (photo)
Calabi-Yau manifolds, 151, 154, 156
mirror symmetry, 152, 157, 158,
160, 163–164, 164, 165,
168–169
Poetry of the Universe (Osserman), 53
Pogorelov, Aleksei, 46, 109
Poincaré, Henri, 46, 117
Poincaré conjecture
solving, 48, 69–75
See also Severi conjecture
Poincaré metric, 86–87, 88
Poincaré recurrence time/theorem,
265–266
Poincaré symmetry, 204
Polchinski, Joe
Calabi-Yau manifolds, 139, 234, 251,
285
cosmic strings, 274
geometry's future, 310–311
string theory, 189, 232, 277
Positive curvature, 27, 27 (fig.)
Positive mass (energy) conjecture
about, 56
proofs, 56–59
Positron, 289
Potential wells, 255
Power series, 153–154
Preissman, Alexandre, 40, 41
Preskill, John, 186, 197
Principle of least energy/action, 49, 152
Ptolemy, 3
Pythagoras, 21, 22
Pythagorean theorem, 21–22, 22 (fig.)
Pythagoreans, 22

Quantization, 165
Quantum cohomology, 164–165
Quantum foam, 314 (fig.), 315 (fig.)

Quantum geometry
about/need for, 308, 311
physics/string theory and, 313–314,
314 (fig.), 315 (fig.), 316–320
Quantum tunneling
about, 255–256
vacuum decay and, 255–256,
264–265, 265 (fig.)
Quarks, 143
Quintic threefold problem, 165–166
See also Schubert problem

Rado, Tibor, 51, 52
Ramanujan, Srinivasa, 300
Randall, Lisa, 232
Randall-Sundrum model, 232, 278, 284
Rees, Martin, 254
Reid, Miles/conjecture, 239, 240, 242,
243, 249
Reshetikhin, Nicolai, 295
Ricci curvature
about, 98–100
Calabi conjecture and, 78, 79, 98, 99,
100, 101
determining, 98
sectional curvature and, 98, 99
zero Ricci curvature description, 78
Ricci-flat manifold, 98
Ricci flow, 70–72, 73–74
Ricci tensor, 99
Richter, Burton, 236
Riemann, Georg Friedrich Bernhard, 8,
18, 29–30, 37, 46, 83, 89
Riemann curvature, 99
Riemann hypothesis, 36–37
Riemann surfaces
about, 83, 84 (fig.), 85, 135, 142,
191, 297, 322
Ahlfors's theorem, 116
Chern classes, 95
conformal property, 83, 84 (fig.), 85

Riemann surfaces (*continued*)
Euler characteristic of, 84 (fig.), 95
Gauss-Bonnet formula, 94
metric, 135
string theory, 83, 84 (fig.)
world sheets and, 152
Riemann zeta function, 37
Riemannian geometry, 8–9, 29, 31–32,
33, 37, 81, 308, 312, 313, 316
Riemannian manifold, 30, 221
Ross, Graham, 145–146
Rotational invariance, 90
Rotational symmetry, 89–90, 133–134
Ruan, Wei-Dong, 177

Salaff, Stephen, 34–35
Salmon, George, 167 (fig.)
Sarason, Donald, 35
Sazhin, Mikhail, 274–275
Scalar field description/example, 228
Scale invariance, 152–153
Scanning tunneling microscopes, 255
Schoen, Richard, 57 (photo)
black holes, 60, 63
positive curvature, 62
positive mass (energy) conjecture,
56–59, 105
Schubert, Hermann, 166, 167 (fig.), 168
(fig.)
Schubert problem
about, 165–166, 167 (fig.), 168
(fig.), 169
mirror symmetry work and,
165–166, 168–173
Schwarz, John
canceling anomalies, 125–126, 212,
213, 214
string theory, 125–126, 127, 128,
212
Schwarz lemma, 116
Schwarzschild, Karl, 184
Schwarzschild black hole, 187

Seiberg, Nathan, 68, 300–301, 316
Seiberg-Witten equation, 68–69
Sen, Ashoke, 155
Severi conjecture, 121, 122
Shandera, Sarah, 279
Shiu, Gary
hidden dimensions, 277, 278–279,
280, 283–284
moduli problem, 228
on verifying string theory, 279, 280
Simon, Leon, 49
Simons, Aaron, 193–194
Singer, I. M., 164, 169, 303
Singularities
about, 19, 183–184
Big Bang as, 184
studying, 183
See also Black holes
Smalley, Richard, 291 (fig.)
Smith, Paul/conjecture, 55
Smoller, Joel, 63–64
Sobolev, Sergei, 117
Soibelman, Yan, 179–180
Spacetime
description, 9–10, 10 (fig.), 31
positive mass (energy) conjecture
and, 59
Riemann's geometry and, 31, 32, 33
Spacetime as five-dimensional
description, 11–12, 12 (fig.)
Kaluza and, 10–11, 12
Sparks, James, 196–197
Standard Model
families of particles, 143, 144, 146,
200
forces included, 202, 203, 204
gauge fields, 204, 205, 206–212
as gauge theory, 203–205
gravity and, 202
heterotic string theory, 204,
205–212
limitations, 200, 201, 202

string theory and, 200–201, 204, 205–212, 212–225
string theory/Calabi-Yau manifolds, 212–225
See also Minimal Supersymmetric Standard Model
Stanford geometry conference (1973), 55–56, 105–106, 118
Statistical mechanics, 188–189, 193, 265–266
Stau particle, 281, 282
Stress energy tensor, 99
String coupling, 142
String description, 14
String phenomenology, 270–272
String theory
   beginnings, 124–129
   compactification, 126–127, 128, 143
   dimension numbers needed, 124–125
   dimensions description, 15 (fig.)
   doubts/questions about (overview), 269, 293–294
   families in particle physics model, 143–147
   four-dimensional look of universe and, 126–127, 142–143
   gauge theories and, 296
   gravity prediction with, 296
   mathematical consistency, 292, 295
   mathematical enrichment examples, 165–166, 169, 173, 296–298, 300–301
   particle physics (Standard Model) and, 212–225, 295–296
   physics/mathematics resources and, 302–304
   supersymmetry and, 93–94, 131–132, 134, 190
   theories on, 147–148, 148 (fig.)
   See also Calabi-Yau manifolds; M-theory; *specific individuals*

String theory landscape
   anthropic arguments, 234–236
   cosmological constant problem and, 237
   debate on, 236
   description, 234, 235 (fig.)
String theory verification
   branes/inflation, 277–278, 279, 280
   bubble collisions evidence, 271–272
   bubble phenomenology, 271–272
   challenges overview, 269–270, 279, 288
   cosmic strings, 272–276, 274 (fig.)
   cosmology and, 270–276, 278
   gravity "weakness" and, 285–287, 287 (photo)
   inflation, 276–280, 281
   Large Hadron Collider and, 134, 281–285, 282 (fig.)
   string phenomenology, 270–272
   supersymmetry and, 281–283
   time frame (for testing), 301–302
   warped throat models, 276–279, 280, 281
Strominger, Andrew, 194 (photo)
   black holes, 184, 186, 187, 189, 192–193, 194–196, 197, 198
   Calabi-Yau manifolds and, 146, 147, 252, 316, 321, 322, 323
   M-theory, 148–149
   mirror symmetry, 297
   on physicists/mathematicians, 128–129, 131–132
   quantum theory and, 13
   string theory, 128–129, 131–132, 134–135, 137, 146, 147, 198
   string theory paper (1985), 135, 137, 142–143, 200, 214–215, 224, 246
   SYZ conjecture, 173, 174, 176 (fig.), 179
   unified field theory, 14
   See also SYZ conjecture

Strominger equations/system (non-Kähler), 246, 247–252
Strømme, Stein Arild, 170
SU(3) holonomy group, 131
Submanifolds, 164–165
    Lagrangian submanifolds, 174–175, 176 (fig.), 178
Sundrum, Raman
    on Calabi-Yau manifolds, 131
    moduli problem, 231
    warped geometry model, 232
    See also Randall-Sundrum model
Superpartners, 133, 134
Supersymmetry
    black holes, 192–193, 194–195
    broken symmetry, 133–134
    Calabi-Yau manifolds, 129, 131–132, 134, 141–142, 190
    covariantly constant spinors, 131
    energy gap and, 134
    energy levels and, 133, 134
    fermions/bosons and, 133–134
    four-dimensional geometry and, 68
    holonomy and, 129
    Standard Model and, 200–201
    string theory and, 93–94, 131–132, 134, 190
    string theory verification, 281–283
    vacuum and, 132
Susskind, Leonard, 237 (photo)
    black holes, 186
    Calabi-Yau manifolds, 150
    string theory, 150
    string theory landscape, 236
Symmetry
    broken symmetries, 206, 207, 207 (fig.)
    heterotic version of string theory, 206–207
Symplectic geometry
    A-branes, 180

    about, 181, 297
    link with algebraic geometry, 180–181
SYZ conjecture
    D-branes, 174–175
    homological mirror symmetry and, 179–180, 297
    K3 surfaces and, 175, 176 (fig.)
    Lagrangian submanifolds, 174–175, 176 (fig.), 178
    mirror symmetry explanation and, 173–175, 176 (fig.), 177–180
    T duality and, 177–179

T duality, 177–179
Tachyons, 134
Tangent bundle, 207, 208, 208 (fig.), 210, 211–212, 213–215, 218
Tau function, 300
Tau lepton, 281
Taubes, Clifford, 68, 69, 304, 306
Tegmark, Max, 304
Telescopes
    invention significance, 1
    See also specific telescopes
Tetrahedron
    description, xvii, xviii (fig.), 327
    See also Platonic solids
Theaetetus, xvii–xviii
Thermal fluctuations and vacuum states, 256–257
Thorne, Kip, 64, 186
Thurston, William, 36, 55, 65
Thurston's geometrization conjecture, 65, 71, 74
Tian, Gang, 316–317
Timaeus (Plato), xvii, xviii, xix, 326
Time Machine, The (Wells), 9
Time-symmetric case, 58
Topology
    four-dimensional topology, 64–69

genus (of surfaces), 5, 6 (fig.), 7,
  64–65
  geometry and, 55
  geometry vs., 5, 6 (fig.), 64–65
  one-dimensional spaces in, 5, 6 (fig.)
  two-dimensional spaces in, 5, 6
    (fig.), 7
Torus
  descriptions, 5, 41, 64–65
  Gauss curvature and, 28 (fig.)
Transcendental numbers, 104–105
Trapped surfaces, 60, 62–63
  See also Black holes
Trivedi, Sandip, 231, 257
  See also KKLT scenario/paper
Trivial fundamental group, 41, 69
Trivial line bundle, 298
Tseng, Li-Sheng, 246, 251
Turner, Michael, 254
Twain, Mark, 310
Tye, Henry
  branes, 278
  cosmic strings, 273–274, 276
  dimension decompactification/
    bubble collisions, 262, 272
  hidden dimensions, 273–274, 276,
    278, 280
  on string theory, 293
Tzeng, Yu-Jong, 300

Uhlenbeck, Karen, 211 (photo)
  DUY theorem, 209, 210
  See also DUY theorem
Uncertainty principle, 255, 284, 307–308
Underwood, Bret, 278–279, 280, 283
Unified field theory, 10, 13, 14, 124
  See also String theory
Universe size, 1–2

Vacuum decay
  about, 254

decompactification, 257–262, 260
  (fig.), 261 (fig.), 264–265, 265
  (fig.), 270
  quantum tunneling and, 255–256,
    264–265, 265 (fig.)
  results, 254–255, 256, 261–262, 263
  thermal fluctuations and, 256–257
  time before, 262–266
Vacuum Einstein equation
  nontrivial solution to, 100
  trivial solution to, 99–100
Vacuum energy/states, 235 (fig.)
"Vacuum States" (Landis), 253
Vafa, Cumrun, 194 (photo)
  black holes, 187, 189, 192–193,
    194–196, 197
  Calabi-Yau manifolds, 321, 322
  on forces, 202–203
  geometry's future, 307, 308, 310
  gravitino, 281
  mirror symmetry, 157, 158, 164
  string theory, 126, 298
Vakil, Ravi, 323
Van de Ven, Anton, 154
Van de Ven, Antonius, 121–122
Velocity curve, 44 (fig.)
Vilenkin, Alexander, 274, 275–276
Virtual particles, 257

Walker, Devin, 283
Wang, Mu-Tao, 60
Warner, Nicholas, 157, 158
Warped throat models, 276–279, 280,
  281
Wave behavior, 82
Wave-particle duality, 140–141
Weil, André, 36–37, 112
Wells, H.G., 9
West, Peter, 127
Weyl, Hermann, 303
Weyl tensor, 99

Wheeler, John, 32, 64, 314 (fig.)
Wijnholt, Martijn, 322
Wilczek, Frank, 254
Wiles, Andrew, 324
Wilkinson Microwave Anisotropy Probe
    (WMAP), 276
Winding number, 177
Wiseman, Toby, 220
Witten, Edward, 149 (photo)
    black holes, 195, 196
    Calabi-Yau manifolds and, 128, 135,
        137, 143–144, 146, 154, 155,
        322–323
    cosmic strings, 273, 274
    DUY theorem and, 212
    four-dimensional geometry, 68–69
    on geometry, 311
    M-theory, 147–148, 148 (fig.), 150,
        151, 180, 195
    on mathematical consistency, 295
    number of dimensions, 124, 135
    positive mass (energy) conjecture,
        58–59, 132
    Seiberg-Witten equations, 300–301
    spatial tears, 318
    string theory duality, 298
    string theory paper (1985), 135,
        137, 142–143, 146, 200,
        214–215, 246
    string theory verification, 288, 296,
        302
World sheet, 152–153

Yang, Chen Ning, 65, 203–204, 205
    (photo), 290
    See also Hermitian Yang-Mills
        equations
Yang-Mills equations, 65, 69, 208, 209,
    290, 296
    See also Hermitian Yang-Mills
        equations

Yang-Mills theory, 69, 205 (fig.), 208,
    212
Yau, Shing-Tung, 136 (photo)
    background, 33–34
    black holes, 60, 63–64
    Bogomolov-Miyaoka-Yau inequality,
        122
    Bourguignon-Li-Yau work, 222–223
    curvature question beginnings, 33,
        36
    Dirichlet problem, 113–114,
        115–116
    education, 34–36, 39–40, 41–43
    flop transition, 316–317, 317 (fig.)
    Hamilton's work and, 71, 72, 73, 74
    Li-Yau inequality, 72–74
    minimal surfaces work/approach, 53,
        54–55, 54 (fig.), 57, 63
    non-Kähler manifolds, 248–252
    partial differential equations, 46–47
    positive mass (energy) conjecture,
        56–59, 105
    Preissman's theorem and, 40–42
    rational curves on K3 surfaces, 297,
        298, 300
    Riemann hypothesis, 36–37
    Schwarz lemma, 116
    Yu-Yun and, 117
    See also DUY theorem; SYZ
        conjecture
Yau, Shing-Tung/Calabi conjecture
    Dirichlet problem and, 113–116
    Minkowski problem and, 108, 109
    skepticism efforts to disprove
        conjecture, 101, 104, 105–106,
        121, 122
    See also Calabi conjecture/proving;
        Calabi-Yau manifolds; Calabi-Yau
        theorem
Yau-Zaslow formula, 297, 298, 300
Yin, Xi, 189–190, 191, 192, 193

Yukawa coupling constant, 218, 219
Yukawa couplings, 156, 168, 219, 224

Zanon, Daniela, 154
Zaslow, Eric
  rational curves/K3 surfaces, 297,
    298, 300

SYZ conjecture, 173, 174, 176 (fig.),
    177, 179
  See also SYZ conjecture
Zeyl, Donald, 325, 326
Zurek, Kathryn, 283